AF539429

ESSENCE OF BUSINESS MATHEMATICS

ESSENCE OF BUSINESS MATHEMATICS

By

Dr. R.K. Rajput

DISCOVERY PUBLISHING HOUSE PVT. LTD.
NEW DELHI-110 002

First Published – 2008

Reprinted – 2026

ISBN: 978-81-8356-339-0

Essence of Business Mathematics

Published by:

DISCOVERY PUBLISHING HOUSE
4383/4B, Ansari Road, Darya Ganj
New Delhi-110 002 (India)
Phone: +91-11-23279245; 23253475; 43596065
Mobile: +91 9811179893 / +91 9871656464
E-mail: discoverybooksindia@gmail.com
orderdphbooks@gmail.com
namitwasan9@gmail.com
web: www.discoverypublishinggroup.com

Printed at:
Infinity Imaging Systems
Delhi

Preface

This book "Essence of Business Mathematics" has been written to provide a comprehensive and updated foundation course in Mathematics and its varied applications in management, commerce and economics.

This book has been written from studies point of view, so that they can easily understand various mathematical concepts, techniques and tool needed for their course. Each new concept and technique is properly supported by suitable solved examples. In solving the questions cave has been taken to explain each step so that students can follow the subject matter themself without even consulting others.

I believe that the book should serve specific need of the students appearing in B.Com. (Hons.), B.A. (Eco. Hons.), M.Com., (M.A.) (Eco.) B.B.A., B.C.A., M.B.A. and other competitive examinations.

Suggestions and comments for the improvement of the book from readers will be thankfully received and duly incorporated in the subsequent editions.

Author

CONTENTS

1

Progressions

ARITHMETICAL PROGRESSION

Quantities $a_1, a_2, a_3, , a_n, \ldots$ are said to be in *Arithmetical Progression* if $a_n - a_{n-1}$ is constant for all integers $n > 1$. The constant quantity $a_n, - a_{n-1}$ is called the *common difference* of the arithmetical progression.

Notation: A.P. stands for an arithmetical progression. Consider the following series

$1, 3, 7, 9, 11, \ldots$

$0, \sqrt{2}, 2\sqrt{2}, 3\sqrt{2}, 4\sqrt{2}, \ldots$

$1, 1/2, 0, -1/2, -1, -3/2, \ldots$

$x + y, x, x - y, x - 2y, \ldots$

$5.3, 5.55, 5.8, 6.05, 6.3, \ldots$

Each of the above series in an A.P. Common differences are respectively $2, \sqrt{2}, -1/2, - y$ and $.25$.

TO FIND THE GENERAL TERM OF A GIVEN ARITHMETICAL PROGRESSION

Let $a_1, a_2, \ldots, a_n, \ldots$ be a given A.P. Let d be their common difference. Then $a_n - a_{n-1} = d$ for all n.

$\Rightarrow a_2 - a_1 = d, a_3 - a_2 = d, a_4 - a_3 = d$ and so on,

$\Rightarrow a_2 = a_1 + d, a_3 = a_2 + d = a_1 + d + d = a_1 + 2d$

$a_4 = a_3 + \ = a_1 + 2d + d = a_1 + 3d$

... ...

$a_{n-1} = a_1 + (n - 2) d$

$a_n = a_{n-1} + \text{d} = a_1 + (n - 2)d + d$

$= a_1 + (n - 1)d.$

Thus nth term, a_n of an arithmetical progression whose first term is a_1 and common difference d, is given by

$$a_n = a_1 + (n - 1)d.$$

Example 1:

Find 16th term of the series 3.75, 3.5, 3.25, ...

Solution:

In this case $a_1 = 3.75$, $a_2 = 3.5$, $a_3 = 3.25$

$$d = a_2 - a_1 = -.25$$

Hence 16th term $= a_{16}$

$$= 3.75 + (16 - 1)(-.25)$$
$$= 3.75 - 15 \times .25$$
$$= 3.75 - 3.75 = 0.$$

Example 2:

Which term of the A.P.

49, 44, 39, ... is 9 ?

Solution:

Let nth term be 9, *i.e.* $a_n = 9$.

Here $a_1 = 49, d = 44 - 49 = -5,$

Thus $a_n = 9 = 49 + (n - 1)(-5)$

$\Rightarrow 9 = 49 - 5n + 5$

or $5n = 54 - 9 = 45$

$n = 9$

Thus 9th term of the given A.P. is 9.

TO FIND THE SEUM OF FINITE NUMBER OF QUANTITIES IN A.P

Let $a_1, a_2, ..., a_n$ be n quantities in A.P., and let the last term a_n be denoted by l. If d is their common difference then

$$a_n = a_1 + (n - 1)d = l.$$

Put $S_n = a_1 + a_2 + ... + a_n$

Thus $S_n = a_1 + (a_1 + d) + (a_2 + 2d) + ... + [a_1 + (n - 1)d]$

$= a_1 + (a_1 + d) + (a_1 + 2d) + ... + (l - d) + l$ ...(1)

Writing, the above series in reverse order, we get

$S_n = l + (l - d) + (l + 2d) + ... + (a_1 + d + a_1$ ...(2)

Adding (1) and (2), we get

$2S_n = (a_1 + l) + (a_1 + l) + ... + (a_1 + l)$, ($n$ times)

$= n(a_1 + l)$

Therefore $S_n = \frac{n}{2}(a_1 + l)$

$= \frac{n}{2}\{a_1 + [a_1 + (n-1)d]\}$

Consequently $S_n = \frac{n}{2}[2a_1 + (n+1)d]$.

Example 3:

Find the sum of $\frac{3}{4}, \frac{2}{3}, \frac{7}{12}, \cdots$ *up to 19 terms.*

Solution:

Here $a_1 = \frac{3}{4}$, $a_2 = \frac{2}{3}$, $n = 19$.

Thus $d = a_2 - a_1 = 2/3 - 3/4 = - 1/12$

i.e., $S_{10} = \frac{19}{2}\left[2 \times \frac{3}{4} + (19-1)\left(-\frac{1}{12}\right)\right]$

$= \frac{19}{2}\left[\frac{3}{2} - \frac{18}{12}\right]$

$= \frac{19}{2}\left(\frac{18-18}{12}\right) = \frac{19}{2} \times 0 = 0.$

Example 4:

How many terms of the following series may be taken so that their sum is 66

− 9, − 6, 3, . . . ?

Solution:

Let $S_n = 66$. Here $a_1 = - 9$, $d = - 6 + 9 = 3$.

or $66 = n/2\ [- 18 + (n - 13)$

$132 = - 18n + 3n^2 - 3n$

i.e. $3n^2 - 21n - 132 = 0$

or $n^2 - 7n - 44 = 0 \quad \Rightarrow \quad (n - 11)(n + 4) = 0$

$\Rightarrow \quad n = 11$ or $n = -4.$

As n is positive integer, second value -4 is rejected.

Thus required number of terms = 11.

ARITHMETIC MEANS

If $a_1, a_2, ..., a_n$ are in A.P., then quantities $a_1, a_3, \ldots, a_{n-1}$ are called *Arithmetic Means* (written as A.M.) between a_1 and a_n.

Thus in the series, 1, 3, 5, 7, 9, 11, 13, 15, . . .

3, 5, are arithmetic means between 1 and 7

9, 11, 13 are arithmetic means between 7 and 15.

TO INSERT *n* ARITHMETIC MEANS BETWEEN TWO GIVEN QUANTITIES

Let a and b be two given quantities and $A_1, A_2, \ldots, A_n$ be the n arithmetic means between them. Then the quntities

$a, A_1, A_2, \ldots, A_n$ b are in A.P.

Let d be their common difference.

Now $\quad$ b = $(n + 2)$th term

$= a + (n + 1)d$

$$\Rightarrow d = \frac{b-a}{n+1}.$$

Further, A_1 = 2nd term

$$= a + d = a + \left(\frac{b-a}{n+1}\right)$$

$$= \frac{na-b}{n+1}.$$

$$= a+2d = a+2\left(\frac{b-a}{n+1}\right)$$

$$= \frac{na-2b-a}{n+1}$$

... ...

$$A_n = a+nd = a+n\left(\frac{b-a}{n+1}\right)$$

$$=\frac{a+nb}{n+1}.$$

Hence $\quad \frac{na+b}{n+1}, \frac{na+2b-a}{n+1}, \ldots, \frac{a+nb}{n+1}.$

are n Arithmetic means between a and b.

Example 5:

Insert 6 Arithmetic means between 1 and 19.

Solution:

Let $\quad A_1, A_2, A_3, A_4, A_5, A_6$, be the required arithmetic means.

Then $\quad 1, A_1, A_2, A_3, A_4, A_5, A_6$. 19 are in A.P.

Let d be their common difference.

Then $\quad 19 = 8$th terms

$= 1 + (8 - 1)d$

$= 1 + 7d$

Thus $\quad d = 18/7$

Hence $\quad A_1 = 2$nd term

$$=1+\frac{18}{7}=\frac{25}{7}$$

$$A_2 = \text{3rd term } = 1+2\times\frac{18}{7} = 1+\frac{36}{7} = \frac{43}{7}$$

$$A_3 = \text{4th term } = 1+3\times\frac{18}{7} = 1+\frac{54}{7} = \frac{61}{7}$$

$$A_4 = \text{5th term } = 1+4\times\frac{18}{7} = 1+\frac{72}{7} = \frac{79}{7}$$

$$A_5 = \text{6th term } = 1+5\times\frac{18}{7} = 1+\frac{90}{7} = \frac{97}{7}$$

$$A_6 = \text{7th term } = 1+6\times\frac{18}{7} = 1+\frac{108}{7} = \frac{115}{7}$$

So, the requiired means are

$$\frac{25}{7}, \frac{43}{7}, \frac{61}{7}, \frac{79}{7}, \frac{97}{7}, \frac{115}{7}.$$

Example 6:

How many terms of the series 4, 2, 0 – 2, 4, – 6 , . . . may be taken to that their sum is 6 ?

Solution:

Let n be the required number of terms.

Here $a_1 = 4, d = 2 - 4 = -2$.

So $6 = S_n = n/2\ [8 + (n - 1)(-2)]$

i.e. $12 = 8n - 2n^2 + 2n$

or $2n^2 - 10n + 12 = 0$

or $n^2 - 5n + 6 = 0$

or $(n - 3)(n - 2) = 0$

Hence $n = 3$ or 2.

We get two answers. As the third term is zero, therefore, sum of the first two terms is ame as the sum of first three terms.

Example 7:

If pth, qth, rth term of an A.P. are a, b, c respectively, show that

$$(q - r)\ a + (r - p)\ b + (p - q)\ c = 0.$$

Solution:

Here pth term $= a = a_1 + (p - 1)\ d$ ...(*i*)

qth term $= b = a_1 + (q - 1)\ d$ ...(*ii*)

rth term $= c = a_1 + (r - 1)\ d$ ...(*iii*)

where a_1 is the first term and d is the common difference of the A.P.

Multiply (*i*) by $q - r$, (*ii*) $r - p$, (*iii*) by $p - q$ and add to obtain $(q-r)\ a + (r - p)\ b + (p - q)\ c$

$$= a_1 (q - r) + a_1 (r - p) + a_1 (p - q)$$
$$+ d\ [(q - 1)(q - r) + (q - 1)(r - p) + (r - 1)(p - q)]$$
$$= a_1[q - r + r - p + p - q.$$
$$+ d\ [pq + r - pr - q + qr + - r + p - pq$$
$$+ rp - rq - p + q] = 0.$$

Example 8:

The sum of n terms of two A.P.'s are in the ration of $7_n + 1 : 4_n + 27$. *Find the ratio of their 11th terms.*

Solution:

Let a_1 and b_1 be the first terms of two A.P.'s and d_1, d_2 be their common difference respectively.

Then $S_n = \frac{n}{2}[2a_1 + (n-1)\,d_1]$

$S'_n = \frac{n}{2}[2b_1 + (n-1)\,d_2]$

So $\frac{S_n}{S'_n} = \frac{2a_1 + (n-1)\,d_1}{2b_1 + (n-1)\,d_2} = \frac{7n+1}{4n+27}$

Putting $n = 21$, we get

$$\frac{2a_1 + 20d_1}{2b_1 + 20d_2} = \frac{148}{111}$$

or $\frac{a_1 + 10d_1}{b_1 + 10d_2} = \frac{148}{111}$

or $\frac{a_{11}}{b_{11}} = \frac{148}{111}$

where a_{11} and b_{11} are 11th terms of two A.P.'s respectively.

$$= \frac{4}{3}.$$

Example 9:

If a^2, b^2, c^2 are in A.P., show that

$\frac{1}{b+c}, \frac{1}{c+a}, \frac{1}{a+b}$ *are also in A.P.*

Solution:

$\frac{1}{b+c}, \frac{1}{c+a}, \frac{1}{a+b}$ are in AP.

$\Leftrightarrow \quad \frac{1}{c+a} - \frac{1}{b+c} = \frac{1}{a+b} - \frac{1}{c+a}$

$\Leftrightarrow \quad \frac{b-a}{(c+a)(b+c)} = \frac{c-b}{(a+b)(c+a)}$

$\Leftrightarrow \quad \frac{b-a}{c+b} = \frac{c-b}{b+a}$

$\Leftrightarrow \quad b^2 - a^2 = c^2 - b^2$

$\Leftrightarrow \quad a^2, b^2, c^2$ are in A.P.

Hence the result follows.

Example 10:

Sum to n terms of three A.P.'s are s_1, s_2 and s_3. Te first term of eaxh of tem is 1 and common differences are 1, 2 and 3 rspectively. Show that s_1, s_2, s_3 are also in A.P.

Solution:

$$S_1 = \frac{n}{2}[2+(n+1)] = \frac{n(n+1)}{2}$$

$$S_2 = \frac{n}{2}[2+(n-1)2] = n^2$$

$$S_3 = \frac{n}{2}[2+(n-1)3] = \frac{n}{2}(3n-1)$$

or $$S_2 - S_1 = n^2 - \frac{n(n+1)}{2} = \frac{n^2+n}{2} = \frac{n(n-1)}{2}$$

$$S_3 - S_2 = \frac{n}{2}(3n-1) - n^2 = \frac{3n^2-n-2n^2}{2}$$

$$= \frac{n^2-n}{2} = \frac{n(n-1)}{2}$$

i.e. $s_2 - s_1 = s_3 - s_2$

Hence s_1, s_2, s_3 are in A.P.

Example 11:

State giving example whether the following statement is true or false.

In a given A.P. let a be the first term. d the common difference, n the number of terms and s their sum. Given any three of a, b, n and s one can always find a unique value of the fourth quantity.

Solution:

The statement is false. Given values of a, d, and s their might be two values of n. Consider the series

12, 9, 6, 3, 0, –3, –6,

It is fo find the number of terms from beginning whose sum is 30.

In this case, $a = 12$, $d = 9 - 12 = -3$, $s = 3$

Since $$s = \frac{n}{2}[2a+(n-1)d]$$

we have $$30 = \frac{n}{2}[24+(n-1)(-3)]$$

or $\qquad 60 = 24n - 3n^2 + 3n$

or $\qquad 3n^2 - 27n + 60 = 0$

$\Leftrightarrow \qquad n^2 - 9n + 20 = 0$

$\Leftrightarrow \qquad (n - 4)(n - 5) = 0$

$\Leftrightarrow \qquad (n = 4)$ or 5.

Example 12:

Find the sum of all numbers between 200 and 400 which are divisible by 7.

Solution:

Since 4 is left as remainder on division of 200 by 7, the least number greater than 200 divisible by 7 is 203. Again if we divide 400 by 7, 1 is left as remainder. This implies that the greatest number less than 400, which is divisible by 7 is 399.

So we are to find the sum of the series

$203 + 210 + 217 + \ldots + 399$

Here $\quad a_1 = 203, \quad d = 7, \quad l = 399$

Let n be the total number of terms in this series.

Then $\quad 399 = 203 + (n - 1)\,7$

$\Rightarrow 7n = 399 - 203 + 7$

$= 406 - 203 = 203$

$\Rightarrow n = 29.$

Hence required sum $\quad \frac{n}{2}(a+l)$

$= \frac{29}{2}(203+399)$

$= \frac{29}{2}(602)$

$= (29)\,(301)$

$= 8729.$

Example 13:

Mr. X arranges to pay off a debt of Rs 9,600 in 48 annual instalments which form an arithmetical sereis. When 40 of these instalments are paid, Mr. X becomes insolvent and his creditor finds that Rs 2,400 still remains unpaid. find the value of each of the first three instalemetns of Mr. S. Ignore interest.

Solution:

Let a, $a + d$, $a + 2d$, $a + 3d$ be the annual instalments.

The sum of this up to n=48 terms is 9600

i. e., $$9600 = \frac{48}{2}[2a+(48-1)d]$$

$\Rightarrow$ $9600 = 24\ (2a + 47d)$

$\Rightarrow$ $2a + 47d = 400$ ...(1)

After 40 instalments are paid, the balance id Rs 2400. In other words in 40 instalments Mt X has paid Rs (9600 – 2400) i. e. Rs 7200.

So the sum of first 40 terms of the abov series is 7200

Thus $$7200 = \frac{40}{2}[2a+40-1)d]$$

$\Rightarrow$ $7200 = 20\ (2a + 39d)$

$\Rightarrow$ $2a + 39d = 360$ ...(2)

Subtractin (2) from (1), we get

$8d = 40 \quad \Rightarrow \quad d = 5$

Then (2) $\Rightarrow$ $2a = 195 = 360$

$\Rightarrow 2a = 165 \quad \Rightarrow \quad a = 82.50$

Hence first instalment of Mr X is Rs 82050, second instalment is Rs (82.50 = 5.00) i. e., Rs 87050 and third instalment is Rs (87.50 = 5.00) i. e. Rs 92.50.

Example 14:

If $\frac{1}{b+c}, \frac{1}{c+a}, \frac{1}{a+b}$ *are ub A.P;*

Prove that a^2, b^2, c^2 *are also in A.P.*

(*C. A., November, 1975*)

Solution:

Simce $\frac{1}{b+c}, \frac{1}{c+a}, \frac{1}{a+b}$ are in A.P.

we have $$\frac{1}{c+a} - \frac{1}{b+c} = \frac{1}{a+b} - \frac{1}{c+a}$$

$$\Rightarrow \frac{(b+c)-(c+a)}{(b+c)(c+a)} = \frac{(c+a)-(a+b)}{(a+b)(c+a)}$$

$$\Rightarrow \frac{b-a}{b+c} = \frac{c-b}{b+a}$$

$\Rightarrow b^2 - a^2 = c^2 - b^2$

$\Rightarrow b^2, b^2, c^2$ are in A.P.

REMARK, This questionwas also solved implicitly in Example 9.

***Example 15*:**

Thc monthly salary of a persom was Rs 320 for each of the first three years. He next got annual increments of Rs 40 per month ofr each of the following successive 12 years. His salary remained stationary till retirement when he found that his average monthiy salary during the service period was rs 698. Find the period of his service.

Solution:

Let n be the total number of year of the person's service.

His total salary = $112n$. 698 Rs

(As his monthly average is Rs (98)

Total salary in first three years of service

= 320 × 3 × 12 = 960 × 12 Rs

In the 4th year, his monthly salary was Rs (320 + 40) = Rs 260 5th year his monthly salary was Rs 400 and so on.

Then for next 12 years, his total salary

= 12 × [360 + 400 + . . . up to 12 terms] Rs

= 12 × 12/2 [2 × 360 + (12 – 1). 40] Rs

= 12 × 6 (720 + 440) Rs

= 12 × 6 × 1160 Rs

= 12 × 6960 Rs

At the end of following 12 years, his monthly salary was

Rs [360 + (12 – 1).40] = *i.e.*, Rs 800

This salary he got for the remaining $(n - 15)$ years. So his total salary for the remaining $(n - 15)$ years was $(n - 15)$ 800 × 12.

Hence his total salary throughout his service period

= 12 [960 + 6960 + 800 $(n - 15)$]

$= 12\ (7920 + 800n - 12000)$

$= 12\ (800n - 4080)$

This must be same as $12n \times 698$

i.e. $12n \times 698 = 12\ (800n - 4080)$

$\Rightarrow \quad 102n = 4080 \Rightarrow n = 40$ years.

Example 16:

The sequence of natural numbers is written as

		1		
	2	3	4	
5	6	7	8	9
...	...	...	...	...
...	...	...	...	...

Find the sum of the numbers in the rth row. *(C.A. May, 1975)*

Solution:

Let $S_1 = 1,\ S_2 = 2 + 3 + 4, \qquad S_3 = 5 + 6 + 7 + 8 + 9$

Let intial term of S_k be t_k

and suppose that $M = 1 + 2 + 5 + \ldots + t_k$

the sum of first terms of S_1, S_2, and S_n

Now $\quad M = 1 + 2 + 5 + 10 + \ldots + t_k,$

Also $\quad M = 1 + 2 + 5 + \ldots + t_{k-2} + t_k$

Subtracting we get

$0 = (1 + 1 + 3 + 5 + \text{up to k term}) - t_k$

$t_k = 1 + [1 + 3 + 5 + \ldots \text{up to (k} - 1) \text{ terms}]$

$$= 1 + \left(\frac{k-1}{2}\right)[2 + (c-2).2]$$

$= 1 + (k - 1)^2 = k^2 - 2k + 2$

Also in S_1 there is one term, in S_2 there are 3 terms and so on, in S_k there will be $(2k - 1)$ term.

Hence we are to find the sum of series

$r^2 - 2r + 2,\ r^2 - 2r + 3,\ r^2 - 2r + 4,$ up to $(2r - r)$ terms

So $\qquad S_r = \dfrac{(2r-1)}{2}[2(r^2 - 2r + 2) - (2r - 2).1]$

$= (2r - 1)(r^2 - 2r + 2 + r - 1)$

$= (2r - 1)(r^2 - r + 1)$

$= 2r^3 - 3r^2 + 2r - 1.$

Example 17:

A lamp lighter has to light 100 gas lamps. He takes l 1/2 minutes to go from one lamp post to the next. Each lamp post burns 10 c.c. of gas per hour. How many c.c. of gas has been burnt by 8.30 P.M. if he lights the first lamp at 6 P.M.

Solution:

Total time from 6 P.M. to 8.30 P.M. is 2.30 hours *i.e,* 150 minutes

g_1 = Gas burnt in 1st lamp $= \dfrac{150 \times 10}{60}$

$= \dfrac{1}{6}(150)$ *c.c*

Second lamp burns 1st (150 – 1.5) minutes

So the gas burnt in 2nd lamp

$= g_2 = \dfrac{1}{6}(150 - 1.5)$

Similarly $g_3 = \dfrac{1}{6}(150 - 2 \times 1.5)$ and so on.

We are to calculate

$S = g_1 + g_2 \quad + \ldots \quad + g_{100}$

$S = \dfrac{1}{6}[150 + (150 - 1.5) + (150 - 2 \times 1.5) + ..]$

The series in side brackedt is an A.P. with first term = 150, common difference = – 1.5 and number of terms = 100

Hence $S = \dfrac{1}{6}\ \dfrac{100}{2}[300 + 99.(-1.5)]$

$= \dfrac{25}{3}(300 - 1.5 \times 99)$

$= 25\ (100 - 1.5 \times 33)$

$= 25\ (100 - 49.5)$

$= 25\ (50.5)$

$= 1262.255$ cc.

Example 18:

Two posts were offered to a man. In one the starting salary was Rs 120 per month and the annual increment was Rs 8; in the other post the salary commenced at Rs 85 per month but the annual increment was Rs 12. The man decided to accept the post which would give him more earnings in the first twenty years of the service. Which post was acceptable to him ? Justify your answer.

Solution:

The totoal earnings of the man in first job

$$= \frac{20}{2}[2\times120+(20-1)8]\times12$$

$= 10\ (240 + 152) \times 12$

$= 120\ (392)$

$= 47040$ Rupees

His total earnings in second job

$$= \frac{20}{2}[2\times85+(20-1)2]\times12$$

$= 120\ (170 + 228)$

$= 120\ (398)$

$= 47760$ Rupees

which is greater than Rs 47040, Hence the second job was accepted by the man.

Example 19:

Find the sum of all natural numbers between 500 and 1000 which are divisible by 13.

Solution:

500 leaves, o division by 13, 6 as remainder so 507 is the least number greater than 500 which is divisible by 13.

Also the remainder, when 1000 is divided by 13, is 12. So the greatest number less than 1000, divisible by 13 is 988.

We are to find the sum 507 + 520 + . . . + 988.

Let n be the number of terms in the series

Then $988 = 507 + (n - 1)\ 13$

$\Rightarrow \quad 76 = 39 + n - 1 = n + 38$

$\Rightarrow \quad n = 76 - 38 = 38.$

Requred sum. $S = \frac{n}{2}[2a + (n-1)d]$

Hence $n = 38, \quad a = 507, \quad d = 13$

Hence $S = \frac{38}{2}[2 \times 507 + (38-1)13]$

$= \frac{38}{2}(1014 + 37 \times 13)$

$= 19\ (1040 + 481)$

$= 19\ (1495)$

$= 28405.$

Example 20:

A firm produced 1000 sets of TV during its first year. The total sum of the firm's production at the end of 10 year's operation is 14,500 sets.

(*i*) *Estimate by how many units, production increased eaxh year, if the increase each year is uniform, and*

(*ii*) *Forecast, based on the estimate of the annual increment in production, the level of output for the 15th year.*

Solution:

Here $a = 1000, \quad S = 14{,}500, \quad n = 10$ and d is to be evaluated.

$14{,}500 = \frac{10}{2}[2 \times 1000 + (10-1)d]$

$14500 = 5\ (2000 + 9d)$

$\Rightarrow 2000 + 9d = 2900$

$\Rightarrow 9d = 900 \quad \Rightarrow \quad d = 100$

Hence 1000 units is increase per annum

$a + 14d = 1000 + 14 \times 100$

$= 1000 + 1400 = 2400.$

EXERCISE

1. Find the 7th and $(a - 4)$th terms of each of the following series.

(*i*) 3, 5, 7, ... , . . . (*ii*) 1, 4, 7, 10, . . . , . . .

(*iii*) $\frac{11}{5}, 1, \frac{-1}{5}, \frac{-7}{5}$ . . . , . . . (*iv*) 5, 11, 17, . . . , . . .

(v) 8, 3, –2, –7, . . . , . . .

2. The third and thirteenth term of an A.P. are respecitvely equal to – 40 and 0. Find the A.P. and its 20th term.

3. Is 203 any term of A.P., 13, 28, 43, 58, 73, . . . ?

4. Find the sum of the following series :

(i) 49, 44, 39, . . . up to 17 terms.

(ii) 3.75, 3.5, 3.25, . . . to 16 terms.

(iii) $2a - b$, $4a - 3a - 3b$, . . . to n terms.

(iv) 5 1/2, 6 3/4, 8, . . . to 28 terms.

(v) $\frac{3}{\sqrt{5}}, \frac{4}{\sqrt{5}}, \sqrt{5}, \ldots$ to 25 terms.

5. Insert 9 Arithmetic Means between 1/4 and –39/4.

6. Insert 10 Arithmetic Means between 2 and 24.

7. The 3rd term of an A.P. is 18 and 7th term is 30. Find the sum of 17 terms.

8. The sum of four integers in A.P. is 24 and their product is 945. Find them.

[**Hint.** Select the numbers to be $a - 3d$, $a - d$, $a + d$ and $a + 3d$.]

9. Sum of three numbers in A.P. is 15 and sum of the squares of its first and third term is 58. Find the numbers.

10. If x, y, z are in A.P., show that

(i) $\frac{1}{xz}, \frac{1}{zx}, \frac{1}{xy}$ are in A.P.

(ii) $x^2(y + z)$, $y^2(z + x)$, $z^2(x + y)$ are in A.P.

(iii) $x\left(\frac{1}{y}+\frac{1}{z}\right), y\left(\frac{1}{z}+\frac{1}{x}\right), z\left(\frac{1}{x}+\frac{1}{y}\right)$ are in A.P.

11. There are n Arithmetic Means between 3 and 54. If 8th Arithmetic Mean bears a ratio 3 : 5 to $(n - 2)$th Arithmetic Mean, find n.

12. The interior angles of a polygon of n sides are in A.P., the smallest angle is of 42° and the common differnece is 33°. Find n.

13. The sum of n terms of the series 2, 5, 8, . . . , is 950; find n.

14. Between two numbers whose sum is 2 1/8, an even number of arithmetic means is inserted; the sum of these exceeds their number by unity; how many means are there ?

15. The sum of n terms of an A.P. is $2n + 3n^2$; fubd tge rth term.

16. The number of terms in an A.P. is even; the sum of the odd terms is 24, of the even terms 30, and the last term exceeds the first by 10 1/2; find the number of terms.

17. Find the relation between x and y in order that the rth mean between x and $2y$, may be the same as the rth mean between $2x$ and y, n means being inseted in eaxh case.

18. Prove that if unity is added to the sum of any number, terms of the A.P. 3, 5, 7, 9, . . . , the resulting sum is a perfect square.

19. If pth term of an A.P. is $1/q$ and qth term is $1/p$, show that the sum of pq terms is $1/2\ (pq + 1)$

20. Find sum of all natural numbers from 100 to 300 which are divisible by 4.

21. If p, q, r, s are any four consecutive terms of an A.P. whow that $p^2 - 3q^2 + 3r^2 - s^2 = 0$

    [**Hint.** Let $p = x - 3\beta$, $q = x - \beta$, $r = x + \beta$, $s = x + 3\beta$]

22. A piece of equipment costs a certain factory Rs 600,000. If it depreciated in value, 15 percent the first year, 13 1/2 percent the next year, 12 percent the third year and so on, what will be its value at the end of 10 year, all percentages applying to the original cost ?

23. The rate of monthly salary of a person is incresed anuually in A.P. It is know that he was drwaing Rs 400 a month during the 11th year of his service, and Rs 760 during the 29th year. Find his starting salary and the rate of annual (increment. What should be his salary at the time of retirement just on the completion of 36 years of service.

24. Mr. X takes a loan of Rs 2000 from Mr Y and agrees to repay in number of instalments, each instalment (beginning with the second)exceeding the previous one ny Rs 10. If the first instalment is Rs 5 find how many instalments will be necessary to wipe out the loan completely ?

25. A moneylender lends Rs 1000 and charges an overall interst of Rs 140. He recovers the loan and interst by 12 monthly instalments each less by Rs 10 than the preceeding one. Find the amount of the first instalment.

26. To verify cash balances, the auditor of a certain bank; employes his assistant to count cash in hand of Rs 4,500. At first he counts quietly at the rate of Rs. 150 per minute for 10 minuters only but at the end of that time he begins to count at the rate of Rs 2 less every minute

than he could count in the previous minute. Ascertain how much time he will take to count this sum of Rs 4,500.

27. Eighty coins are placed in a line on the ground. The distance between any tow consecutive coins is 10 meters. How far must a person travel to bring them, one by on, to a basket placed 10 miters behind the first coin.

28. A man saved Rs 16,500 in 10 years. In each year after the first he saved Rs 100 more than he did in the preceeding year. How much did he save in the first year ?

29. An emterprise produced 690 units in the 3rd year of its existance and 700 units in its 7th year, what was the initial production in the first year ?

30. A man secures an interest free loan of Rs 14500 from a friend and agrees to repay it in ten intalements. He pays Rs 1000 as first instalment and then increases each instalment by eaqual amount over the preceding instalment. What will be his last instalment ?

ANSWERS

1. (i) 15, 27, $2n - 7$, (ii) 19, 37, $3n - 14$

   (iii) -5, $-12\frac{1}{5}, 8\frac{1}{5} - \frac{6n}{5}$, (iv) 41, 77, $6n - 25$

   (v) $-22, -52, 33 - 5n$.

2. $-48, -44, -40, \ldots ; 28$.

3. No.

4. (i) 153; (ii) 30

   (iii) $(n + 1)a - n^2b$; (iv) 626 1/2

   (v) $75\sqrt{5}$.

5. $-\frac{3}{4}, -\frac{7}{4}, -\frac{11}{4}, \frac{-15}{4}, \frac{-19}{4}, \frac{-23}{4}, \frac{-27}{4}, \frac{-31}{4}, \frac{-35}{4}$.

6. 4, 6, 8, 10, 12, 14, 16, 18, 20, 22. 7. 612.

8. 3, 5, 7, 9. 9. 3, 5, 7.

11. 16. 12. 5.

13. 25. 14. 12.

15. $6r - 1$. 16. 8.

17. $ry - (n + 1 - r)x$. 20. 10200.

22. Rs 105000.
23. Starting salary Rs 200; annal increment Rs 20 and salary at the time of retirement, Rs 900.
24. 20.
25. Rs 150.
26. 34 minutes.
27. 64800 meters.
28. Rs 1200.
29. Rs 550.
30. Rs 1900

GEOMETRICAL PROGRESSION

Non–zero quantities $a_1, a_2, a_3, ..., a_n, ...,$ each term of which is equal to the product of preceding term and a constant number are called to form a *Geometrical Progression* (written as G.P.).

Thus all the following quantities are in G.P.

(*i*) 1, 2, 4, 8, 16, . . .

(*ii*) $3, -1, \frac{1}{3}, \frac{-1}{9}, \frac{1}{27}, ...,$

(*iii*) $1, \sqrt{2}, 2, 2\sqrt{2}, . . . ,$ where $a \neq 0, b \neq 0.$

(*iv*) $1, \frac{1}{5}, \frac{1}{25}, \frac{1}{125}, ...$

The constant number is termed as the *common ratio* of the G.P.

TO FIND THE *n*TH TERM OF A G.P

Let first term be a and r, the common ration. By definition the G.P. is $a, ar, ar^2, . . .$

1st term $a = ar^0 = ar^{1-1}$ 2nd term $ar = ar^1 = ar^{2-1}$

... ...

In general, nth term $= ar^{n-1}$.

In examples above, we compute 5th, 7th, 3rd, 11th and 8th term of (*i*), (*ii*), (*iii*), (*iv*) and (*v*) respectively.

In (*i*) 1st term is 1 and common ration = 2.

Hence 5th term $= ar^4 = 1.2^4 = 16.$

In (*ii*) $a = 3, r = \frac{-1}{3}$, hence 7th term $= ar^6 = 3\left(\frac{-1}{3}\right)^6 = \frac{1}{243}.$

In (*iii*) $a = 1, r = \sqrt{2}$, hence 3rd term $= ar^2 = 2.$

In (*iv*) 1st term = a, $r = \frac{1}{b}$, hence 11th term $= ar^{10}\, \frac{a}{b^{10}}$.

In (*v*) $a = 1$, $r = \frac{1}{5}$, hence 8th term $= ar^{2}\, \frac{1}{5^{7}} = \frac{1}{78125}$.

TO FIND THE SUM OF FIRST *n* OF A G.P

Let $a, ar, ar^2, \ldots$ be a given G.P. and let S_n be the sum of its first n terms.

Then $S_n = a + ar + ar^2 + \ldots + ar^{n-1}$

This gives that $r\,S_n = ar + ar^2 + \ldots + ar^{n-1} + ar^n$

Subtracting, we get $S_n - r\,S_n = a - ar^n = a\,(1 - r^n)$

In case $r \neq 1$, $S_n = \frac{a(1-r^n)}{(1-r)}$

In case $r = 1$, $S_n = a + a + a + \ldots + a$ (n times)

Thus, sum of n terms of a G.P. is $\frac{a(1-r^n)}{1-r}$ provided $r \neq 1$.

In case $r = 1$, sum of G.P. is na.

Example 21:

Find the sum of first 14 terms of a G.P.

3, 9, 27, 81, 243, 729, . . .

Solution:

In this case $a = 3$, $r = 3$, $n = 14$.

So $S_n = \frac{a(1-r^n)}{1-r} = \frac{3(1-3^{14})}{1-3} = \frac{3}{2}(3^{14} - 1)$.

Example 22:

Find the sum of first 11 term of a G.P. given by $1, -\frac{1}{2}, \frac{1}{4}, -\frac{1}{8}, \ldots, \ldots$

Solution:

Here $a = 1$, $r = -1/2$, $n = 11$.

So $$S_n = \frac{a(1-r^n)}{1-r} = \frac{1\left[1-\left(-\frac{1}{2}\right)^{11}\right.}{1+\frac{1}{2}}$$

$$=\frac{2^{11}+1}{3\times 2^{10}}-\frac{683}{1024}.$$

TO FIND THE SUM TO INFINTY OF A GEOMETRIC PROGRESSION WHOSE COMMON RATIO IS LESS THAN 1

Let a, ar, ar^3, . . . be a G.P. with $r < 1$.

Now $r < 1 \Rightarrow r^2 < r,\ r^3 < r^2, \ldots$

Thus as power of r goes on increasing, the corresponding term in G.P. decreases in value. So, we can assume that as n becomes indefinitely large, r^n becomes indefinitely small i.e., $r^n \to 0$.

Now $\quad S_n=\frac{a(1-r^n)}{1-r}=\frac{a}{1-r}-\frac{ar}{1-r}.$

So, $\quad$ as $n \to \infty$, $S\infty$, $\quad =\frac{a}{1-r}.$

Example 23:

Find the sum of the following series up to infinity

$$1+\frac{3}{7}+\frac{9}{49}+\frac{27}{343}+\frac{81}{2401}+\ldots,\ldots$$

Solution:

Here $\quad a = 1,\quad r=\frac{3}{7}<1.$

So, $\quad S\infty=\frac{a}{1-r}=\frac{1}{1-\frac{3}{7}}=\frac{7}{4}.$

Example 24:

Evaluate the recurring decimal 17.

Solution:

Now $\quad .17 = .1 + .07 + .007 + .0007 + \ldots$

$$=\frac{1}{10}+\frac{7}{10^2}+\frac{7}{10^3}+\ldots$$

$$=\frac{1}{10}+\frac{7}{10^2}\left\{1+\frac{1}{10}+\frac{1}{10^2}+\ldots+\ldots\right\}$$

$$=\frac{1}{10}+\frac{7}{10^2}\frac{1}{1-\frac{1}{10}}$$

$$= \frac{1}{10} + \frac{7}{100} \frac{10}{9}$$

$$= \frac{1}{10} + \frac{7}{90} = \frac{16}{90} = \frac{8}{45}.$$

GEOMETRIC MEAN

If α, β, γ are in G.P. then β is called a *Geometric Mean* between α and γ (written as G.M.).

If $a_1, a_2, \ldots, a_n$ are in G.P., then $a_2, \ldots, a_{n-1}$, are called *Geometric Means* between a_1 and a_n.

Thus 3, 9, 27 are three geomtric means between 1 and 81.

TO FIND *N* GEOMERIC MEANS BETWEEN TWO GIVEN NUMBERS *a* AND *b*

Let $G_1, G_3, \ldots G_n$, be n Geometric Means between a and b.

Thus $a, G_1, G_2, \ldots G_n, b$ is a G.P., n being $(n + 2)$th term $= ar^{n+1}$ where r is the common ratio of G.P.

Thus $b = ar^{n+1} \quad \Rightarrow r = \left(\frac{b}{a}\right)^{\frac{1}{n+1}}$

So, $G_1 = ar = a = \left(\frac{b}{a}\right)^{\frac{1}{n+1}} = \left(a^n b\right)^{\frac{1}{n+1}}$

$$G_2 = ar^2 = a\left(\frac{b}{a}\right)^{\frac{2}{n+1}} = \left(a^{n-1}b^2\right)^{\frac{1}{n+1}}$$

... ...

$$G_n = ar^{n-1} = a\left(\frac{b}{a}\right)^{\frac{n-1}{n+1}} = \left(a^2b^{n-1}\right)^{\frac{1}{n+1}}$$

Example 25:

Find 7 G.M. s. between 1 and 256.

Solution:

Let $G_1, G_3, \ldots G_7$, be 7 G.M's between 1 and 256.

Then 256 = 9th term of G.P.,

$= 1.r^8$ where r is the common ratio of the G.P.

This gives that $r^8 = 256 \Rightarrow r = 2$.

$G_1 = ar = 1.2 = 2$

$G_2 = ar^2 = 1.4 = 4$

$G_3 = ar^3 = 1.8 = 8$

$G_4 = ar^1 = 1.16 = 16$

$G_5 = ar^5 = 1.32 = 32$

$G_6 = ar^6 = 1.64 = 64$

$G_7 = ar^7 = 1.128 = 128$

Hence required G.M.'s are 2, 4, 8, 16, 32, 64, 128.

Example 26:

Sum the series $1 + 3x + 5x^2 + 7x^2 + \ldots$ *up to term* $x \neq 1$.

Solution:

Note that nth term of this series $= (2n - 1)\, x^{n-1}$.

Let $\quad S_n + 1 + 3x + 5x^2 + \ldots + (2n - 1)\, x^{n-1}$.

Then $\quad xS_n = x + 3x^2 + \ldots + (2n - 3)\, x^{n-1} + (2n - 1)\, x^n$.

Subtracting, we get

$$S_n(1 - x) = 1 + 2x + 2x^2 + \ldots + 2x^{n-1} - (2n - 1)\, x^n$$

$$= 1 + 2x.\frac{1 - x^{n-1}}{1 - x} - (2n - 1)x^n$$

$$= \frac{1 - x + 2x - 2x^n - (2n - 1)x^n(1 - x)}{1 - x}$$

$$= \frac{1 + x - 2x^n - (2n - 1)x^n + (2n - 1)x^{n+1}}{1 - x}$$

$$= \frac{1 + x - (2n + 1)x^n + (2n - 1)x^{n+1}}{1 - x}$$

Hence $\quad S = \dfrac{1 + x - (2n + 1)x^n + (2n - 1)x^{n+1}}{(1 - x)^2}$

Example 27:

If in a G.P. $(p + q)$*th term* $= m$ *and* $(p - q)$*th term* $= n$, *them find its pth and qth terms.*

Solution:

Suppose that the given G.P. be $a, ar, ar^2, ar^3 \ldots$

By hypothesis, $(p + q)$th term $= m\ ar^{p+q-1}$

$(p + q)$th term $= n = ar^{p-q-1}$.

Then $\frac{m}{n} = r^{2q} \Rightarrow r = \left(\frac{m}{n}\right)^{1/2q}$

Hence $m = a\left(\frac{m}{n}\right)^{(p+q-1)/2q'} \Rightarrow a = m^{(q-p+1)/2q} n^{(p+q-1)/2q}$.

Thus pth term $ar^{p-1} = m\frac{1}{2}\ n\frac{1}{2}$

$= \sqrt{mn}$

qth term $= ar^{q-1} = m\frac{2q-p}{2q}\ n\frac{p}{2q}$

Example 28:

Sum the series 5 + 55 + 555 + . . . up to n terms.

Solution:

Let $S_n = 5 + 55 + 555 + \ldots.$

$S_n = 5\ (1 + 11 + 111 + \ldots.)$

$= \frac{5}{9}(9+99+999+\ldots)$

$= \frac{5}{9}[(10-1)+(100-1)+(1000-1)+\ldots]$

$= \frac{5}{9}[(10+10^2+10^3+\ldots+10^n)-(1+1+\ldots n \text{ terms})]$

$= \frac{5}{9}[(10+10^2+10^3+\ldots+10^n)-n]$

$= \frac{5}{9}\left[\frac{10(1-10^n)}{1-10}-n\right]$

$= \frac{5}{9}\left[\frac{10(10^n-1)}{9}-n\right]$

$= \frac{50}{81}(10^n-1)-\frac{5n}{9}.$

Example 29:

If a, b, c, d are in G.P. prove that $a^2 - b^2$, $b^2 - c^2$ and $c^2 - d^2$ are also in G.P.

Solution:

Since $\frac{b}{a}=\frac{c}{b}=\frac{d}{c}=k(\text{say})$

we have $b = ak$, $c = bk$, $d = ck$

i.e. $b = ak$, $c = ak^2$, $d = ak^3$.

Now $(b^2 - c^2)^2 = (a^2k^2 - a^2k^4)^2$

$= a^4k^4 (1 - k^2)^2$

Hence $(b^2 - c^2) - \rightarrow = (a^2 - b^2)(c^2 - d^2)$.

This gives that $a^2 - b^2$, $b^2 - c^2$, $c^2 - d^2$ are in G.P.

Example 30:

There numbers are in G.P. Their product is 64 and sum is $\frac{124}{5}$. *Find them.*

Solution:

Let the numbers be $\frac{a}{r}$ a, ar.

Since $=\frac{a}{r}+a+ar\frac{124}{5}$ and $\frac{a}{r}.a\ ar=64$

we have $a^3 = 64 \Rightarrow a = 4$.

This gives that $\frac{4}{r}+4+4r=\frac{124}{5}$

$\Rightarrow \frac{1}{r}+1+r=\frac{31}{5}$

$\Rightarrow \frac{r^2+1}{r}=\frac{26}{5}$

$\Rightarrow 5r^2 + 5 = 26r$

$\Rightarrow r=\frac{1}{5}$ or 5

In either case, the numbers are 4/5, 4, and 20.

Example 31:

If a, b, c are in G.P. and $a^x = b^y = c^z$, *prove that*

$\frac{1}{x}+\frac{1}{z}=\frac{2}{y}$.

Solution:

a, b, c are in G.P., $b^2 = ac$

But $b^y = a^x \Rightarrow a = b^{y/x}$

and $b^y = c^z \Rightarrow c = b^{y/z}$

So we get $b^2 = b^{y/x}.\ b^{y/z}$

$$= b^{y\left(\frac{1}{x}+\frac{1}{z}\right)}$$

$$\Rightarrow 2 = y\left(\frac{1}{x}+\frac{1}{z}\right)$$

$$\Rightarrow \frac{1}{x}+\frac{1}{z}=\frac{2}{y}.$$

Example 32:

Sum to n terms the series

.7 + .77 + .777 + . . .

Solution:

Given series

$= .7 + .77 + 777 + \ldots$ up to n terms

$= 7\,(.1 + .11 + .111 + \ldots$,, ,, ,, $)$

$$=\frac{7}{9}(9+.99+.999+\ldots ,, ,, ,,)$$

$$=\frac{7}{9}\left[\left(1-\frac{1}{10}\right)+\left(1-\frac{1}{10^2}+\right)+\left(1-\frac{1}{10^3}\right)+\ldots ,, ,,\right]$$

$$=\frac{7}{9}\left[n-\left(\frac{1}{10}+\frac{1}{10^2}+\ldots \text{ up to } n \text{ terms}\right)\right]$$

$$=\frac{7}{9}\left[n-\frac{1/10(1-1/10^n}{1-1/10}\right]$$

$$=\frac{7}{9}\left[n-\frac{1}{9}\left(1-\frac{1}{10^n}\right)\right]$$

$$=\frac{7}{9}\left[n-\frac{1}{9}\left(1-\frac{1}{10^n}\right)\right].$$

Example 33:

The sum of three numbers in G.P. is 35 and their product is 1000. Find the numbers.

Solution:

Let the numbers be, $\frac{\alpha}{r}, \alpha, \alpha r$

Their product $\alpha^3 = 1000$

$\Rightarrow \alpha = 10$

So the numbers are $\frac{10}{r}, 10, 10r$

The sum of these numbers = 35

$\Rightarrow \frac{10}{r} + 10 + 10r = 35$

$\Rightarrow \frac{2}{r} + 2r = 5$

$\Rightarrow 2r^2 - 5r + 2 = 0.$

$\Rightarrow (2r - 1)(r - 2) = 0.$

$\Rightarrow r = 2 \quad \text{or} \quad \frac{1}{2}$

$r = 2$ gives the numbers as 5, 10, 20

$r = \frac{1}{2}$ gives the numbers as 20, 10, 5, the same as the first set.

Hence the required numbers are 5, 10 and 20.

Example 34:

The sum of the first eight terms of a G.P. (of real terms) is five time the sum of the first four terms. Find the common ration.

Solution:

Let the G.P. be $a, ar, ar^2, \ldots$

S_8 = Sum of first eight terms $= \frac{a(1-r^8)}{1-r}$

S_4 = Sum of first four terms $= \frac{a(1-r^4)}{1-r}$

By hypothesis $S_8 = 5s_4 \Rightarrow \frac{a(1-r^8)}{1-r} = \frac{5a(1-r^4)}{1-r}$

$\Rightarrow 1 - r^3 = 5(1 - r^4)$

$\Rightarrow (1 - r^4)(1 - r^1) = 5(1 - r^4)$

In case $r^4 - 1 = 0$ we get $(r^2 - 1) = 0 \Rightarrow r = \pm 1$

(note that $r^2 + 1 = 0 \Rightarrow r$ is imaginary)

Now $r = 1 \Rightarrow$ the given series in $a + a + a + \ldots$

but then $S_8 = 8a$ and $S_4 = 4a$. So $S_8 \neq 4S_4$

In case $r = -1$, we get $S_8 = 0$ and $S_4 = 0$ hence the hypothesis is satisfied.

Suppose now $r^4 - 1 \neq 0$ then $1 + r^4 = 5$

$\Rightarrow r^4 = 4 \Rightarrow r^2 = 2$ $(r^2 \neq -2)$

Hence $r = -1$ or $\pm\sqrt{2}$.

Example 35:

If S is the sum, P the product and R the sum of reciprocls of n terms in G. P. prove that

$F^2 R^n = S^n$.

Solution:

Let $a, ar, ar^2, \ldots$ be the given G.P.

Then $S = a + ar + ar^2 + \ldots$ up to n terms

$$\frac{a(1-r)}{1-r} \qquad \ldots(1)$$

$$P = a\ ar\ ar^2 \ldots ar^{n-1}$$

$$= a^n r^{1+2+3+\ldots+(n-1)}$$

$$\frac{(n-1)^{(1+n-1)}}{2}$$

$$a^n r$$

$$a^n r^{\left(\frac{n-1}{2}\right)^t} \qquad \ldots(2)$$

$$R = \frac{1}{a} + \frac{1}{ar} + \frac{1}{ar^2} + \ldots \text{ up to } n \text{ terms}$$

$$= \frac{\frac{1}{a}\left(1 - \frac{1}{r^n}\right)}{1 - \frac{1}{r}} = \frac{r}{a}\frac{(r^n - 1)}{(r-1)r^n}$$

$$= \frac{(1-r^n)}{a(1-r)r^{n-1}} \qquad ...(3)$$

By (2) and (3), $P^2 R^n = a^{2n} r^{n(n-1)} \dfrac{(1-r^n)^n}{a^n(1-r)^n r^{n(n-1)}}$

$$= \frac{a^n(1-r^n)^n}{(1-r)^n} = S^n \text{ by (1).}$$

Example 36:

The ratio of the 4th to the 12th term of a G.P. with positive comnton ratio is 1/256. If the sum of the two terms is 61.68 find the sum of series to 8 terns.

Solution:

Let the series be $a, ar, ar^2, \ldots,$

T_4 = 4th term = ar^3

T_{12} = 12th term = ar^{11}

By hypothesis $\dfrac{T_4}{T_{12}} = \dfrac{1}{256}$

i.e., $\dfrac{ar^3}{ar^{11}} = \dfrac{1}{256}$

$$\frac{1}{r^3} = \frac{1}{256}$$

$\Rightarrow \qquad r^3 = 256$

$\Rightarrow \qquad r = \pm 2$

Since r is given to be positive, we reject negative sign.

Again it is given that

$T_4 + T_{12} = 61.68$

i.e., $a(r^3 + r^{11}) = 61.68$

$a(8 + 2048) = 61.68$

$$a = \frac{61.68}{2056} = .03$$

Hence S_8 = sum to eight terms

$$= \frac{a(1-r^2)}{1-r} = \frac{a(r^3-1)}{r-1}$$

$$= \frac{(.03)(256-1)}{(2-1)}$$

$= .03 \times 255$

$= 7.65$

Example 37:

A manufacturer reckons that the value of a machine which costs him Rs 18750 will dipreciate each year by 20%. Find the estimated value at the end of 5 years.

Solution:

At the end of first year the value of machine

$$= 18750 \times \frac{80}{100}$$

$$= \frac{4}{5}(18750)$$

At the end of 2nd year it is equal to $\left(\frac{4}{5}\right)^2$ (18750); proceeding in this manner, the estimated value of machine at the end of 5 years is $\left(\frac{4}{5}\right)^5$ (18750)

$$= \frac{64 \times 16}{125 \times 25} \times 18750$$

$$= \frac{1024}{125} \times 750$$

$$= (1024) \times 6$$

$= 6144$ Rupees.

Example 38:

Show that a given sum of money accumulated at 20 percent per annum more than doubles itself in 4 years at compound interst.

Solution:

Let the given sum be a repees. After 1 year it becomes $6a/5$ (it is increased by $a/5$).

At the end of two years it becomes $\frac{6}{5}\left(\frac{6a}{5}\right) = \left(\frac{6}{5}\right)^2 a.$

Proceeding in the manner, we get that at the end of 4th year, the amount will be $\left(\frac{6}{5}\right)^4 a = \frac{1296}{625} a$

Now $\frac{1296}{625}a - 2a\frac{46}{625}a$, a + ve quantity, so the amount after 4 year is more than double of the original amount.

Example 39:

If $x = a + \frac{a}{r} + \frac{a}{r^2} + \ldots \infty$

$y = b - \frac{b}{r} + \frac{b}{r^2} + \ldots \infty$

and $z = c + \frac{c}{r^2} + \frac{c}{r^4} + \ldots \infty$

Show that $\frac{xy}{z} = \frac{ab}{c}$

Solution:

Clearly $x = \frac{a}{1 - \frac{1}{r}} = \frac{ar}{r-1}$,

$y = \frac{b}{1-(-1/r)} = \frac{br}{r+1}$

and $z = \frac{c}{1 - \frac{1}{r^2}} = \frac{cr^2}{r^2 - 1}$

Now $\frac{xy}{z} = \frac{ab\, r^2}{(r^2-1)} / \left(\frac{cr^2}{r^2-1}\right)$

$= \frac{ab}{c}$.

Example 40:

If $a^2 + b^2$, $ab + bc$ and $b^2 + c^2$ are in G. P. prove that a, b, c are also in G.P.

Solution:

Since $a^2 + b^2$, $ab + bc$ and $b^2 + c^2$ are in G.P. we get

$(ab + bc)^2 = (a^2 + b^2)(b^2 + c^2)$

$b^2(a^2 + 2ac + c^2) = a^2b^2 + a^2c^2 + b^4 + b^2c^2$

$\Rightarrow 2ab^2c^2 = a^2c^2 + b^4$

$\Rightarrow a^2c^2 - 2ab^2c^2 + b^4 = 0$

$\Rightarrow (ac - b^2)^2 = 0$

$\Rightarrow ac = b^2$

$\Rightarrow a, b, c$ are in G.P.

EXERCISES

1. Find 6th and 8th terms in each of the following series:

   (*i*) 4, 12, 36, ...

   (*ii*) 1, 3. 9, – 27, ...

   (*iii*) –21, 14, –28/3, ...

   (*iv*) $1, \frac{1}{2}, \frac{1}{2^2}, \frac{1}{2^3}, \ldots \quad \ldots \quad \ldots$

   (*v*) $\frac{1}{\sqrt{2}}, -2, \frac{8}{\sqrt{2}}, \ldots \quad \ldots \quad \ldots$

2. Find the sum of following series :

   (*i*) $\frac{1}{2}+\frac{1}{3}+\frac{2}{9}, \ldots \quad \ldots \quad \ldots$ up to 7 terms.

   (*ii*) 16.2 + 5.4 + 1.8 + ... up to 9 terms.

   (*iii*) $-\frac{1}{3}+\frac{1}{2}-\frac{3}{4}+ \ldots \quad \ldots \quad \ldots$ up to 10 terms.

   (*iv*) $3 + \sqrt{3} + 1 + \ldots \quad \ldots \quad \ldots$ up to 4 terms and infinity.

   (*v*) 1.665 + (– 1.1) + .74 + ... up to infinity.

3. (*i*) Insert 5 geometric means between $\frac{32}{9}$ and $\frac{81}{2}$.

   (*ii*) Insert 3 geometric means between $\frac{9}{4}$ and $\frac{4}{9}$.

4. (*a*) Sum the following series:

   (*i*) $1 + 4x + 7x^2 + 10x^2 + \ldots \quad \ldots \quad \ldots$ up to infinity, $x < 1$.

   (*ii*) 3 + 33 + 333 + ... up to 8 terms.

   (*b*) Evaluate

   (*i*) .132 (*ii*) .178.

5. If *p*th, *q*th term of a G.P. are respectively equal to *a, b, c*, then prove that

$$a^{q-r} \; b^{r-p} \; c^{p-q} = 1.$$

6. If a, b, c, d are in G.P., show that

   (*i*) $(a-b)^2, (b-c)^2$ and $(c-d)^2$ are in G.P.

   (*ii*) $a^2+b^2, ab+bc$ and b^2+c^2 are in G.P.

   (*iii*) $(b-c)^2+(c-a)^2+(d-b)^2=(a-d)^2$

7. If $a = 1 + c^a + c2a + c3a + \ldots \quad \ldots \quad \ldots$ up to infinity (0 < c < r) and $b = 1 + c^b + c^{ab} + c^{3b} + \ldots \quad \ldots \quad \ldots$ up to infinity (0 < c < 1) then prove that $c=\left(\frac{a'-1}{a}\right)^{1/a}=\left(\frac{b'-1}{b'}\right)^{1/b}$

8. The sum of first six terms of a G.P. is 9 times the sum of first three terms. Find the common ration.

9. The sum of three numbers in a G.P. is 38 and their product is 1728. Find them.

10. There are n terms in G.P., show that nth root of the product is equal to the square root of the product of its first and last terms.

11. The sum of an infinite numer of terms of a G.P. is 4 and the sum of their cubes is 192, find the series.

12. Prove that the $(n+1)^{\text{th}}$ term of a G.P. of which the first term is a and the third term is b, is equal to the $(2n+1)^{\text{th}}$ term of a G.P. of which the first term is a and the fifth term is b.

13. Find the sum of the infinite series.

    $1+(1+b)r+(1+b+b^2)r^2+(1+b+b^2+b^3)r^3 \ldots \quad \ldots$

    where $0 < b, \_ r < 1$.

14. If the arithmetic mean between a and b $(a > b)$ is twice as great as the geometric mean, show that either $a/b = (2+\sqrt{3})$ or $(2-\sqrt{3})$.

15. The sum of $2n$ terms of a G.P. whose first teerm is a and common ratior is equal to the sum of n terms of G.P. whose first term is b and common ratlo r^2. Prove that b is equal to the sum of the first two terms of the first series.

16. Find the sum of the serics.

$$\frac{1}{2}+\frac{1}{3^2}+\frac{1}{2^3}+\frac{1}{3^1}+\frac{1}{2^5}+\frac{1}{3^6}+\ldots\text{to } \infty$$

17. If $x - 1 + a + a^2 + \ldots \quad \ldots \quad \ldots \infty, \; y - 1 + b + b^2 + \ldots \infty$

    then $1+ab+a^2b^2+\ldots = \frac{xy}{x+y-1}$

18. If a, b, c, d are in G.P., prove that

(*i*) $\frac{ab-cd}{b^2+c^2}=\frac{a+c}{b}$

(*ii*) $(ab + bc + cd)^2 = (a^2, b^2, c^2)(b^2, c^2, d^2)$

19. At 10 percent per annum compound interst, a sum of money accumulates to Rs 8650 in 5 years. Find the sum invested initially.

20. If the population of a town increases 25 per thousand per year of the population at the beginning of a year and the present population is 26,24000, what will be the population in three year's time? What was it a year ago ?

ANSWERS

1. (i) 972, 8748; (ii) – 243, – 2187;
   (iii) $2\frac{62}{81}, 1\frac{167}{729}$; (iv) $\frac{1}{32}, \frac{1}{128}$;
   (v) – 128, –1024.
2. (i) $1\frac{601}{1458}$; (ii) $242\frac{80}{81}$;
   (iii) $7\frac{853}{1536}$; (iv) $\frac{8}{\sqrt{3}(\sqrt{3}-1)}, \frac{3\sqrt{3}}{\sqrt{3}-1}$,
   (v) .999
3. (i) 5 1/2, 8, 12, 18, 27
   (ii) 1 1/2, 1, 2/3.
4. (a)(i) $\frac{1+2x}{(1-x)^2}$, (*ii*) $\frac{10}{27}(108-1)-\frac{8}{3}$.
   (*b*)(*i*) $\frac{131}{990}$. (*ii*) $\frac{161}{9000}$.

8. .2 9. 8, 12, 18.

11. 6, –3, 1 1/2, ... 13. $\frac{1}{(1-r)(1-br)}$

16. $\frac{19}{24}$ 19. Rupees 5908.07

20. 2757840, 2560000.

HARMONICAL PROGRESSION

Non–zero quantities whose reciprocals are in A.P. are said to be in *Harmonical Progression* (written as H.P.)

Consider the following examples:

1. $1, \frac{1}{3}, \frac{1}{5}, \frac{1}{7}, \ldots\ldots$
2. $\frac{1}{2}, \frac{1}{5}, \frac{1}{8}, \frac{1}{11}, \ldots\ldots$
3. $2, \frac{5}{2}, \frac{10}{8}, \frac{1}{3}, \ldots$
4. $\frac{1}{a}, \frac{1}{a+b}, \frac{1}{a+2b}, \ldots\ldots\ a, b > 0.$
5. $5, \frac{55}{9}, \frac{55}{7}, 11, \ldots\ldots$

It can be easily checked that in each case the series obtained by aking reciprocal of each of the term is an A.P.

HARMONIC MEAN

If a, b, c are in H.P., then b is called a *Harmonic Mean* between a and c (written as H.M.)

TO INSERT n HARMONIC MEAMS BETWEEN a AND b

Let $H_1, H_2, H_3, \ldots, H_n$ be the requires harmonic means.

Then $\quad a, H_1, H_2, \ldots, H_n\ b$ are in H.P.

i.e. $\quad \frac{1}{a}, \frac{1}{H_1} \frac{1}{H_2}, \ldots, \frac{1}{H_n}, \frac{1}{b}$ are in A.P.

Then $\frac{1}{b}$ = $(n + 2)$th term of an A.P.

$= \frac{1}{a} + (n + 1)\, d$, where d is the common difference of A.P.

This gives $\quad d = \frac{a-b}{(n+1)ab}$.

Now $\quad \frac{1}{H_1} = \frac{1}{a} + d = \frac{1}{a} + \frac{a-b}{(n+1)ab}$

$$= \frac{nb+b+a-b}{(n+1)ab} = \frac{a+nb}{(n+1)ab}$$

So $$\frac{1}{H_1}\frac{a-nb}{(n+1)ab}$$

$$\Rightarrow \quad H_1=\frac{(n+1)ab}{a+nb}.$$

Again $$\frac{1}{H_2}=\frac{1}{a}+2d=\frac{1}{a}+\frac{2(a-b)}{(n+1)ab}$$

$$=\frac{nb+b+2a-2b}{(n+1)ab}=\frac{2a-b+nb}{(n+1)ab}$$

$$\Rightarrow \quad H_2=\frac{(n+1)ab}{2a-b+nb}$$

Similarly, $$\frac{1}{H_3}=\frac{1}{a}+3b=\frac{3a-2b+nb}{(n+1)ab}$$

$$\Rightarrow \quad H_3=\frac{(n+1)ab}{3a-2b+nb}$$

and so on,

$$\frac{1}{H_n}=\frac{1}{a}+nd=\frac{1}{a}+\frac{n(a-b)}{(n+1)ab}$$

$$=\frac{nb+b+na-nb}{(n+1)ab}$$

$$=\frac{na+b}{(n+1)ab} \quad \Rightarrow \quad H_n=\frac{(n+1)ab}{na+b}.$$

Examle 41:

Find the 5th term of 2, $2\frac{1}{2}, \; 3\frac{1}{3}, \ldots \quad \ldots$

Solution:

Let 5th term be x. Thrn $1/x$ is 5th term of corresponding A.P.

$\frac{1}{2}, \frac{2}{5}, \frac{3}{10}, \ldots \quad \ldots$

Then $$\frac{1}{x}=\frac{1}{2}+4\left(\frac{2}{5}-\frac{1}{2}\right)=\frac{1}{2}+4\left(\frac{-1}{10}\right)$$

$$\Rightarrow \quad \frac{1}{x}=\frac{1}{2}-\frac{2}{5}=\frac{1}{10} \quad \Rightarrow \quad x=10.$$

Example 42:

Insert two harmonic means between 1/2 *and* 4/17 .

Solution:

Let H_1, H_2 be two harmonic means between 1/2 and 4/17.

Then $2, \frac{1}{H_1}, \frac{1}{H_2}, \frac{17}{4}$ are in A.P. Let d be their common difference.

Then $\frac{17}{4} = 2 + 3d$

$\Rightarrow \quad 3d = \frac{9}{4} \Rightarrow d = \frac{3}{4}.$

Thus $\frac{1}{H_1} = 2 + \frac{3}{4} = \frac{11}{4} \Rightarrow H_1 = \frac{4}{11}$

$\frac{1}{H_2} = 2 + 2.\frac{3}{4} = \frac{7}{2} \Rightarrow H_2 = \frac{2}{7}.$

Required harmonic means are $\frac{4}{11}, \frac{2}{7}.$

EXERCISES

1. Find 5th and 12th terms of the following series:

   (*i*) $\frac{2}{9}, \frac{4}{17}, \frac{1}{4}, \quad \ldots.$

   (*ii*) $\frac{1}{3}, \frac{8}{23}, \frac{4}{11}, \quad \ldots.$

2. Insert 4 harmonic means between 4 and 1/4.

3. Insert 7 harmonic means between 5 and – 11.

ANSWERS

1. (*i*) $\frac{2}{7}, \frac{4}{7}$ (*ii*) $\frac{2}{5}, \frac{8}{13}.$

2. $1, \frac{4}{7}, \frac{2}{5}, \frac{4}{13}.$

3. $6\frac{1}{9}, 7\frac{6}{7}, 11, 18\frac{1}{3}, 55, -55, -11\frac{1}{3}.$

SUM OF FIRST n NATURAL NUMBERS

We wish to find the sum of the series 1 + 2 + 3 + . . . + n .

This is an A.P. with first term 1 and common difference 1.

Thus $S_n = \frac{n}{2}[2.1+(n-1).1] = \frac{n(n+1)}{2}.$

Hence sum of first n natural numbers is $\frac{n(n+1)}{2}$.

SUM OF THE SQUARES OF FIRST n NATURAL NUMBERS

Let $S = 1^2 + 2^2 + 3^2 + \ldots + n^2$.

Now $n^3 - (n-1)^3 = 3n^2 - 3n + 1$

Putting successively $n = 1, 2, \ldots, \ldots, \ldots n$, we get

$1^3 - 0^3 = 3.1^2 - 3.1 + 1$

$2^3 - 1^3 = 3.1^2 - 3.2 + 1$

... ...

$n^3 - (n-1)^3 = 3n^2 - 3.n + 1.$

Adding, we obtain

$n^3 = 3(1^2 + 2^2 + \ldots + n^2) - 3(1 + 2 + \ldots + n) + n$

Then $3(1^2 + 2^2 + \ldots + n^2) = n^3 + (1 + 2 + \ldots + n) - n$

$$= n^3 + \frac{3n(n+1)}{2} - n$$

$$= \frac{n}{2}[2n^2 + 3n + 3 - 2]$$

$$= \frac{n}{2}[2n^2 + 3n + 1]$$

i.e. $3S = \frac{n}{2} = [(n+1)(2n+1)]$

Hence $S = \frac{n}{6}(n+1)(2n+1).$

SUM OF THE CUBES OF FIRST n NATURAL NUMBERS

Let $S = 1^3 + 2^3 + 3^3 \ldots + n^3$.

Now $n^4 - (n-1)^4 = 4n^3 - 6n^2 + 4n - 1.$

Putting succcessively, $n = 1, 2, \ldots, n$, we get

$1^4 - 0^4 = 4.1^3 - 6.1^2 + 4.1 - 1$

$2^4 - 1^4 = 4.2^3 - 6.2^2 + 4.2 - 1$

$3^4 - 2^4 = 4.3^3 - 6.3^2 + 4.3 - 1$

... ...

$n^4 - (n-1)^4 = 4.n^3 - 6.n^2 + 4.n - 1.$

Adding we obtain

$n^4 = 4\ (1^3 + 2^3 + \ldots + n^3) - 6\ \ (1^2 + 2^2 + 3^2 + \ldots + n^2)$

$+ 4\ (1 + 2 + 3 \ldots + n) - n$

Then $4S = n^4 + 6\ \ (1^2 + 2^2 + \ \ldots + n^2) - 4\ (1 + 2 + \ldots + n) + n$

$= n^4 + \dfrac{6n(2n+1)(n+1)}{6} - \dfrac{4n(n+1)}{2} + n$

$= n^4 + n\ (2n^2 + 3n + 1) - 2 - 2n^2 + n$

$= n^4 + 2n^3 + 3n^2 + n - 2n^2 - 2n + r$

$= n^4 + 2n^3 + n^2$

$= n^2\ \ (n+1)^2$

$\Rightarrow S = \left[\dfrac{n(n+1)}{2}\right]^2.$

Example 43:

a and b are two positive real numbers and A,G, H arre their A. M., G. M. and H.M. respeclively. If G is to be taken positive, then prove that A ≥ G ≥ H and they are in G.P.

Solution:

Clearly $A = \dfrac{a+b}{2}, G = \sqrt{ab}$ and $H = \dfrac{2ab}{a+b}.$

Now $A - G = \dfrac{a+b}{2} - \sqrt{ab} = \dfrac{a+b-2\sqrt{ab}}{2}$

$= \dfrac{(\sqrt{a}-\sqrt{b})^2}{2} \geq 0$

$\Rightarrow \quad A \geq G$

Further $G - H = \sqrt{ab} - \dfrac{2ab}{a+b} = \sqrt{ab}\,\dfrac{a+b-2\sqrt{ab}}{a+b}$

$= \sqrt{ab}\,\dfrac{(\sqrt{a}-\sqrt{b})^2}{a+b} \geq 0$

$\Rightarrow G \geq H.$

Hence $\quad A \geq G \geq H.$

Again $G^2 = ab$ and $AH = \frac{a+b}{2} \cdot \frac{2ab}{a+b} = ab$ imply that $G^2 = AH$ In other words A, G, H are in G.P.

Example 44:

Find the sum of the series

$$1+\frac{2}{5}+\frac{3}{5^2}+\ldots \text{ upto } n \text{ terms.}$$

Solution:

Let $S = 1+\frac{2}{5}+\frac{3}{5^2}+\frac{4}{5^2}+\ldots+t_n$

where t_n = nth term of the series $= \frac{n}{5^{n-1}}$. (Note this step)

Thus $S = 1+\frac{2}{5}+\frac{3}{5^2}+\ldots+\frac{n}{5^{n-1}}$.

Multiplying each side by $\frac{1}{5}$ we get.

$$\frac{1}{5}S = \frac{1}{5}+\frac{2}{5^2}+\ldots.+\frac{n-1}{5^{n-1}}+\frac{n}{5^n}.$$

On subtraction, we obtain

$$\frac{4}{5}S = 1+\frac{1}{5}+\frac{2}{5^2}+\ldots+\frac{1}{5^{n-1}}-\frac{n}{5^n}$$

$$= \frac{1\left(1-\frac{1}{5^n}\right)}{1-\frac{1}{5}} - \frac{n}{5^n} = \frac{5^{n-1}}{4.5^{n-1}} - \frac{n}{5^n}.$$

$$\Rightarrow \frac{4}{5}S = \frac{5^{n+1}-5-4n}{4.5^n} \Rightarrow \qquad S = \frac{5^{n+1}-5-4n}{16.5^{n-1}}.$$

Example 45:

The number of terms in an A.P. is even; the sum of the odd terms is 24, of the even terms 30, and the last term exdeeds the first term by 10 1/2. Find the number of terms.

Solution:

Let a be the first term and d be the common difference of the given A.P., the odd terms are a, $a + 2d$, $a + 4d$ . . . ., $a + (2n - 2)d$ and the even terms are $a + d$, $a + 3d$, . . . $a + (2n - 1)d$.

By hypothesis

$24 = a + (a + 2d) + (a + 4d) + \ldots + a + (2n - 2)d$ ...(1)

$30 = (a + d) + (a + 3d) + (a + 5d) + \ldots + a + (2n - 1)d$ ...(2)

and $[a + (2n - 1)d] - a = 10\ 1/2 = 21/2$ ...(3)

Simplyfying (1), (2) and (3) we get

$$24 = \frac{n}{2}[2a + (n-1)2d] \Rightarrow 24 = n(a + nd - d) \quad ...(4)$$

$$30 = \frac{n}{2}[2a + 2d + (n-1)2d] \Rightarrow 30 = n(a + nd) \quad ...(5)$$

and $(2n - 1)d = 21/2.$ ...(6)

Subtracting (4) from (5) we obtain $nd = 6.$ ...(7)

Then (6) implies $12 - d = 21/2 \Rightarrow d = 3/2.$

Finally (7) gives that $n = 4$ and so $2n = 8.$

Example 46:

If $H_1, H_2, \ldots, H_n$ are n harmonic means between two given numbers then show that

$$H_1H_2 + H_2H_3 + \ldots + H_{n-1}H_n = (n-1)\,H_1H_n.$$

Solution:

Let a and b be the two given numbers. Let d be the common difference of the series $\frac{1}{a}, \frac{1}{H_1}, \frac{1}{H_2}, \ldots \frac{1}{H_n}, \frac{1}{b}.$

Now $\frac{1}{H_2} - \frac{1}{H_1} = d \Rightarrow H_1 - H_2 = d(H_1H_2)$

Similarly $\frac{1}{H_3} - \frac{1}{H_2} = d \Rightarrow H_2 - H_3 = d(H_2H_3)$

... ...

... ...

Finaly $\frac{1}{H_n} - \frac{1}{H_{n-1}} = d \Rightarrow H_{n-1} - H_n = d(H_{n-1}H_n)$

Adding, we get $H_1 - H_2 = d\,(H_1H_2 + H_2H_3 + \ldots + H_{n-1}H_n)$

or $H_1H_2 + H_2H_3 + \ldots + H_{n-1}H_n = \frac{H_1 - H_n}{d}.$

Now $\frac{1}{H_n}$ is nth term of the A.P. with first term as H_1 and common difference d.

So $$\frac{1}{H_n}=\frac{1}{H_1}+(n-1)d$$

$$\Rightarrow \frac{1}{H_n}-\frac{1}{H_1}=(n-1)d$$

$$\Rightarrow H_1-H_n=(n-1)d\,(H_1H_n)$$

Hence $$H_1H_2+H_2H_3+\ldots H_{n-1}H_n=(n-1)H_1H_n.$$

Example 47:

A_1, A_2 ; G_1, G_2 and H_1, H_2 are A.M.'s, G.M's and H.M's respectively between two given numbers. Prove that

$$G_1G_2 : H_1H_2 = A_1 + A_2 : H_1 + H_2.$$

Solution:

Let the given numbers be x and y.

Then x, A_1, A_2, y; x, G_1, G_2, y ; x, H_1, H_2, y are in A.P., G.P, and H.P. respectively.

This implies that $A_1=\frac{y+2x}{3}$, $A_2=\frac{2y+x}{3}$; $G_1=x^{2/3}y^{1/3}$,

$G_2=x^{1/3}y^{2/3}$ and $H_1=\frac{3xy}{x+2y}$, $H_2=\frac{3xy}{2x+y}$.

Now $$\frac{G_1G_2}{H_1H_2}=\frac{(xy)(x+2y)(2x+y)}{9x^2y^2}=\frac{(x+2y)(2x+y)}{9xy} \qquad \ldots(1)$$

While $$\frac{A_1+A_2}{H_1+H_2}=\frac{\frac{y+2x}{3}+\frac{x+2y}{3}}{\frac{9xy}{x+2y}+\frac{3xy}{2x+y}}$$

$$=\frac{(x+y)(x+2y)(2x+y)}{3xy(3x+3y)}$$

$$=\frac{(x+2y)(2x+y)}{9xy} \qquad \ldots(2)$$

(1) and (2) give the required result.

Example 48:

If $a^x = b^y = c^z = \ldots$ etc and a, b, c . . . are in G.P., then show that x, y, z, . . . are in H.P.

Solution:

Let $a^x = b^y = c^z = \ldots = k$

Then $a = k^{1/x},\ b = k^{1/y},\ c = k^{1/z}, \ldots$ etc.

This in turn implies that $\frac{b}{a} = k^{1/y - 1/x},\ \frac{c}{b} = k^{1/z - 1/y}, \ldots,$

But $a, b, c, \ldots$ are in G.P., So $\frac{b}{a} = \frac{c}{b} = \ldots$

$\Rightarrow \frac{1}{y} - \frac{1}{x} = \frac{1}{z} - \frac{1}{y} = \ldots$

$\Rightarrow \frac{1}{x}, \frac{1}{y}, \frac{1}{z}, \ldots$ are in A.P.

$\Rightarrow x,\ y,\ z, \ldots$ are in H.P.

Note. See also Example 31.

Example 49:

If the ration of H.M. and G.M. between two numbers is 12 : 13 then prove that the ratio between the numbers will be 9 : 4.

Solution:

Let the numbers be a and b. If G and H stand for the geometric and harmonic means between a and b respectively, then

$$\frac{H}{G} = \frac{12}{13} \Rightarrow \frac{\frac{2ab}{a+b}}{\sqrt{ab}} = \frac{12}{13} \Rightarrow \frac{2\sqrt{ab}}{a+b} = \frac{12}{13}$$

$$\Rightarrow \frac{a+b-2\sqrt{ab}}{a+b+2\sqrt{ab}} = \frac{13-12}{13+12} = \frac{1}{25} \Rightarrow \frac{\sqrt{a}-\sqrt{b}}{\sqrt{a}+\sqrt{b}}$$

$$= \frac{1}{5} \Rightarrow \frac{\sqrt{a}}{\sqrt{b}} = \frac{6}{4}$$

$$\Rightarrow \qquad \frac{\sqrt{a}}{\sqrt{b}} = \frac{3}{2} \Rightarrow \frac{a}{b} = \frac{9}{4}.$$

Example 50:

If 12 and 9 3/5 are the geometric and harmonic means, respecitvely, between two numbers, find them.

Solution:

Let the two nimhers be x and y. Then $\sqrt{xy} = 12$ and $\frac{2xy}{x+y} = \frac{48}{5}$. Since $xy = 144$, we get $x + y = \frac{5}{24} xy = 30$. Solving the two equations, we get $x = 6$, $y = 24$ or $x = 24$, $y = 6$. Hence the two numbers are 6 and 24.

Example 51:

Sum to n terms the series

$$12 + (1^2 + 2^2) + (1^2 + 2^2 + 3^2) + \ldots$$

Solution:

t_r = rth term of the given series

$$= 1^2 + 2^2 + 3^2 + \ldots + r^2$$

$$= \frac{r(r+1)(2r+1)}{6}$$

$$= \frac{r(2r^2+3r+1)}{6}$$

$$= \frac{r^2}{3} + \frac{r^2}{2} + \frac{r}{6}$$

So $$t_1 = \frac{1^3}{3} + \frac{1^2}{2} + \frac{1}{6}$$

$$t_2 = \frac{2^3}{3} + \frac{2^2}{2} + \frac{2}{6}$$

$$t_3 = \frac{3^3}{3} + \frac{3^2}{2} + \frac{3}{6} \text{ and so on}$$

$$t_n = \frac{n^3}{3} + \frac{n^2}{2} + \frac{n}{6}$$

Hence required sum $= t_1 + t_2 + \ldots + t_n$

$$= \frac{1}{3}(1^3 + 2^3 + \ldots + n^3) + \frac{1}{2}(1^2 + 2^2 + \ldots + n^2) + \frac{1}{6}(1 + 2 + \ldots + n)$$

$$= \frac{1}{3}\frac{n^2(n+1)^2}{4} + \frac{1}{2}\frac{n(n+1)(n+2)}{6} + \frac{1}{6}\frac{n(n+1)}{2}$$

$$= \frac{n(n+1)}{12} \quad [n(n+1)+2n+1+1]$$

$$= \frac{n(n+1)}{12}(n^2+3n+2) = \frac{n(n+1)^2(n+2)}{12}.$$

Example 52:

Sum the series

$1^2 + 3^2 + 5^2 + 7^2 + \ldots$ *up to n terms*

Solution:

nth term of the series 1, 3, 5, 7, . . ., is $2n - 1$

So the nth term t_n of the given series

$= (2n - 1)^2 = 4n^2 - 4n + 1$

Thus $\quad t_1 = 4.1^2 - 4.1 + 1$

$t_2 = 4.2^2 - 4.2 + 1$

$t_3 = 4.3^2 - 4.3 + 1$

. . . . and so on

$t_n = 4.n^2 - 4.n + 1$

Hence $\quad S = t_1 + t_2 + \ldots + t_n$

$= 4\,(1^2 + 2^2 + \ldots + n^2) - 4\,(1 + 2 + \ldots + n) + n$

$$= 4\frac{n(n+1)(2n+1)}{6} - 4\frac{n(n+1)}{2} + n$$

$$= \frac{2n(n+1)(2n+1)}{3} - 2n(n+1) + n$$

$$= \frac{n(n+1)}{3}(4n+2-6) + n$$

$$= \frac{n(n+1)(4n-4)}{3} + n$$

$$= \frac{4n(n^2-1)}{3} + n$$

$$= \frac{n}{3}(4n^2-4+3) = \frac{n}{3}(4n^2-1).$$

Example 53:

Sum the series to n terms

4 + 14 + 30 + 52 + 80 + . . .

Solution:

Let $S = 4 + 14 + 30 + 52 + 80 + \ldots + t_n$ ...(1)

where t_n is the nth term of the given series.

Also $S = 4 + 14 + 30 + 52 + \ldots + t_{n-1} + t_n$ ...(2)

Subtracting (2) frin (1) we get

$0 = 4 + 10 + 16 + 22 + 28 + \ldots$ up to n terms $- t_n$

$\Rightarrow t_n = 4 + 10 + 16 + \ldots$ up to n terms

$= \frac{n}{2}\,[\,8 + (n-1)\,6]$

Hence $S_n = \Sigma t_n = \Sigma\,(3n^2 + n)$

$3\Sigma\, n^2 + \Sigma n$

$$= 3\frac{n(n+1)(2n+1)}{6} + \frac{n(n+1)}{2}$$

$$= \frac{n(n+1)}{2}(2n+1+1)$$

$= n\,(n + 1)\,(n + 1)$

$= n\,(n + 1)^2$

Example 54:

Sum the series to n terms

2.4.6 + 4.6.8 + 6.8.10 + . . .

Solution:

Evidently nth terrms of the given

series $= t_n = 2n\,(2n + 2)\,(2n + 4)$

$= 8n\,(n + 1)\,(n + 2)$

$= 2\,(n^3 + 3n^2 + 2n)$

Hence $S = \Sigma t_n = 8\Sigma^3 + 24\Sigma^2 + 16\Sigma n$

$$= 8\frac{n^2(n+1)^2}{4} + 24\frac{n(n+1)(2n+1)}{6} + 16\frac{n(n+1)}{2}$$

$= 2n^2\,(n + 1)^2 + 4n\,(n + 1)\,(2n + 1) + 8\,n\,(n + 1)$

$= 2n\,(n + 1)\,[\,n\,(n + 1) + 2\;\;(2n + 1) + 4]$

$= 2n\;\;(n + 1)\;\;(n^2 + 5n + 6)$

$= \;2n\;\;(n + 1)\,(n + 2)\,(n + 3)$

Example 55:

Find the sum of n terms of the series whose nth term is $3\ (4^3 + 2n^2)$ -- $4n^3$.

Solution:

Here $= t_n = 3\ (4^n + 2n^2) - 4n^3$

So $S = \Sigma t_n = \Sigma\ [3\ (4^n + 2n^2 - 4n^3]$

$= 3\Sigma\, 4^n + 6\Sigma n^2 - 4\Sigma n^3$

$$= 3\frac{4(1-4^n)}{1-4} + 6\frac{n(n+1)(2n+1)}{6} - 4\frac{n^2(n+1)^2}{4}$$

$$= 4^{n+1} - 4 + n(n+1)[2n+1-n(n+1)]$$

$$= 4^{n+1} - 4 + n(n+1)[2n+1-n(n+1)]$$

$$= 4^{n+1} - 4 + n(n+1)(-n^2+n+1)$$

$$= 4^{n+1} - 4 + n(n+1)(n^2-n-1).$$

EXERCISES

1. Sum the following series:

   (*i*) $2^2 + 4^2 + 6^2 + \ldots$, *up to n terms.*

   (ii) $1.2 + 2.3 + 3.4 + 4.5 + \ldots \quad \ldots$ up to n terms.

   (*iii*) $1.3.5 + 3.5.7 + 5.7.9 + \ldots \quad \ldots$ *up to n terms.*

2. Find the sum of n terms of the series whose rth term given by $2^{r-1} + 8r^3 - 6r^2$.

3. If pth, qth, rth terms of a H.P. are a, b, c respectively, Prove that $(q-r)\ bc + (r-p)\ ca + (p-q)\ ab = 0$.

4. If l, m, n are three numbers in G.P. prove that the first of an A.P. whose ith, mth and nth terms are in H.P. is to the common difference as $m + 1$ to 1.

5. Show that a, b, c are in A.P. or in G.P. or in H.P. according as

   $$\frac{a-b}{b-c} = \frac{a}{a} \text{ or } \frac{a}{b} \text{ or } \frac{a}{c} \text{ respectively.}$$

6. If H.M. and G.M. of two numbers are $14\frac{2}{5}$ and 24 respectively, find the numbers.

7. If a, b, c are in A.P., x, y, z in H.P, ax, by, cz are in G.P. prove that

$$\frac{x}{z}+\frac{z}{x}=\frac{a}{c}+\frac{c}{a}.$$

8. If $2(y-a)$ is H.M. between $(y-x)$ and $(y-z)$, then prove that $(x-a)$, $(y-a)$ and $(z-a)$ are in A.P.

9. If a, b, c are in H.P. then prove that

$$\frac{a}{b+c}, \frac{b}{c+a}, \frac{c}{a+b} \text{ are also in H.P.}$$

10. If a, b, c are in H.P. then prove that

$$a^2 + c^2 > 2b^2.$$

11. Find nth term and the sum to n terms the following series

 (*i*) $4 + 6 + 9 + 13 + 18 + \ldots\ldots$

 (*ii*) $11 + 23 + 59 + 167 + \ldots\ldots$

 (*iii*) $1 + 5 + 12 + 22 + \ldots\ldots$

 (*iv*) $1.2.4 + 2.3.5 + 3.4.6 + \ldots\ldots$

12. Find the sum of n terms of the series whose nth is

 (*i*) $n^3 + \frac{2}{3}n$

 (*ii*) $3^n - 2^n$

 (*iii*) $3n^2 - n$.

ARITHMETIC-GEOMETRIC SERIES

A series each term of which is the product of the corresponding terms of an A.P. and a G.P. is called an Arithmetico-Geometric series:

For example the following series is one of this type:

$a + (a + d)\, r + (a + 2d)\, r^2 + (a + 3d)\, r^3 + \ldots$

$+ [\, a + (n - 1)\, d]\, b)\, r^{n-1} + \ldots.$

To find the sum of such a series upto n terms.

Let $\quad S = a + (a + d)\, r + (a + 2d)\, r^2 + \ldots + [\, a + (n - 1)\, r^{n-1}$

then $\quad rS = ar + (a + d)\, r^2 + \ldots + [\, a + (n - 2)\, d]\, r^{n-1}$

$+ [\, a + (n - 2)\, d]\, r^n$

$\therefore S\,(1 - r) = a + (dr + dr^2 + \ldots + dr^{n-1}) - [\, a + (n - 1)\, r^n$

$= a + dr\,(1 + r + .. + r^{n-2}) - [\, a + (n - 1)\, d]\, r^n$

$$= a + dr\frac{1-r^{n-1}}{r-1} - [a+(n-1)d]r^n$$

$$\therefore S = \frac{a}{1-r} + \frac{dr}{(1-r)^2} - \frac{dr^n}{(1-r)^2} - \frac{[a+(n-1)d]r^n}{(1-r)}$$

Note.

If $r < 1$ numerically and the number of terms is infinite, then in the limit $r^n = 0$, so that the sum of an infinite, Arithmetico-Geometric series is

$$S = \frac{a}{1-r} + \frac{dr}{(1-r)^2}.$$

Example 1:

Sum theseries 1.2 + 2.4 + 3.8 + 4.16 + ... to n terms,

Solution:

Here nth term = (nth term of 1, 2, 3, 4...) × (nth term of 2, 4, 8, 16...)

$= n.\ 2^n$.

$S = 1.\ 2 + 2.4 + 3.8 + ...\ n\ .\ 2^n$

Subtracting $\dfrac{2S = 1.4 + 2.8 + ... + (n-1)2^n + n2^{n+1}}{-S1.2 - 1.4 + 1.8 + 1.16 + ... 1.2^n - n.2^{n+1}}$

or $-S = 2\dfrac{2^n - 1}{2-1} - n2^{n+1} = 2^{n+1} - 2 - n.n2^{n+1}$

$\therefore\ S = 2 + (n-1)2^{n+1}$.

Example 2:

Find the sum of the series $1 + 4x + 7x^2 + ...$ to n terms: hence write down the sum to infinity when $x < 1$ numerically,

Solution:

$$S = 1 + 4x + 7x^2 + ... + (3n-5)x^{n+2} + (3n-2)x^{n-1}$$

$$xS = \quad x + 4x^2 + ... + (3n-5)x^{n-1} + (3n-2)x^n$$

$$\therefore\ S(1-x) = 1 + 3x + 3x^2 + ... + 3x^{n-1}(3n-2)x^n$$

$$= 1 + \frac{3x(1-x^{n-1})}{1-x} - (3n-2)x^n$$

$$\therefore\ S=\frac{1}{1-x}+\frac{3x(1-x)^{n-1}}{(1-x)^2}-\frac{(3n-2)x^n}{1-x}$$

Sum of infinity $=\frac{1}{1-x}+\frac{3x}{(1-x)^2}$.

If the sum upto infinity is only required, we proceed as follows:

$$S = 1 + 4x + 7x^2 + 10x^3 + ...$$

$$xS = x+4x^2+7x^2 \quad +...$$

$$\therefore\ S(1-x)=1+3x+3x^2+3x^3+...$$

$$=1+3x(1+x+x^2+...)=1+3x\frac{1}{1-x}$$

$$\therefore\ S=\frac{1}{1-x}+\frac{3x}{(1-x)^2}.$$

Example 3:

Sum to infinity $1^2 + 2^2x + 3^2x^2 + 4^2x^3 + ...$ when $x < 1$ numerically.

Solution:

Though this is not an Arithmetico-Geometric series but a similar method can be applied.

$$S = 1 + 4x + 9x^2 + 16x^3 + ...$$

$$xS = \quad x+4x^2+9x^3+...$$

$$\therefore\ S(1-x)=1+3x-5x^2+7x^3+...$$

This is an Arithmetico-Geometric series.

Multiplying by x again, we get

$$Sx(1-x)=x+3x^2-5x^3+...$$

$$S(1-x)^2=1+2x+2x^2+2x^3...$$

$$=+\frac{2x}{1-x}=\frac{1+x}{1-x}$$

$$S=\frac{1+x}{(1-x)^3}$$

Example 4:

Sum to (2 n + 1) terms the series 1 - 2 + 3 - 4...

Solution:

For the series 1, 3, 5, ... n th terms is $(2n - 1)$

and for the series 2, 4, 6, ... n th term is $2n$

Therefore the sum of the given series upto $2n$ terms

= (1 + 3 + 5 + ...to n terms) - (2 + 4 + 6 ...to n terms)

$$\frac{n}{2}(1+2n-1)-\frac{n}{2}(2+2n)=-n$$

and the $(2n + 1)$th term of the given series is $2n + 1$.

hence total sum = $- n + 2n + 1 = n + 1$.

Example 5:

Show that $1 - 3 + 5 - 7$ ... to n terms = $(-1)^{n+1} n$.

Solution:

$S = 1 - 3 + 5 - 7 ... - (-1)^{n+1}(2n - 1)$

$S = 1 - 3 + 5 + ... + (-1)^{n}(2n - 3) + (-1)^{n+1}(2n - 1)$

adding $2S = 1 - 2 + 2 - 2 ... + (-1)^{n}(-2) + (-1)^{n+1}(2n - 1)$

If n be even, last but one term of the series 2S is - 2 and the sast term is $- 2n - 1$,

then sum of the series - 2 + 2 - 2 + ...+ 2 - 2 is - 2

hence $2S = 1 - 2 - 2n - 1 = 2n \therefore S = -n = (-1)^{n+1} n$.

If n be odd, last but one term of 2S is + 2 and the sast term is $2n - 1$ hence the sum of the series

- 2 + 2 + ... + 2 is 0

so that $2S\ 1 + 2n - 1 = 2n \therefore S = -n = (-1)^{n+1} n$.

Hence in either case $S = (-1)^{n+1} n$.

Example 6:

Find the sum to n terms of the series $8 - 11 + 14 - 17$ ...

Solution:

$S = 8 - 11 + 14 - 17 ... + (-1)^{n+1}(3n + 5)$

$$\underline{S = \quad 8-11+14...+(-1)^{n}(3n+2)+(-1)^{n+1}(3n+5)}$$

$$\therefore 2S = 8-3+3-3...+(-1)^{n}(-3)+(-1)^{n+1}(3n+5)$$

When n is even, sum of the series - 3 + 3 ...is - 3

so that $2S = 8 - 3 - 3n - 5\ \ 3n$

$$\therefore S = -\frac{3}{2}n$$

When n is odd, sum of the series - 3 + 3 ... is 0

$\therefore \quad 2S = 8 + 3n + 5 + 13 + 3n$

$\therefore S = \frac{13}{2} + \frac{3}{2}n.$

Example 7:

Sum to n terms $1 + 3.\ 6 + 5.\ 6^2 + 7\ .\ 6^3 + ...$

Solution:

$Tn = \{1 + (n - 1)\ 2\}6^{n-1} = (2n - 1)\ 6^{n-1}$

$S_n = 1 + 3 \cdot 6 + 5 \cdot 6^2 + ... + (2n - 1)\ 6^{n-1}$

$6S_n = 1.6 + 3.6^2 + ... + (2n-3)6^{n-1} + (2n-1)6^n$

$-5S_n = 1 + 2\cdot 6 + 2\cdot 6^2 + ... \quad + 2\cdot 6^{n-1} - (2n-1)6^n$

$= 1 + 2 \cdot 6\ \{1 + 6 + ...\text{to } (n - 1) \text{ terms}\} - (2n - 1)6^n$

$= 1 + 2 \cdot 6 \cdot \frac{6^{n-1} - 1}{6-1} - (2n-1)6^n$

$\therefore S_n = \frac{(10_n - 7)6^n + 7}{25}.$

EXERCISES

1. Find the sum n terms of the following series:

(1) $1 + 2x + 2x^2 + 4x^3 + ...$

(2) $1 + 3x + 5x^2 + ...$

(3) $1 + \frac{4}{5} + \frac{7}{5^2} + \frac{10}{5^3} + ...$

(4) $2 \cdot 1 + 3 \cdot 2 + 4 \cdot 4 + 5 \cdot 8 + ...$

2. Sum up to infinity the following series:

(1) $1 + 3x + 5x^2 + 7x^3 + ...$ when $x < 1$

(2) $1 + \frac{2}{3} + \frac{1}{3} + \frac{4}{27} + ...$

(3) $\frac{3}{2} - \frac{5}{6} + \frac{7}{18} - ...$

(4) $1 - \frac{2}{5} + \frac{3}{5^2} - \frac{4}{5^3} + ...$

(5) $1 - x + 2x^2 - 3x^3 + ...$

(6) $1^2 + 3x^3 + 5^2x^2 + 7^2x^3 + ...$

SIGMA-NOTATION

When terms of the same type, with the same co-efficient occur, the symbol Σ (sigma) affixed before the general term signifies the sum of all such terms.

Thus $\Sigma n = 1 + 2 + 3 + ... + n$

$\Sigma 2^n = 2^1 + 2^2 + 2^3 + ... + 2n.$

$$\Sigma\left(n^2+\frac{1}{n}\right)=(1^2+1)(2^2+\frac{1}{2})+(3^2+\frac{1}{3}+...+\left(n^2+\frac{1}{n}\right).$$

TO FIND THE SUM OF THE FIRST n NATUREAL NUMBERS

$S = \Sigma n = 1 + 2 + 3 + ... + n$

It is an A.P. whose first term is 1 and common difference is 1.

$$S=\frac{n}{2}[2\times1+(n-1)\times1]=\frac{1}{2}n(n+1)$$

TO FIND THE SUM OF THE SQUARES OF THE FIRST n NATUREAL NUMBERS

$S = \Sigma 2^n = 1^2 + 2^2 + 3^2 + ... + n^2.$

We know that $x^3 - (x - 1)^3 = 3x^2 - 3x + 1.$

Put for x successively the values 1, 2, 3, ..., $(n - 1)$, n.

Then $1^3 - 0^3 = 3 . 1^2 - 3 . 1 + 1$

$2^3 - 1^3 = 3 . 2^2 - 3 . 2 + 1$

$3^3 - 2^3 = 3 . 3^2 - 3 . 3 + 1$

... ... = ...

... ... = ...

$(n - 1)^3 - (n - 2)^3 = 3 (n - 1)^2 - 3 (n - 1) + 1$

$n^3 - (n - 1)^3 = 3n^2 - 3n + 1$

Adding both sides, we get

$n^3 = 3 (1^2 + 2^2 + ... + n^2) - 3 (1 + 2 + ... + n) + n$

$$=3\Sigma n^2-3.\frac{n}{2}(n+1)+n$$

$$\therefore =3\Sigma n^2=n^3+\frac{2}{3}n(n+1)-n=n(n^2-1)-\frac{2}{3}n(n+1).$$

$$= n(n+1)[n-1+\frac{2}{3}] = \frac{1}{2}n(n+1)(2n+1)$$

$$\sum n^2 = \frac{1}{6}n(n+1)(2n+1).$$

TO FIND THE SUM OF THE CUBES OF THE FIRST n NATURAL NUMBERS

$$S = \sum n^3 = 1^3 + 2^3 + 3^3 + ... + n^3$$

We know that $x^4 - (x - 1)^4 = 4x^3 - 6x^2 + 4x - 1$.

Put for x successively the values 1, 2, 3, ... $(n - 1)$, n.

Then $\quad 1^4 - 0^4 = 4 . 1^3 - 6 . 2^2 + 4.1 - 1$

$$2^4 - 1^4 = 4 . 2^3 - 6 . 2^2 - 4 . 2 - 1$$

$$3^4 - 2^4 = 4 . 3^3 - 6 . 3^2 - 4 . 3 - 1$$

... ... = ...

... ... = ...

$$(n - 1)^4 - (n - 2)^4 = 4(n - 1)^3 - 6(n - 1)^2 + 4(n - 1) - 1$$

$$n^4 - (n - 1)^4 = 4.n^3 - 6.n^2 + 4.n - 1$$

Adding both sides, we get

$$n^4 = 4.\sum n^3 - 6\sum n^2 + 4\sum n - n$$

$$\therefore 4\sum n^3 = n^4 + n + 6\sum n^2 - 4\sum n$$

$$4\sum n^3 = n^4 + n + 6 \cdot \frac{1}{6} n\ (n+1)(2n + 1) - 4 \cdot \frac{1}{2} n(n + 1)$$

$$= n(n+1)(n^2 - n + 1) + n(n+1)(2n + 1) - 2n(n+1)$$

$$= n(n+1)(n^2 + n)\ \ n^2 (n+1)^2.$$

$$\sum n^3 = \left[\frac{n(n+1)}{2}\right]^2$$

TO FIND THE SUM OF n TERMS OF A SERIES WHOSE nTH TERM IS $an^3 + bn^2 + cn + d$

$$S = \sum(an^3 + bn^2 + cn + d) = a\sum n^3 + b\sum n^2 + c\sum n + d\sum 1$$

$$= a.\left[\frac{n(n+1)}{2}\right]^2 + b.\frac{1}{6}n(n+1)(2n+1)$$

$$+ c.\frac{1}{2}n(n-1) + dn.$$

Example 1:

Sum to n terms the series $2 \cdot 3 + 4 \cdot 5 + 6 \cdot 7 + \ldots$

Solution:

nth term = (nth term of 2, 4, 6...) (nth term of 3, 5, 7...)

$= [2 + (n - 1) \cdot 2][3 + (n - 1) \cdot 2]$

$= 2n\,(2n + 1) = 4n^2 + 2n$

$$\therefore\ S = 4\sum n^2 + 2\sum n = 4 \cdot \frac{1}{6} n(n+1)(2n+1)$$

$$+2 \cdot \frac{1}{2} n(n+1)(2n+1)$$

$$= \frac{n(n+1)(4n+5)}{3}.$$

Example 2:

Sum the series

$(1 + 2) = (1 + 2 + 3) + \ldots$ to n terms.

Solution:

$$T_n = 1 + 2 + 3 + \ldots \text{ to } n \text{ terms } \frac{1}{2} n(n+1) = \frac{n^2}{2} + \frac{n}{2}$$

$$\square \therefore \qquad S_n = \frac{1}{2}\sum n^2 + \frac{1}{2}\sum n = \frac{1}{6} n(n+1)(n+2).$$

Example 3:

Find the sum of $1^2 + 2^2 + 3^2 - 4^2 \ldots$ to $2n$ terms.

Solution:

$S = (1^2 + 3^2 + 5^2 + \ldots$ to n terms$) - (2^2 + 4^2 + \ldots$ to n terms$)$

$= \Sigma(2n - 1)^2 - \Sigma 4n^2 = \Sigma(4n^2 - 4n + 1 - 4n^2)$

$$= -4\sum n + \sum 1 = -4\frac{n}{2}(n+1) + n = -2n(n+1) + n$$

$= -\,n\,(2n + 1).$

Example 4:

Sum the series

$1 + 2 - 3 + 4 + 5 - 6 + 7 + 8 - 9 \ldots$ to $(3n + 1)$ terms.

Solution:

nth term of 1, 4, 7 ... = 1 + (n - 1) 3 = $3n - 2$

nth term of 2, 5, 8 ... = $3n - 1$

nth term of 3, 6, 9 ... = $3n$

$\therefore$ sum of the corresponding three terms of the given series

$= (3n - 2) + (3n - 1) - 3n = 3n - 3$

$\therefore$ sum upto $3n$ terms $= 3\Sigma n - 3\Sigma 1$

$$= 3\frac{n}{2}(n+1) - 3n = \frac{3}{2}n(n-1)$$

the $(3n + 1)$th term $= 3n + 1$

$$\therefore \text{total sum } \frac{3}{2}n(n-1) + 3n + 1 = \frac{1}{2}(3n^2 + 3n + 2).$$

Example 5:

Sum the series

$n \cdot 1 + (n - 1) \cdot 2 + (n - 2) \cdot 3 + ... + 1 \cdot n$

Solution:

$T_r = r\,(n - r + 1) = (n - 1)\,r - r^2$

$$\therefore \text{sum } = (n+1)\sum_{r=1}^{n} r - \sum_{r=1}^{n} r^2$$

$$= (n+1)\frac{n}{2}(n+1) - \frac{1}{6}n(n+1)(2n+1)$$

$$\frac{n}{6}(n+1)(n+2).$$

Example 6:

Sum the series

$1^2 + (1^2 + 2^2) + (1^2 + 2^2 + 3^2) + ...$ to n terms.

Solution:

$$T_n = 1^2 + 2^2 + ... + n^2 = \frac{1}{6}n(n+1)(2n+1)$$

$$= \frac{1}{3}n^3 + \frac{1}{2}n^2 + \frac{1}{6}n$$

$$= \frac{1}{3}n^3 + \frac{1}{2}n^2 + \frac{1}{6}n$$

$$\therefore S_n = \frac{1}{2}\Sigma n^3 + \frac{1}{2}\Sigma n^2 + \frac{1}{6}n$$

$$= \frac{1}{2}\frac{n^2(n+1)^2}{4} + \frac{1}{2}\frac{n(n+1)(2n+1)}{6} + \frac{1}{6}.\frac{n(n+1)}{2}$$

$$= \frac{n(n+1)^2(n+2)}{12}.$$

METHOD OF DIFFERENCES

If $T_1, T_2, T_3, \ldots T_n$ be the terms of a series and the series formed by the differences, $T_2 - T_1, T_3 - T_2, T_4 - T_3$ and so on is an A.P. or G.P., such a series can be summed up as shown below.

Example 1:

Find the *n*th term and sum to *n* terms of the series

$3 + 5 + 9 + 15 + 23 + \ldots$

Solution:

$S = 3 + 5 + 9 + 15 + 23 + \ldots + T_n.$

Also $$\frac{S = 3+5+9+15+\ldots+T_{n-1}+T_n}{0 = 3+24+6+8+\ldots+(T_n - T_{n-1}) - T_n}$$

Subtracting, $T_n = 3 + [2 + 4 + 6 + 8 + \ldots$ to $(n - 1)$ terms]

$$3+\frac{n-1}{2}.[4+(n-2)\cdot 2]$$

$T_n = 3 + n(n - 1) = 3 - n + n^2$

$\therefore S\ \Sigma(3 - n + n^2) = 3\ \Sigma 1 - \Sigma n + \Sigma n^2$

$$3\cdot n - \frac{1}{2}.n(n+1) + \frac{1}{6}n(n+1)(2n+1) = \frac{n^3+8n}{3}.$$

Example 2:

$S = 1 + 5 + 13 + 29 + \ldots$ to n terms,

$\therefore$ Subtracting $$\frac{S = 1+5+13+\ldots+T_{n-1}+T_n}{0 = 1+4+8+16+\ldots\text{to } n\text{ -1 terms } -T_n}$$

$\therefore T_n = 1 + (4 + 8 + 16 + \ldots$ to $n - 1$ terms)*

$$1+4\cdot\frac{2^{n-1}-1}{2-1}1+4(2^{n-1}-1) = 2^{n+1}-3.$$

$$\therefore S = \sum 2^{n+1} - 3\sum 1$$

$$= 2^2 + 2^3 + ...2^{n+1} - 3n = \frac{2^2 + 2^n}{2-1} - 3n$$

$$= 2^{n+2} - 4 - 3n.$$

Example 3:

Find the sum of n terms of the series

1.3..5 + 3.5.7 +5.7..9 + ...

Solution:

nth term is the prlduct of the nth term of each of the three arithmetic series 1, 3, 5, ..., 3, 5, 7, ...; 5, 7, 9....

$\therefore$ nth term = $(2n - 1)(2n + 1)(2n + 3)$

$= 8n^3 + 12n^2 - 2n - 3$

$$\therefore\ S = 8\sum n^3 + 12\sum n^2 - 2\sum n - 3\sum 1$$

$$= 8\cdot\frac{n^2(n+1)^2}{4} + 12\cdot\frac{1}{6}n(n+1)(2n+1) - 2\cdot\frac{1}{6}n(n+1) - 3n$$

$$= 2n^4 + 8n^3 + 7n^2 - 2n.$$

Example 4:

Sum to n terms the series $\frac{1}{1.3} + \frac{1}{3.5} + \frac{1}{5.7} + ...$

Solution:

Here $T_n = \frac{1}{(2n-1)(2n+1)} = \frac{1}{2}\left(\frac{1}{2n-1} - \frac{1}{2n+1}\right)$

$$\therefore\ T_1 = \frac{1}{2}\left(\frac{1}{1} - \frac{1}{3}\right)$$

$$T_2 = \frac{1}{2}\left(\frac{1}{3} - \frac{1}{5}\right)$$

$$T_3 = \frac{1}{2}\left(\frac{1}{5} - \frac{1}{7}\right)$$

... ...

$$T_n = \frac{1}{2}\left(\frac{1}{2n-1} - \frac{1}{2n+1}\right)$$

Adding both sides, we get

$$S = \frac{1}{2}\left(-\frac{1}{2n+1}\right) = \frac{n}{(2n+1)}$$

If the series proceeds upto infinity then the sum is given by

$$S = \lim_{n \to \infty} \frac{n}{2n+1} \lim_{n \to \infty} \frac{1}{2+\frac{1}{n}} = \frac{1}{2+0} = \frac{1}{2}.$$

Example 5:

Sum to n terms

$$1 + \frac{1}{2+1} + \frac{1}{1+2+3} + \frac{1}{1+2+3+4} + \ldots$$

Solution:

$$T_n = \frac{1}{\frac{n}{2}(n-1)} = 2\left(\frac{1}{n} - \frac{1}{n+1}\right)$$

$$T_1 = 2\left(1 - \frac{1}{2}\right)$$

$$T_2 = 2\left(\frac{1}{2} - \frac{1}{3}\right)$$

$$T_3 = 2\left(\frac{1}{3} - \frac{1}{4}\right)$$

....

....

$$T_n = 2\left(\frac{1}{n} - \frac{1}{n+1}\right)$$

$$S_n = 2\left(1 - \frac{1}{n+1}\right) = \frac{2n}{n+1}.$$

Example 6:

Sum to n terms

$$\frac{1}{1.3.5} + \frac{1}{3.5.7} + \frac{1}{5.7.9} + \ldots$$

Solution:

$$T_n = \frac{1}{(2n-1)(2n+1)(2n+3)}$$

$$= \frac{1}{2}\left(\frac{1}{2n-1} - \frac{1}{2n+1}\right)\frac{1}{(2n+3)}$$

$$=\frac{1}{2}\frac{1}{(2n-1)(2n+3)}-\frac{1}{2}\frac{1}{(2n+1)(2n+3)}$$

$$=\frac{1}{8}\left(\frac{1}{2n-1}-\frac{2}{2n+1}+\frac{1}{2n+3}\right)$$

$$T_1=\frac{1}{8}\left(\frac{1}{3}-\frac{2}{5}+\frac{1}{7}\right)$$

$$T_2=\frac{1}{8}\left(\frac{1}{3}-\frac{2}{5}+\frac{1}{7}\right)$$

$$T_3=\frac{1}{8}\left(\frac{1}{5}-\frac{2}{7}+\frac{1}{9}\right)$$

... ...

... ...

$$T_n=\frac{1}{8}\left(\frac{1}{2n-1}-\frac{1}{2n+1}+\frac{1}{2n+3}\right)$$

$$S=\frac{1}{8}\left(1-\frac{1}{3}-\frac{2}{2n+1}+\frac{1}{(2n+3)}\right)$$

$$=\frac{n(n+3)}{3(2n+1)(2n+3)}.$$

Example 7:

Find the sum of n terms of the series

$$1+\left(1+\frac{1}{2}\right)+\left(1+\frac{1}{2}+\frac{1}{2^2}\right)+\ldots$$

Solution:

$$T_n=1+\frac{1}{2}+\frac{1}{2^2}+\ldots+\frac{1}{2^{n-1}}=\frac{1-\frac{1}{2^n}}{1-\frac{1}{2}}=-\frac{1}{2^{n-1}}$$

$$T_1=2-\frac{1}{1}$$

$$T_2=2-\frac{1}{2}$$

$$T_3=2-\frac{1}{2^2}$$

... ...

... ...

$$T_n = 2 - \frac{1}{2^{n-1}}$$

$$S_n = 2n - \left(1 + \frac{1}{2} + \frac{1}{2^2} + ... + \frac{1}{2^{n-1}}\right)$$

$$= 2n - \frac{1 - \frac{1}{2^n}}{1 - \frac{1}{2}} = 2n - 2\left(1 - \frac{1}{2_n}\right)$$

$$= 2n - 2 - \frac{1}{2^{n-1}}.$$

EXERCLSES

1. Sum the following series:

(1) $1 - \frac{3}{2} + \frac{5}{4} + \frac{7}{8} ...$ to infinity.

(2) $1 - \frac{3}{4} + \frac{7}{16} + \frac{15}{64} + \frac{31}{256} + ...$ to infinity.

(3) $1 + 4x + 9x^2 + 16x^3 + ...$ to infinity ($x < 1$).

(4) $1 + 3x + 6x^2 + 10x^3 + ...$ to infinity ($x < 1$).

2. Find the sum up to n terms of a series whose nth term is

(1) $(n + n^2)$.

(2) $3n^2 + 2n$.

(3) $(4n^3 + 6n^2 + 2n$.

(4) $(2n + 1) 2^n$

(5) $\frac{n}{2^{n-1}}$ (infinite sum).

3. Find the sum upto n terms of the following series :

(1) $1^2 + 3^2 + 5^2 + ...$

(2) $2^2 + 4^2 + 6^2 + ...$

(3) $1^3 + 3^3 + 5^3 + ...$

4. Find the sum of the infinite series :

$1 + (1 + b) r + (1 + b + b^2) r^2 + ...$

5. Sum the following series upto n terms :

(1) 1.3 + 3.5 + 5.7 + ...

(2) 1.2.4 + 2.3.5 + 3.4.6 + ...

(3) $1.2^2 + 2.3^2 + 3.4^2 + ...$

(4) 2.5.8 + 5.8.11 + 8.11.14 + ...

6. Find then nth term and the sum of n terms of the series :

(1) 1 + 4 + 9 + 16 + ...

(2) 3 + 15 + 35 + 63 + ...

(3) 12 + 16 + 24 + 40 + ...

(4) 5 + 7 + 11 + 19 + ...

(5) 4 + 14 + 30 + 52 + 80 + ...

(6) 1 + 5 + 13 + 29 + 61 + ...

(7) 1 + 3 + 7 + 15 + 31 + ...

7. Find the sum of n terms of the series :

$$1+\left(1+\frac{1}{3}\right)+\left(1+\frac{1}{3}+\frac{1}{3^2}\right)+...$$

8. Sum to n terms the following series :

(1) $\frac{1}{3.7}+\frac{1}{7.11}+\frac{1}{11.15}+...$

(2) $\frac{1}{5.8}+\frac{1}{8.11}+\frac{1}{11.14}+...$

(3) $\frac{1}{1.2.3}+\frac{1}{2.3.4}+\frac{1}{3.4.5}+...$

2

Determinants

INTRODUCTION

Consider the following two equations:

$$a_1x + b_1y = 0$$

$$a_2x + b_2y = 0$$

$$\therefore \qquad -\frac{a_1}{b_1} = \frac{y}{x} = -\frac{a_2}{b_2}$$

Eliminating x and y we get $-\frac{a_1}{b_1} = -\frac{a_2}{b_2}$

or $a_1b_2 - a_2b_1 = 0$

We shall experts the above eliminate in the form of

$$\begin{vmatrix} a_1 & b_1 \\ a_2 & b_2 \end{vmatrix} = 0 \qquad ...(1)$$

(1) is called a determinant of second order and its value is $a_1b_2 - a_2b_1 = 0$

Let us non eliminate x, y, z from the following equations:

$$a_1x + b_1y + c_1 = 0$$

$$a_2x + b_2y + c_2 = 0$$

$$a_3x + b_3y + c_3 = 0$$

Solving eqn. (2) and (3) by cross multiplication, we have

$$\frac{x}{b_2c_3 - b_3c_2} = \frac{y}{(a_2c_3 - a_3c_2)} = \frac{y}{a_2c_3 - a_3b_2}$$

Substituting the value of x_1 and z in (1), we get

$$K[a_1(b_2c_3 - b_3c_2) - b_1(a_2c_3 - c_3a_2) + c_1(a_2b_3 - a_3b_2)] = 0$$

∴ The eliminant is $a_1(b_2c_3 - b_3c_2) - (a_2c_3 - c_3a_2) + c_1(a_2b_3 - a_3b_2) = 0$

$$\begin{vmatrix} a_1 & b_1 & c_1 \\ a_2 & b_2 & c_2 \\ a_3 & b_3 & c_3 \end{vmatrix} = 0 \qquad ...(2)$$

eqn. (2) is called the determinant of order (3).

COFACTOR OF AN ELEMENT

Definition: *If in the expansion of a determinant* $| a_{ij} |$*, all the terms containing* a_{ij} *as a factor are collected and their sum be denoted by* $a_{ij} C_{ij}$ *then the factor* C_{ij} *is defined as the cofactor of the element* a_{ij}.

From the above definition we find that if $[a_{ij}]$ be the n × n matrix whose determinant is $| a_{ij} |$ then if from $[a_{ij}]$ the element of its ith row and jth column are removed, the terms of C_{ij} are then composed of elements from the remaining (n – 1) × (n – 1) sub-matrix M_{ij} of $[a_{ij}]$.

Hence, the determinant $| a_{ij} |$ can be expressed as a function of the elements of ith row, by collecting all the terms containing $a_{i1}, a_{i2}, a_{i3}, \ldots,$ a_{in} and finding their sum.

i.e., $| a_{ij} | = a_{i1} C_{i1} + a_{i2} C_{i2} + \ldots + a_{in} C_{in}$

$$= \sum_{j-1}^{n} a_{ij} C_{ij},$$

which is the expansion of the determinant $| a_{ij} |$ by the elements of the ith row and their cofactors.

In a similar manner we can expand the determinant $| a_{ij} |$ by the elements of the kth column and their cofactor and write as $a_{ij} | = \sum_{j-k}^{n} a_{ik} C_{ik}$

Or

The determinant that is left by cancelling the row and columns intersecting at a particular constituent, when the particular constituent is brought in the top left hand corner of a determinant is called its cofactor and is denoted by corresponding capital letter.

PROPERTIES OF DETERMINANTS

Prop. I:

If $A = [a_{ij}]$ is an $n \times n$ matrix then $|A'| = |A|$, where A' is the transpose of the matrix A.

Or

The value of a determinant is not altered by changing the rows into column and column into rows.

Proof:

If $A = [a_{ij}]$, then $A' = [a'_{ij}]$, where $a'_{ij} = a_{ij}$.

Now the product of the elements of the principal diagonal of

$$A' = a'_{11}\, a'_{22}\, a'_{33} \ldots a'_{nn}.$$

Operating on the row subscripts of the elements of this product by the permutation $p = \begin{pmatrix} 1 & 2 & 3 & n \\ i_1 & i_2 & i_3 \ldots i_n \end{pmatrix}$, where $i_1, i_2, i_3, \ldots$ are 1, 2, 3, ... in some order, we have $\pm\, a'_{i11}\, a'_{i22}\, a'_{i33} \ldots a'_{inn}$ as a term of $|A'|$ plus or minus sign to be taken according as p is even or odd.

Now as $a'_{ij} = a_{ji}$, so we have

$a'_{i11}\, a'_{i22}\, a_{i33} \ldots a'_{inn} = a_{1/1}\, a_{2/2}\, a_{3/3} \ldots a_{n/n}$

The term $a_{1/1}\, a_{2/2} \ldots a_{n/n}$ can be obtained from the term $a_{i1/1}\, a_{i2/2}\, a_{i3/3} \ldots a_{in/n}$ by operating on its row subscripts the permutation $p' = \begin{pmatrix} i_1 & i_2 & i_3 \ldots i_n \\ 1 & 2 & 3 & n \end{pmatrix} = p^{-1}$.

Hence p' is even or odd according as p is even or odd. Therefore the term $\pm a'_{i11}\, a'_{i22}\, a'_{i33} \ldots a'_{inn}$ of $|A'|$ is also a term of $|A|$.

We can thus prove that every one of the n! terms of $|A'|$ is a term of $|A|$. Hence the property.

For example, If $|A| = \begin{vmatrix} a_1 & a_2 & a_3 \\ b_1 & b_2 & b_3 \\ c_1 & c_2 & c_3 \end{vmatrix}$

$$= a_1 \begin{vmatrix} b_2 & b_3 \\ c_2 & c_3 \end{vmatrix} - a_2 \begin{vmatrix} b_1 & b_3 \\ c_1 & c_3 \end{vmatrix} + a_3 \begin{vmatrix} b_1 & b_2 \\ c_1 & c_2 \end{vmatrix}$$

$$= a_1b_2c_3 - a_1b_3c_2 - a_2b_1c_3 + a_2b_3c_1 + a_3b_1c_2 - a_3b_2c_1.$$

Then $|A'| = \begin{vmatrix} a_1 & b_1 & c_1 \\ a_2 & b_2 & c_2 \\ a_3 & b_3 & c_3 \end{vmatrix}$

$$= a_1 \begin{vmatrix} b_2 & c_2 \\ b_3 & c_3 \end{vmatrix} - b_1 \begin{vmatrix} a_2 & c_2 \\ a_3 & c_3 \end{vmatrix} + c_1 \begin{vmatrix} a_2 & b_2 \\ a_3 & b_3 \end{vmatrix}$$

$$= a_1b_2c_3 - a_1b_3c_2 - a_2b_1c_3 + a_3b_1c_2 + a_2b_3c_1 - a_3b_2c_1.$$

$\therefore \quad |A'| = |A|.$

Prop. 2:

If B is obtained from A by interchanging two raws (or columns) then $|B| = \; |A|$.

Or

It two adjacent rows or columns of a determinant are interchanged the sign of the determinant is changed whereas its numerical value remarks the same.

Proof:

Let us suppose that sth and rth columns of the determinant A are interchanged where $s < 1$.

Let $A = \begin{bmatrix} a_{11} & a_{12} & \cdots & a_{1s} & \cdots & a_{1t} & \cdots & a_{1n} \\ a_{21} & a_{22} & \cdots & a_{2s} & \cdots & a_{2t} & \cdots & a_{2n} \\ \cdots & \cdots & \cdots & \cdots & \cdots & \cdots & \cdots & \cdots \\ a_{i1} & a_{i2} & \cdots & a_{is} & \cdots & a_{it} & \cdots & a_{in} \\ \cdots & \cdots & \cdots & \cdots & \cdots & \cdots & \cdots & \cdots \\ a_{n1} & a_{n2} & \cdots & a_{ns} & \cdots & a_{nt} & \cdots & a_{nn} \end{bmatrix}$

Then the trace of A *i.e.,* the product of the elements of the principal diagonal of A $= a_{11}\, a_{22} \ldots a_{ss} \ldots a_{tt} \ldots a_{nn}$.

If the sth and rth columns are interchanged the product of the elements of the principal diagonal of B

$= a_{11}\, a_{22} \ldots a_{ss} \ldots a_{tt} \ldots a_{nn}$ **(Note)**

In order to have a term of $|A|$, let us operate on the row subscripts of the trace of A by the permutation

$$p = \begin{pmatrix} 1 & 2 & \ldots & s & \ldots & t & \ldots & n \\ i_1 & i_2 & \ldots & i_s & \ldots & i_t & \ldots & i_n \end{pmatrix}$$

Then we have $a_{i11}\ a_{i22}\ \ldots\ a_{iss}\ \ldots\ a_{itt}\ \ldots\ a_{inn}$.

This term can also be obtained from the trace of B by operating on its row subscripts by the permutation

$$p' = \begin{pmatrix} 1 & 2 & \ldots & s & \ldots & t & \ldots & n \\ i_1 & i_2 & \ldots & i_t & \ldots & i_s & \ldots & i_n \end{pmatrix}$$

Here we observe that $p' = p\,(i_s\, i_t)$, since $(i_s\, i_t)$ is a transposition. Therefore p' is odd or even according as p is even or odd. Hence the term $\pm a_{i11}\ a_{i22}\ \ldots\ a_{iss}\ \ldots\ a_{itt}\ \ldots\ a_{inn}$ is also a term of | B | but with its sign changed. Thus every one of the n! terms of | A | is a term of | B | but with sign changed.

Here | B | = – | A |.

Example:

$$\text{Let } |A| = \begin{vmatrix} a_1 & a_2 & a_3 \\ b_1 & b_2 & b_3 \\ c_1 & c_2 & c_3 \end{vmatrix}$$

Let the determinant | B | be formed by interchanging second and third columns.

$$\text{Then } |B| = \begin{vmatrix} a_1 & a_2 & a_3 \\ b_1 & b_2 & b_3 \\ c_1 & c_2 & c_3 \end{vmatrix}$$

Expanding | B | by the elements of its first row, we get

$$|B| = a_1 \begin{vmatrix} b_3 & b_2 \\ c_3 & c_2 \end{vmatrix} - a_3 \begin{vmatrix} b_1 & b_2 \\ c_1 & c_2 \end{vmatrix} + a_2 \begin{vmatrix} b_1 & b_3 \\ c_1 & c_3 \end{vmatrix}$$

$$= a_1(b_3c_2 - b_2c_3) - a_3(b_1c_2 - b_2c_1) + a_2(b_1c_3 - b_3c_1)$$

$$= -\ [a_1(b_2c_3 - b_3c_2) - a_2(b_1c_3 - b_3c_1) + a_3(b_1c_2 - b_2c_1)$$

$$= -\left[a_1 \begin{vmatrix} b_2 & b_3 \\ c_2 & c_3 \end{vmatrix} - a_2 \begin{vmatrix} b_1 & b_3 \\ c_1 & c_3 \end{vmatrix} + a_3 \begin{vmatrix} b_1 & b_2 \\ c_1 & c_2 \end{vmatrix}\right]$$

$$= - \begin{vmatrix} a_1 & a_2 & a_3 \\ b_1 & b_2 & b_3 \\ c_1 & c_2 & c_3 \end{vmatrix} = -|A|.$$

Prop. 3:

If the elements of ith row (or ith column) of the determinant $|a_{ij}|$ *are multiplied by a scalar c then the resulting determinant is* $c\,|a_{ij}|$.

Proof:

We can write $| a_{ij} | = \sum_{j=1}^{n} a_{ij} . C_{ij} \ldots$

In this case, the resulting determinant (when the elements of ith row are multiplied by c)

$$= \sum_{j=1}^{n} ca_{ij} C_{ij} = c \sum_{j=1}^{n} a_{ij} C_{ij} = c \, | a_{ij} |.$$

Similarly we can prove that statement when elements of kth column are multiplied by c.

Prop. 4:

If two rows or two columns of a determinant | A | are identical then | A | = 0.

Proof:

In Prop. 2 above we have proved that if any two rows or columns of a determinant are interchanged then the value of the determinant changes in sign only.

Thus, if the two *identical* columns (or rows) of a determinant are interchanged, then the determinant does not change but its sign only changes.

Hence | A | = – | A | *i.e.,* | A | + | A | = 0 *i.e.,* | A | = 0.

Example:

Evaluate $\begin{vmatrix} a_1 & a_2 & a_1 \\ b_1 & b_2 & b_1 \\ c_1 & c_2 & c_1 \end{vmatrix}$

Expanding the determinant with respect to the first row, we have the given determinant

$$= a_1 \begin{vmatrix} b_2 & b_1 \\ c_2 & c_1 \end{vmatrix} - a_2 \begin{vmatrix} b_1 & b_1 \\ c_1 & c_1 \end{vmatrix} + a_1 \begin{vmatrix} b_1 & b_2 \\ c_1 & c_2 \end{vmatrix}$$

$$= a_1(b_2c_1 - b_1c_2) - a_2(b_1c_1 - b_1c_1) + a_1(b_1c_2 - b_2c_1)$$

$$= a_1b_2c_1 - a_1b_1c_2 - a_2(0) + a_1b_1c_2 - a_1b_2c_1 = 0.$$

AN IMPORTANT PROPERTY OF THE DETERMINANT

(a) *If A_r, the ith row of a determinant $| A | = | a_{ij} |$ of order n be replaced by $A_j + cA_k$, where c is a scalar and A_k denotes the kth row of the determinant | A |, then the value of the determinant remains unaltered.*

Proof:

The determinant $|A| = \sum_{j=1}^{n} a_{ij} C_{ij}$.

Replacing A_i by $A_i + c A_k$ we get the new determinant $|B|$, say

$$\text{Then} \quad |B| = \sum_{j=1}^{n} (a_{ij} + c\, a_{kj}) C_{ij}$$

$$= \sum_{j=1}^{n} a_{ij} C_{ij} + c \sum_{j=1}^{n} a_{kj} C_{ij}$$

$$= |A| + c.0$$

$$\text{i.e.,} \quad |B| = |A|.$$

(b) *If C_i, the ith column of determinant $|A| = |a_{ij}|$ of order n be replaced by $C_i + \lambda C_k$ where λ is a scalar and C_k denotes the kth column of $|A|$, then the value of the determinant remains unaltered.*

Proof is similar to part (a) above.

In the following examples R_1. R_2, R_3, . . . stand for first, second third, . . . rows and C_1, C_2,C_3, . . . stand for first, second, third . . . columns.

DETERMINANT OF ORDER TWO

Let us consider a square matrix $A = \begin{bmatrix} a_{11} & a_{12} \\ a_{21} & a_{22} \end{bmatrix}$ of order 2 × 2.

Then 2! permutations on two symbols 1 and 2 are

$$I = \begin{pmatrix} 1 & 2 \\ 1 & 2 \end{pmatrix}, \; p = \begin{pmatrix} 1 & 2 \\ 2 & 1 \end{pmatrix}.$$

The product of the elements of the principal diagonal are $a_{11}\, a_{22}$.

Operating on the row subscripts of $a_{11}\, a_{22}$ by the permutation I we get $+ a_{11}\, a_{22}$, prefixing + sign as the permutation I is even and by the permutation p we get $-a_{21}\, a_{12}$, prefixing – sign as the permutation is odd. **(Note)**

$$\text{Hence} \quad \begin{vmatrix} a_{11} & a_{12} \\ a_{21} & a_{22} \end{vmatrix} = a_{11}\, a_{22} - a_{21}\, a_{12}$$

DETERMINANT OF ORDER THREE

Let us consider a square matrix $A = \begin{bmatrix} a_{11} & a_{12} & a_{13} \\ a_{21} & a_{22} & a_{23} \\ a_{31} & a_{32} & a_{33} \end{bmatrix}$, of order 3 × 3.

Then 3! permutations on three symbols 1, 2, and 3 are

$$I = \begin{pmatrix} 1 & 2 & 3 \\ 1 & 2 & 3 \end{pmatrix}; \ p_1 = \begin{pmatrix} 1 & 2 & 3 \\ 1 & 3 & 2 \end{pmatrix}; \ p_2 = \begin{pmatrix} 1 & 2 & 3 \\ 2 & 3 & 1 \end{pmatrix};$$

$$p_3 = \begin{pmatrix} 1 & 2 & 3 \\ 2 & 1 & 3 \end{pmatrix}; \ p_4 = \begin{pmatrix} 1 & 2 & 3 \\ 3 & 2 & 1 \end{pmatrix}; \text{ and } p_5 = \begin{pmatrix} 1 & 2 & 3 \\ 3 & 2 & 1 \end{pmatrix};$$

Operating on the row subscripts of $a_{11}\ a_{22}\ a_{33}$ by the permutation I, p_1, p_2, . . ., p_5 we have successively

(a) $+a_{11}\ a_{22}\ a_{33}$, prefixing + as permutation I is even,

(b) $-a_{11}\ a_{32}\ a_{23}$, prefixing – as permutation p_1 is odd,

(c) $+a_{21}\ a_{32}\ a_{13}$, prefixing + sign as permutation p_2 is even,

(d) $-a_{21}\ a_{12}\ a_{33}$, prefixing + sign as permutation p_3 is odd,

(e) $+a_{31}\ a_{12}\ a_{23}$, prefixing + sign as permutation p_4 is even,

(f) $-a_{31}\ a_{22}\ a_{13}$, prefixing – sign as permutation p_5 is odd.

Hence we have

$$\begin{bmatrix} a_{11} & a_{12} & a_{13} \\ a_{21} & a_{22} & a_{23} \\ a_{31} & a_{32} & a_{33} \end{bmatrix} = a_{11}\ a_{23}\ a_{33} - a_{11}\ a_{32}\ a_{23} + a_{21}\ a_{32}\ a_{13} - a_{21}\ a_{12}\ a_{33}$$

$$+ a_{31}\ a_{12}\ a_{23} - a_{31}\ a_{22}\ a_{13}$$

$$= a_{11}\ (a_{22}\ a_{33} - a_{23}\ a_{32}) - a_{13}\ (a_{21}\ a_{33} - a_{23}\ a_{21}) + a_{13}\ (a_{21}\ a_{22} - a_{22}\ a_{31})$$

$$= a_{11} \begin{vmatrix} a_{22} & a_{23} \\ a_{32} & a_{33} \end{vmatrix} - a_{12} \begin{vmatrix} a_{21} & a_{23} \\ a_{31} & a_{33} \end{vmatrix} + a_{13} \begin{vmatrix} a_{21} & a_{22} \\ a_{31} & a_{32} \end{vmatrix}.$$

DETERMINANT OF A SQUARE MATRIX

Let us consider a square matrix A of order n × n given by

$$A = \begin{bmatrix} a_{11} & a_{12} & a_{13} & \dots & \dots & a_{1n} \\ a_{21} & a_{22} & a_{23} & \dots & \dots & a_{2n} \\ a_{31} & a_{32} & a_{33} & \dots & \dots & a_{3n} \\ \dots & \dots & \dots & \dots & \dots & \dots \\ a_{n1} & a_{n2} & a_{n3} & \dots & \dots & a_{nn} \end{bmatrix}$$

The product of the elements in the principal diagonal is

$$a_{11}\ a_{22}\ a_{33}\ .\ .\ .\ a_{nn}.$$

This is also called the *trace of the matrix.*

Now obtain n! terms of the above type by operating on the row-subscripts of the elements of the above expression by n! permutations

$p = \begin{pmatrix} 1 & 2 & 3 & \dots & n \\ i_1 & i_2 & i_3 & \dots & i_n \end{pmatrix}$, where $i_1, i_2, i_3, \dots i_n$ are one of the n! permutations of the integers 1, 2, 3, . . ., n.

The sum of n! signed terms thus obtained is defined as the *determinants of the matrix* A and is denoted by | A | or | a_{ij} |.

Therefore the determinant of the square matrix A = $[a_{ij}]$ of order n × n is given by

$$| a_{ij} | = \sum \pm \; a_{\alpha 1} \, a_{\beta 2} \, a_{\gamma 3} \, a_{\delta 4} \dots a_{k_n}$$

where + or – sign is taken where α, β, γ, δ, . . ., k is an even or odd permutation of 1, 2, 3, . . ., n, and the summation extends over n! permutations of the row subscripts 1, 2, 3, ...

Note: The determinant of a square matrix of order n is known as a determinant of order n.

MINOR OF AN ELEMENT

Definition:

If M_{ij} be the (n – 1) × (n – 1) sub-matrix of the matrix A = $[a_{ij}]$ obtained by removing the ith row and jth column, then the determinant | M_{ij} | is defined as the minor of the element a_{ij} in the determinant | a_{ij} | of order n.

Theorem 1:

The cofactor C_{ij} of the element a_{ij} in the determinant | a_{ij} | is given by $C_{ij} = (-1)^{i+j} \,|\, M_{ij} \,|$.

Proof:

Let us first of all prove the case $C_{11} = (-1)^2 \,|\, M_{11} \,|$ *i.e.*, $C_{11} = |\, M_{11} \,|$.

The terms in C_{11} are composed of elements taken from the (n – 1) × (n – 1) sub-matrix M_{11} of A.

The general term of $a_{11}\, C_{11} = \pm\, a_{11}\, a_{i22}\, a_{i33} \dots a_{inn}$, where $i_2, i_3, \dots i_n$ are 2, 3, . . ., n is some order.

This term can also be obtained from the trace of matrix A *i.e.*, the product of the elements of the diagonal of the matrix A *i.e.*, $a_{11}\, a_{22}\, a_{33} \dots a_{nn}$, by operating on its row subscripts by the permutation $p = \begin{pmatrix} 1 & 2 & 3 & \dots & n \\ i_1 & i_2 & i_3 & \dots & i_n \end{pmatrix}$,

where $i_2, i_3, \dots, i_n$ are defined as above.

Thus the permutation p may be regarded as a permutation on the symbols 2, 3, . . ., n. Hence all the terms of a_{11} C_{11} can be obtained by running p through the (n – 1)! permutations on the symbols 2, 3, . . ., n keeping 1 fixed.

Thus, the terms of C_{11} can be obtained by operating on the row subscripts of the elements of the product a_{22} a_{33} . . . a_{nn}, which is the product of the elements of the diagonal of M_{11} *i.e.,* the trace of M_{11}.

Hence $C_{11} = | M_{11} |$.

Now let us prove $C_{ij} = (-1)^{i+j} | M_{11} |$.

Move the jth column of the matrix A to the first column by performing (j – 1) successive interchanges of adjacent columns and move the ith row of the matrix A to the first row by performing (i – 1) successive interchanges of adjacent rows. Then the element a_{ij} is in the first row and first column of the resulting matrix B, say. The sub-matrix of B obtained by removing the first row and first column is the sub-matrix M_{ij} of the matrix A. Hence a_{ij} $| M_{ij} |$ is the term of | B | containing a_{ij}.

Also we know that, if two rows or two columns of a determinant | A | are interchanged, the new determinant = – | A |.

$\therefore$ We have $| B | = (-1)^{i-1+j-i} | A |$ **(Note)**

$= (-1)^{i+j} (-1)^{-2} | A |$

$= (-1)^{i+j} | A |,$ $\because$ $(-1)^{-2} = 1$

$\Rightarrow$ $| A | = (-1)^{i+j} | B |$ **(Note)**

Equating the coefficient of a_{ij} from both sides, we have

$C_{ij} = (-1)^{i+j} | M_{ij} |$.

Theorem 2:

If C_{ij} is the cofactor of a_{ij} in the determinant $| A | = | a_{ij} |$ of order n, then (i) the sum of the products of the elements of the ith row with the cofactors of the corresponding elements of the kth row is zero provided $i \neq k$.

(ii) Also the sum of the products of the elements of the jth column with the cofactors of the corresponding elements of the kth column is zero provided $j \neq k$,

i.e., (i) $$\sum_{j=1}^{n} c_{ij} C_{kj} = 0, \text{ if } i \neq k$$

and (ii) $$\sum_{j=1}^{n} c_{ij} C_{ik} = 0, \text{ if } j \neq k$$

Proof:

(i) The given determinant $|A| = \sum_{j=1}^{n} a_{kj}C_{kj}$.

Now replace the kth row by ith row, then we have the new determinant $= \sum_{j=1}^{n} a_{ij} \cdot C_{kj}$.

But the kth and ith rows of the new determinant are identical, hence its values is zero.

$$\sum_{j=1}^{n} a_{ij} C_{kj} = 0$$

(ii) The given determinant $|A| = \sum_{i=1}^{n} a_{ik} C_{ik}$

Now replace the kth column by jth column, then we have the new determinant $= \sum_{i=1}^{n} a_{ij} C_{ik}$.

But the kth and jth columns of the new determinant are identical, hence its value is zero.

$$\therefore \quad \sum_{i=1}^{n} a_{ij} C_{ik} = 0.$$

SOLVED EXAMPLES

Example 1:

Evaluate $\begin{vmatrix} 1^2 & 2^2 & 3^2 & 4^2 \\ 2^2 & 3^2 & 4^2 & 5^2 \\ 3^2 & 4^2 & 5^2 & 6^2 \\ 4^2 & 5^2 & 6^2 & 7^2 \end{vmatrix}$

Solution:

We have

$$= \begin{vmatrix} 1 & 4 & 9 & 16 \\ 4 & 9 & 16 & 25 \\ 9 & 16 & 25 & 36 \\ 16 & 25 & 36 & 49 \end{vmatrix}$$

$$= \begin{vmatrix} 1 & 0 & 0 & 0 \\ 4 & -7 & -20 & -39 \\ 9 & -20 & -56 & -108 \\ 16 & -39 & -108 & -207 \end{vmatrix}$$, replacing C_2, C_3, C_4 by $C_2 - 4C_1, C_2 - 9C_1$ and $C_4 - 16C_1$ respctively.

$$= (-1)^2 \begin{vmatrix} 7 & 20 & 39 \\ 20 & 56 & 108 \\ 39 & 108 & 207 \end{vmatrix},$$

$$= \begin{vmatrix} 7 & -1 & -1 \\ 20 & -4 & -4 \\ 39 & -9 & -9 \end{vmatrix}$$, replacing C_2 and C_3 by $C_2 - 3C_1$ and $C_3 - 2C_3$ respctively.

= 0, as two columns are identical. **Ans.**

Example 2:

Show that $\begin{vmatrix} 1 & a & bc \\ 1 & b & ca \\ 1 & c & ab \end{vmatrix} = (b - c)\ (c - a)\ (a - b)$

Solution:

The given determinant

$$= \begin{vmatrix} 1 & a & bc \\ 0 & b - a & ca - bc \\ 0 & c - a & ab - bc \end{vmatrix}$$, replacing R_2 and R_3 by $R_2 - R_1$ and $R_3 - R_1$ respective.

$$= \begin{vmatrix} b - a & -c\,(b - a) \\ c - a & -b\,(c - a) \end{vmatrix}$$, expanding with respect to C_1

$$= (b - a)\ (c - a) \begin{vmatrix} 1 & -c \\ 1 & -c \end{vmatrix}$$, taking common factors out

$= (b - a)\ (c - a)\ (-b + c) = (a - b)\ (b - c)\ (c - a)$ **Hence proved.**

Example 3:

Evaluate $\begin{vmatrix} a & x & y & a \\ x & 0 & 0 & y \\ y & 0 & 0 & x \\ a & y & z & a \end{vmatrix}$

Solution:

The given determinant

$$= \begin{vmatrix} 0 & x-y & y & a \\ x-y & 0 & 0 & y \\ y-x & 0 & 0 & x \\ a & y+x & x & a \end{vmatrix},$$ replacing C_1 and C_2 by $C_1 - C_4$ and $C_2 + C_3$ respectively.

$$= (x-y)(x+y) \begin{vmatrix} 0 & 1 & y & a \\ 1 & 0 & 0 & y \\ -1 & 0 & 0 & x \\ 0 & 1 & x & a \end{vmatrix},$$ taking out $x - y$ and $x + y$ common from C_1 and C_2 respectively.

$$= (x^2 - y^2) \begin{vmatrix} 0 & 1 & y & a \\ 1 & 0 & 0 & y \\ 0 & 0 & 0 & x+y \\ 0 & 0 & x-y & a \end{vmatrix},$$ replacing R_3 and R_4 by $R_2 + R_3$ and $R_4 - R_1$ respectively.

$$= -(x^2 - y^2) \begin{vmatrix} 1 & y & a \\ 0 & 0 & x+y \\ 0 & x-y & 0 \end{vmatrix},$$ expanding with respect to C_1

$$= -(x^2 - y^2) \begin{vmatrix} 0 & x+y \\ x-y & 0 \end{vmatrix}$$

$$= -(x^2 - y^2)[-(x+y)(x-y)] = (x^2 - y^2)^2.$$

Example 4:

Evaluate $\begin{vmatrix} 1 & bc & a(b+c) \\ 1 & ca & b(c+a) \\ 1 & ab & c(a+b) \end{vmatrix}$

Solution:

The given determinant

$$= \begin{vmatrix} 1 & bc & a(b+c) \\ 0 & c(a-b) & c(b-a) \\ 0 & b(a-c) & b(c-a) \end{vmatrix},$$ replacing R_2 and R_3 by $R_2 - R_1$ and $R_3 - R_1$ respectively.

$= \begin{vmatrix} c(a-b) & c(b-a) \\ b(a-c) & b(c-a) \end{vmatrix}$, expanding with respect to C_1

$= (a-b)(a-c) \begin{vmatrix} c & -c \\ b & -b \end{vmatrix}$, taking out (a – b) and (a – c) common from R_1 and R_2 respectively

$= -(a-b)(a-c) \begin{vmatrix} c & c \\ b & b \end{vmatrix} = 0$, two columns belong identical. **Ans.**

Example 5:

Evaluate $\begin{vmatrix} 1 & 1 & 1 \\ a & b & c \\ a^2 & b^2 & c^2 \end{vmatrix}$

Solution:

The given determinant

$= \begin{vmatrix} 1 & 0 & 0 \\ a & b-a & c-a \\ a^2 & b^2-a^2 & c^2-a^2 \end{vmatrix}$, replacing R_2 and R_3 by $R_3 - R_1$ and $R_3 - R_1$ respectively.

$= \begin{vmatrix} b-a & c-a \\ b^2-a^2 & c^2-a^2 \end{vmatrix}$, expanding with respect to R_1

$= (b-a)(c-a) \begin{vmatrix} 1 & 1 \\ b+a & c+a \end{vmatrix}$, taking common factors out

$= (b-a)(c-a)[(c+a)-(b+a)]$

$= (a-b)(c-a)(b-c).$ **Ans.**

Example 6:

Prove that

$$\begin{vmatrix} 1 & a^2+bc & a^3 \\ 1 & b^2+ca & b^3 \\ 1 & c^2+ab & c^3 \end{vmatrix} = -(b-c)(c-a)(a-b)(a^2+b^2+c^2)$$

Solution:

The given determinant

$$= \begin{vmatrix} 1 & a^2 & a^3 \\ 1 & b^2 & b^3 \\ 1 & c^2 & c^3 \end{vmatrix} + \begin{vmatrix} 1 & bc & a^3 \\ 1 & ca & b^3 \\ 1 & ab & c^3 \end{vmatrix},$$ breaking into two determinante **(Note)**

$$= \begin{vmatrix} 1 & 1 & 1 \\ a^2 & b^2 & c^2 \\ a^3 & b^3 & c^3 \end{vmatrix} + \begin{vmatrix} 1 & 1 & 1 \\ bc & ca & ab \\ a^3 & b^3 & c^3 \end{vmatrix},$$ interchanging rows and columns **(Note)** ...(i)

The first determinant

= (a – b) (b – c) (c – a) (ab + bc + ca) ...(ii)

The second determinant

$$= \begin{vmatrix} 1 & 0 & 0 \\ ba & ca - bc & ab - bc \\ a^2 & b^3 - a^3 & c^3 - a^3 \end{vmatrix},$$ replacing C_2 and C_3 by $C_2 - C_1$ and $C_3 - C_1$

$$= \begin{vmatrix} c(a-b) & b(a-c) \\ (b-a)(b^2 + ab + a^2) & (c-a)(c^2 + ac + a^2) \end{vmatrix},$$

expanding with respect to R_1

$$= (a-b)(a-c) \begin{vmatrix} c & b \\ -(b^2 + ab + a^2) & -(c^2 + ac + a^2) \end{vmatrix},$$

taking out common factors from C_1 and C_2

$$= (a-b)(a-c) \begin{vmatrix} c-b & b \\ -(b^2 - c^2 + ab - ac) & -(c^2 + ac + a^2) \end{vmatrix},$$

replacing C_1 by $C_1 - C_2$

$$= (a-b)(a-c) \begin{vmatrix} (c-b) & b \\ (c-b)(c + b + a) & -(c^2 + ac + a^2) \end{vmatrix}$$

$$= (a - b)\ (a - c)\ (c - b) \begin{vmatrix} 1 & b \\ a + b + c & -(c^2 + ca + a^2) \end{vmatrix},$$

taking (c – b) common from C_1.

$= (a - b)\ (b - c)\ (c - a)\ [-(c^2 + ca + a^2) - b\ (a + b + c)]$

$= -(a - b)\ (b - c)\ (c - a)\ (a^2 + b^2 + c^2 + ab + bc + ca)$ ...(iii)

Substituting values from (ii) and (iii) in (i), we find the given determinant

$= (a - b)\ (b - c)\ (c - a)\ (ab + bc + ca)$

$- (a - b)\ (b - c)\ (c - a)\ (a^2 + b^2 + c^2 + ab + bc + ca)]$

$= (a - b)\ (b - c)\ (c - a)\ (a^2 + b^2 + c^2)$. **Ans.**

Example 8:

Evaluate $\begin{vmatrix} 1 & 1 & 1 \\ a^2 & b^2 & c^2 \\ a^3 & b^3 & c^3 \end{vmatrix}$

Solution:

We have

$$= \begin{vmatrix} 1 & 0 & 0 \\ a^2 & b^2 - a^2 & c^2 - a^2 \\ a^3 & b^3 - a^3 & c^3 - a^3 \end{vmatrix},$$ replacing C_2 and C_3 by $C_2 - C_1$ and $C_3 - C_1$ respectively.

$$= \begin{vmatrix} b^2 - a^2 & c^2 - a^2 \\ b^3 - a^3 & c^3 - a^3 \end{vmatrix},$$ expanding with respect to R_1

$$= (b - a)\ (c - a) \begin{vmatrix} b + a & c + a \\ b^2 + ab + a^2 & c^2 + ac + a^2 \end{vmatrix},$$

taking out common factors of C_1, C_2

$$= (b - a)\ (c - a) \begin{vmatrix} b - c & c + a \\ b^2 + ab - c^2 - ca & c^2 + ca + a^2 \end{vmatrix},$$

replacing C_1 by $C_1 - C_2$

$$= (b - a)(c - a)\begin{vmatrix} b-c & c+a \\ (b^2-c^2)+a(b-c) & c^2+ca+a^2 \end{vmatrix}$$

$$= (b - a)(c - a)\begin{vmatrix} 1 & c+a \\ b+c+a & c^2+ca+a^2 \end{vmatrix}$$

$= -(a - b)(b - c)(c - a)\,[1\,(c^2 + ca + a^2) - (c + a)(b + c + a)$

$= -(a - b)(b - c)(c - a)$

$[c^2 + ca + a^2 - cb - c^2 - ac - ab - ac - a^2]$

$= -(a - b)(b - c)(c - a)\,[-ab - bc - ca]$

$= (a - b)(b - c)(c - a)(ab + bc + ca).$ **Ans.**

Example 9:

Show that $\begin{vmatrix} a & b & c \\ a^2 & b^2 & c^2 \\ a^3 & b^3 & c^3 \end{vmatrix} = abc\,(a - b)(b - c)(c - a)$

Solution:

The given determinant

$= abc\begin{vmatrix} 1 & 1 & 1 \\ a & b & c \\ a^2 & b^2 & c^2 \end{vmatrix}$, taking out a, b and c common from C_1, C_2 and C_3 respectively.

Ans. $abc\,(a - b)(b - c)(c - a)$

Example 10:

Evaluate $\begin{vmatrix} 1 & 2 & 3 & 4 \\ 2 & 3 & 4 & 1 \\ 3 & 4 & 1 & 2 \\ 4 & 1 & 2 & 3 \end{vmatrix}$

Solution:

The given determinant

$= \begin{vmatrix} 10 & 2 & 3 & 4 \\ 10 & 3 & 4 & 1 \\ 10 & 4 & 1 & 2 \\ 10 & 1 & 2 & 3 \end{vmatrix}$, replacing C_1 by $C_1 + C_2 + C_3 + C_4$

$$= 10 \begin{vmatrix} 1 & 2 & 3 & 4 \\ 1 & 3 & 4 & 1 \\ 1 & 4 & 1 & 2 \\ 1 & 1 & 2 & 3 \end{vmatrix}$$, taking out 10 common from C_1

$$= 10 \begin{vmatrix} 1 & 2 & 3 & 4 \\ 0 & 1 & 1 & -3 \\ 0 & 2 & -2 & -2 \\ 0 & -1 & -1 & -1 \end{vmatrix}$$, replacing R_2, R_3, R_4 by $R_2 - R_1$, $R_3 - R_1$ and $R_4 - R_1$ respectively

$$= 10 \begin{vmatrix} 1 & 1 & -3 \\ 2 & -2 & -2 \\ -1 & -1 & -1 \end{vmatrix}$$, expanding with respect to C_1

$$= -20 \begin{vmatrix} 1 & 1 & -3 \\ 1 & -1 & -1 \\ 1 & 1 & 1 \end{vmatrix}$$, taking out 2 common from R_2 and -1 common from R_3

$$= -20 \begin{vmatrix} 1 & 1 & -3 \\ 0 & -2 & 2 \\ 0 & 0 & 4 \end{vmatrix}$$, replacing R_2 and R_3 by $R_2 - R_1$ and $R_3 - R_1$ respectively

$$= -20 \begin{vmatrix} -2 & 2 \\ 0 & 4 \end{vmatrix}$$, expanding with respect to C_1

$= -20\ [(-2).4 - 0.2] = -20\ [-8] = 160.$ **Ans.**

Example 11:

Evaluate $\begin{vmatrix} 1 & 1 & 1 \\ 1 & 1+x & 1 \\ 1 & 1 & 1+y \end{vmatrix}$

Solution:

We have

$$= \begin{vmatrix} 1 & 1 & 1 \\ 0 & x & 0 \\ 0 & 0 & y \end{vmatrix}$$, replacing R_3 by $R_2 - R_1$ and R_3 by $R_3 - R_1$

$$= 1 \times \begin{vmatrix} x & 0 \\ 0 & y \end{vmatrix}$$, expanding with respect to the first column.

$= xy.$ **Ans.**

Example 12:

Evaluate $\begin{vmatrix} a & -a & -a & -a \\ b & -b & -b & -b \\ c & -c & -c & -c \\ d & -d & -d & -d \end{vmatrix}$

Solution:

Since three columns of the given determinant are identical, so the value of the determinant is zero. **Ans.**

Example 13:

Evaluate $\begin{vmatrix} -4 & 1 & 1 & 1 & 1 \\ 1 & -4 & 1 & 1 & 1 \\ 1 & 1 & -4 & 1 & 1 \\ 1 & 1 & 1 & -4 & 1 \\ 1 & 1 & 1 & 1 & -4 \end{vmatrix}$

Solution:

We have

$$= \begin{vmatrix} 0 & 1 & 1 & 1 & 1 \\ 0 & -4 & 1 & 1 & 1 \\ 0 & 1 & -4 & 1 & 1 \\ 0 & 1 & 1 & -4 & 1 \\ 0 & 1 & 1 & 1 & -4 \end{vmatrix}, \text{ replacing } C_1 \text{ by } C_1 + C_2 + C_3 + C_4 + C_5$$

= 0, expanding with respect to elements of first column. **Ans.**

Example 14:

Evaluate $\begin{vmatrix} 3 & 2 & 1 & 4 \\ 15 & 29 & 2 & 14 \\ 16 & 19 & 3 & 17 \\ 33 & 39 & 8 & 38 \end{vmatrix}$

Solution:

The given determinant

$$= \begin{vmatrix} 0 & 0 & 1 & 0 \\ 9 & 25 & 2 & 6 \\ 7 & 13 & 3 & 5 \\ 9 & 23 & 8 & 6 \end{vmatrix},$$ replacing C_1, C_2 and C_4 by $C_1 - 3C_3$, $C_2 - 2C_3$, $C_4 - 4C_3$ respectively.

$$= \begin{vmatrix} 9 & 25 & 6 \\ 7 & 13 & 5 \\ 9 & 23 & 6 \end{vmatrix}, \text{ expanding with respect to } R_1$$

$$= \begin{vmatrix} 0 & 2 & 0 \\ 7 & 13 & 5 \\ 9 & 23 & 6 \end{vmatrix}, \text{ replacing } R_1 \text{ by } R_1 - R_3$$

$$= -2 \begin{vmatrix} 7 & 5 \\ 9 & 6 \end{vmatrix}, \text{ expanding with respect to } R_1$$

$= -2\ [42 - 45] = (-2) \times (-3) = 6.$ **Ans.**

Example 15:

Prove that

$$\begin{vmatrix} b+c & c+a & a+b \\ q+r & r+p & p+q \\ y+z & z+x & x+y \end{vmatrix} = 2 \begin{vmatrix} a & b & c \\ p & q & r \\ x & y & z \end{vmatrix}$$

Solution:

We have

$$\begin{vmatrix} b+c & c+a & a+b \\ q+r & r+p & p+q \\ y+z & z+x & x+y \end{vmatrix}$$

$$= \begin{vmatrix} b & c+a & a+b \\ q & r+p & p+q \\ y & z+x & x+y \end{vmatrix} + \begin{vmatrix} c & c+a & a+b \\ r & r+p & p+q \\ z & z+x & x+y \end{vmatrix} \quad \text{(Note)}$$

$$= \begin{vmatrix} b & c+a & a \\ q & r+p & p \\ y & z+x & x \end{vmatrix} + \begin{vmatrix} b & c+a & b \\ q & r+p & q \\ y & z+x & y \end{vmatrix} + \begin{vmatrix} c & c & a+b \\ r & r & p+q \\ z & z & x+y \end{vmatrix} + \begin{vmatrix} c & a & a+b \\ r & p & p+q \\ z & x & x+y \end{vmatrix}$$

$$= \begin{vmatrix} b & c & a \\ q & r & p \\ y & z & x \end{vmatrix} + \begin{vmatrix} b & a & a \\ q & p & p \\ y & x & x \end{vmatrix} + \begin{vmatrix} c & a & a \\ r & p & p \\ z & x & x \end{vmatrix} + \begin{vmatrix} c & a & b \\ r & p & q \\ z & x & y \end{vmatrix},$$

second and third deter minants vanish as two coumns in each are identical.

$$= \begin{vmatrix} b & c & a \\ q & r & p \\ y & z & x \end{vmatrix} + \begin{vmatrix} c & a & b \\ r & p & q \\ z & x & y \end{vmatrix},$$ second and 'hird deter minants vanish as two columns in each are identical.

$$= -\begin{vmatrix} b & a & c \\ q & p & r \\ y & x & z \end{vmatrix} - \begin{vmatrix} a & c & b \\ p & r & q \\ x & z & y \end{vmatrix},$$ interchanging C_2 and C_3 in first determin ant and C_1 and C_2 in second.

$$= \begin{vmatrix} a & b & c \\ p & q & r \\ x & y & z \end{vmatrix} + \begin{vmatrix} a & b & c \\ p & q & r \\ x & y & z \end{vmatrix},$$ interchanging C_1 and C_2 in first and C_2 and C_3 in second determinat.

$$= 2\begin{vmatrix} a & b & c \\ p & q & r \\ x & y & z \end{vmatrix}$$

Hence proved.

Example 16:

Evaluate $\begin{vmatrix} 5 & 7 & 10 & 14 \\ 2 & 3 & 7 & 6 \\ 3 & 3 & 6 & 9 \\ 5 & 6 & 11 & 20 \end{vmatrix}$

Solution:

The given determinant

$$= \begin{vmatrix} 0 & 1 & -1 & -6 \\ 2 & 3 & 7 & 6 \\ 1 & 0 & -1 & 3 \\ 5 & 6 & 11 & 20 \end{vmatrix},$$ replacing R_1 and R_3 by $R_1 - R_4$ and $R_3 - R_2$ respectively.

$$= \begin{vmatrix} 0 & 0 & -1 & 0 \\ 2 & 10 & 7 & 24 \\ 1 & -1 & -1 & 5 \\ 5 & 17 & 11 & 56 \end{vmatrix},$$ replacing C_2 and C_4 by $C_2 + C_3$ and $C_4 + 6C_2$ respectively.

Now expand with respect to 1st row and proceed. **Ans.** 96

Example 17:

Show that

$$\begin{vmatrix} a + b + 2c & a & b \\ c & b + c + 2a & b \\ c & a & c + a + 2b \end{vmatrix} = 2\,(a + b + c)^2$$

Solution:

The given determinant

$$= \begin{vmatrix} 2a + 2b + 2c & a & b \\ 2a + 2b + 2c & b + c + 2a & b \\ 2a + 2b + 2c & a & c + a + 2b \end{vmatrix},$$ replacing C_1 by $C_1 + C_2 + C_3$

$$= (2a + 2b + 2c)\begin{vmatrix} 1 & a & b \\ 1 & b + c + 2a & b \\ 1 & a & c + a + 2b \end{vmatrix},$$ taking out $2a + 2b + 2c$ common

$$= 2(a + b + c)\begin{vmatrix} 1 & a & b \\ 0 & b + c + a & b \\ 0 & a & c + a + b \end{vmatrix},$$ replacing R_2 and R_3 by $R_2 - R_1$ and $R_3 - R_1$ respectively

$$= 2(a + b + c)\begin{vmatrix} b + c + a & 0 \\ 0 & c + a + b \end{vmatrix},$$ expanding with respect to Ist column.

$= 2(a + b + c)\ [(b + c + a)\ (c + a + b)]$

$= 2\ (a + b + c)^3.$

Example 18:

Evaluate $\begin{vmatrix} 1 & 1 & 1 & 1 & 1 \\ 1 & 2 & 3 & 4 & 5 \\ 1 & 3 & 6 & 10 & 15 \\ 1 & 4 & 10 & 20 & 35 \\ 1 & 5 & 15 & 35 & 69 \end{vmatrix}$

Solution:

We have

$$= \begin{vmatrix} 1 & 0 & 0 & 0 & 0 \\ 1 & 1 & 2 & 3 & 4 \\ 1 & 2 & 5 & 9 & 14 \\ 1 & 3 & 9 & 19 & 34 \\ 1 & 4 & 14 & 34 & 68 \end{vmatrix},$$ replacing C_2, C_3, C_4 and C_5 by $C_2 - C_1$, $C_3 - C_1$, $C_4 - C_1$ and $C_5 - C_1$ respectively

$$= \begin{vmatrix} 1 & 2 & 3 & 4 \\ 2 & 5 & 9 & 14 \\ 3 & 9 & 19 & 34 \\ 4 & 14 & 34 & 68 \end{vmatrix},$$ expanding with respect to first row.

$$= \begin{vmatrix} 1 & 0 & 0 & 0 \\ 2 & 1 & 3 & 6 \\ 3 & 3 & 10 & 22 \\ 4 & 6 & 22 & 52 \end{vmatrix},$$ replacing C_2, C_3 and C_4 by $C_2 - 2C_1$, $C_3 - 3C_1$ and $C_4 - 4C_1$ respectively.

$$= \begin{vmatrix} 1 & 3 & 6 \\ 3 & 10 & 22 \\ 6 & 22 & 52 \end{vmatrix},$$ expanding with respect to first row

$$= \begin{vmatrix} 1 & 0 & 0 \\ 3 & 1 & 4 \\ 6 & 4 & 16 \end{vmatrix},$$ replacing C_2 and C_3 by $C_3 - 3C_1$ and $C_3 - 6C_1$ respectively.

$$= \begin{vmatrix} 1 & 4 \\ 4 & 16 \end{vmatrix},$$ expanding with respect to first row.

$= 1\ (16) - 4\ 4 = 0.$ **Ans.**

Example 19:

Show that $= \begin{vmatrix} 1 & 1 & 1 & 1 \\ \alpha & \beta & \gamma & \delta \\ \beta+\gamma & \gamma+\delta & \delta+\alpha & \alpha+\beta \\ \delta & \alpha & \beta & \gamma \end{vmatrix} = 0$

Solution:

We have

$$= \begin{vmatrix} 1 & 1 & 1 & 1 \\ \alpha & \beta & \gamma & \delta \\ \beta+\gamma & \gamma+\delta & \delta+\alpha & \alpha+\beta \\ \alpha+\beta+\gamma+\delta & \alpha+\beta+\gamma+\delta & \alpha+\beta+\gamma+\delta & \alpha+\beta+\gamma+\delta \end{vmatrix},$$

replacing R_4 by $R_2 + R_3 + R$

$$= (\alpha + \beta + \gamma + \delta)\begin{vmatrix} 1 & 1 & 1 & 1 \\ \alpha & \beta & \gamma & \delta \\ \beta + \gamma & \gamma + \delta & \delta + \alpha & \alpha + \beta \\ 1 & 1 & 1 & 1 \end{vmatrix},$$

taking out $(\alpha + \beta + \gamma + \delta)$ common from R_4

$= 0$, since two row are identical. **Hence provec**

Example 20:

Evaluate $\begin{vmatrix} a - b - c & 2a & 2a \\ 2b & b - c - a & 2b \\ 2c & 2c & c - a - b \end{vmatrix}$

Solution:

The given determinant

$$= \begin{vmatrix} a + b + c & b + c + a & a + b + c \\ 2b & b - c - a & 2b \\ 2c & 2c & c - a - b \end{vmatrix},$$ replacing R_1 by $R_1 + R_2 + R_3$

$$= (a + b + c)\begin{vmatrix} 1 & 1 & 1 \\ 2b & b - c - a & 2b \\ 2c & 2c & c - a - b \end{vmatrix},$$ taking out $(a + b + c)$ common.

$$= (a + b + c)\begin{vmatrix} 1 & 0 & 0 \\ 2b & - a - b - c & 0 \\ 2c & 0 & - a - b - c \end{vmatrix},$$ replacing C_2 and C_3 by $C_2 = C_1$ and $C_3 - C_1$ respectively

$$= (a + b + c)\begin{vmatrix} - a - b - c & 0 \\ 0 & - a - b - c \end{vmatrix},$$ expanding with respect to first row.

$= (a + b + c)(- a - b - c)(- a - b - c) = (a + b + c)^2$. **Ans**

Example 21(a):

Evaluate $\begin{vmatrix} y+z & x & y \\ z+x & z & x \\ x+y & y & z \end{vmatrix}$

Solution:

We have

$$= \begin{vmatrix} 2x+2y+2z & x+y+z & x+y+z \\ z+x & z & x \\ x+y & y & z \end{vmatrix}, \text{ replacing } R_1 \text{ by } R_1 + R_2 + R_3$$

$$= (x+y+z)\begin{vmatrix} 2 & 1 & 1 \\ z+x & z & x \\ x+y & y & z \end{vmatrix},$$

$$= (x+y+z)\begin{vmatrix} 0 & 1 & 1 \\ 0 & z & x \\ x-y & y & z \end{vmatrix}, \text{ replacing } C_1 \text{ by } C_1 - C_2 - C_3$$

$$= (x+y+z)(x-z)\begin{vmatrix} 1 & 1 \\ z & x \end{vmatrix}, \text{ expanding with respect to first column.}$$

$$= (x+y+z)(x-z) = (x+y+z)(x-z)^2$$ **Ans.**

Example 21(b):

Evaluate $\begin{vmatrix} b+c & a+b & a \\ c+a & b+c & b \\ a+b & c+a & c \end{vmatrix}$

Solution:

We have

$$= \begin{vmatrix} 2a+2b+2c & 2a+2b+2c & a+b+c \\ c+a & b+c & b \\ a+b & c+a & c \end{vmatrix}, \text{ replacing } R_1 \text{ by } R_1 + R_2 + R_3$$

$$= (a+b+c)\begin{vmatrix} 2 & 2 & 1 \\ c+a & b+c & b \\ a+b & c+a & c \end{vmatrix}, \text{ taking out } (a+b+c) \text{ common.}$$

$$= (a+b+c)\begin{vmatrix} 0 & 2 & 1 \\ a-b & b+c & b \\ b-c & c+a & c \end{vmatrix}, \text{ replacing } C_1 \text{ by } C_1 - C_2$$

$$= (a+b+c)\begin{vmatrix} 0 & 0 & 1 \\ a-b & c-b & b \\ b-c & a-c & c \end{vmatrix}, \text{ replacing } C_2 \text{ by } C_2 - 2C_2$$

$= (a + b + c) [(a - b) (a - c) - (b - c) (c - b)]$

$= (a + b + c) [a^2 - ac - ba + bc + b^2 + c^2 - 2bc]$

$= (a + b + c) (a^2 + b^2 + c^2 - ab - bc - ca)$

$= a^2 + b^2 + c^2 - 3abc.$ **Ans.**

Example 21(c):

Evaluate $\begin{vmatrix} a & 1 & 1 & 1 \\ 1 & a & 1 & 1 \\ 1 & 1 & a & 1 \\ 1 & 1 & 1 & a \end{vmatrix}$

Solution:

We have

$$= \begin{vmatrix} a+3 & 1 & 1 & 1 \\ a+3 & a & 1 & 1 \\ a+3 & 1 & a & 1 \\ a+3 & 1 & 1 & a \end{vmatrix}, \text{ replacing } C_1 \text{ by } C_1 + C_2 + C_3 + C_4$$

$$= (a+3)\begin{vmatrix} 1 & 1 & 1 & 1 \\ 1 & a & 1 & 1 \\ 1 & 1 & a & 1 \\ 1 & 1 & 1 & a \end{vmatrix}, \text{ taking out } (a+3) \text{ common from } C_1$$

$$= (a+3)\begin{vmatrix} 1 & 1 & 1 & 1 \\ 0 & a-1 & 0 & 0 \\ 0 & 0 & a-1 & 0 \\ 0 & 0 & 0 & a-1 \end{vmatrix}$$, replacing R_2, R_3, R_4 by $R_2 - R_1$, $R_3 - R_1$ and $R_4 - R_1$ respectively.

$$= (a+3)\begin{vmatrix} a-1 & 0 & 0 \\ 0 & a-1 & 0 \\ 0 & 0 & a-1 \end{vmatrix}$$, expanding with respect to C_1

$$= (a+3)(a-1)\begin{vmatrix} a-1 & 0 \\ 0 & a-1 \end{vmatrix}$$, expanding with respect to R_1

$$= (a+3)(a-1)[(a-1)(a-1) = 0.0]$$

$$= (a+3)(a-1)^3.$$ **Ans.**

Example 22:

Prove that

$$\begin{vmatrix} 1+a_1 & a_2 & a_3 & a_4 \\ a_1 & 1+a_2 & a_3 & a_4 \\ a_1 & a_2 & 1+a_3 & a_4 \\ a_1 & a_2 & a_3 & 1+a_4 \end{vmatrix} = 1 + a_1 + a_2 + a_3 + a_4$$

Solution:

We have

$$= \begin{vmatrix} 1+a_1+a_2+a_3+a_4 & a_2 & a_3 & a_4 \\ 1+a_1+a_2+a_3+a_4 & 1+a_2 & a_3 & a_4 \\ 1+a_1+a_2+a_3+a_4 & a_2 & 1+a_3 & a_4 \\ 1+a_1+a_2+a_3+a_2 & a_2 & a_3 & 1+a_4 \end{vmatrix}$$, replacing C_1 by $C_1 + C_2 + C_3 + C_4$

$$= (1+a_1+a_2+a_3+a_4)\begin{vmatrix} 1 & a_2 & a_3 & a_4 \\ 1 & 1+a_2 & a_3 & a_4 \\ 1 & a_2 & 1+a_3 & a_4 \\ 1 & a_2 & a_3 & 1+a_4 \end{vmatrix}$$, taking out $(1 + a_1 + a_2 + a_3 + a_4)$ common from C_1

$$= (1 + a_1 + a_2 + a_3 + a_4)\begin{vmatrix} 1 & a_2 & a_3 & a_4 \\ 0 & 1 & 0 & 0 \\ 0 & 0 & 1 & 0 \\ 0 & 0 & 0 & 1 \end{vmatrix}$$, replacing R_2, R_3 and R_4 by $R_2 - R_1$, $R_3 - R_1$ and $R_4 - R_1$ respectively.

$$= (1 + a_1 + a_2 + a_3 + a_4)\begin{vmatrix} 1 & 0 & 0 \\ 0 & 1 & 0 \\ 0 & 0 & 1 \end{vmatrix}$$, expanding with respect to C_1

$$= (1 + a_1 + a_2 + a_3 + a_4)\begin{vmatrix} 1 & 0 \\ 0 & 1 \end{vmatrix}$$, expanding with respect to R_1

$= (1 + a_1 + a_2 + a_3 + a_4)$. **Hence proved.**

Example 23:

Evaluate $\begin{vmatrix} 1+a & 1 & 1 & 1 \\ 1 & 1+a & 1 & 1 \\ 1 & 1 & 1+a & 1 \\ 1 & 1 & 1 & 1+a \end{vmatrix}$

Solution:

We have

$$= \begin{vmatrix} 4+a & 1 & 1 & 1 \\ 4+a & 1+a & 1 & 1 \\ 4+a & 1 & 1+a & 1 \\ 4+a & 1 & 1 & 1+a \end{vmatrix}$$, replacing C_1 by $C_1 + C_2 + C_3 + C_4$

$$= (4+a)\begin{vmatrix} 1 & 1 & 1 & 1 \\ 1 & 1+a & 1 & 1 \\ 1 & 1 & 1+a & 1 \\ 1 & 1 & 1 & 1+a \end{vmatrix}$$, taking out $(4 + a)$ common from first column

$$= (4 + a)\begin{vmatrix} 1 & 0 & 0 & 0 \\ 1 & a & 0 & 0 \\ 1 & 0 & a & 0 \\ 1 & 0 & 0 & a \end{vmatrix},$$ replacing C_1, C_2 and C_4 by $C_2 - C_1$, $C_3 - C_1$ and $C_4 - C_1$ respectively.

$$= (4 + a)\begin{vmatrix} a & 0 & 0 \\ 0 & a & 0 \\ 0 & 0 & a \end{vmatrix},$$ expanding with respect to R_1.

$$= (4 + a)\begin{vmatrix} a & 0 \\ 0 & a \end{vmatrix},$$ expanding with respect to R_1.

$= (4 + a)\ a\ (a^2) = (4 + a)\ a^2$. **Ans.**

Example 24(a):

Evaluate $\begin{vmatrix} x & a & a & a \\ a & x & a & a \\ a & a & x & a \\ a & a & a & x \end{vmatrix}$

Solution:

We have

$$= \begin{vmatrix} x + 3a & a & a & a \\ x + 3a & x & a & a \\ x + 3a & a & x & a \\ x + 3a & a & a & x \end{vmatrix},$$ replacing C_1 by $C_1 + C_2 + C_3 + C_4$

$$= (x + 3a)\begin{vmatrix} 1 & a & a & a \\ 1 & x & a & a \\ 1 & a & x & a \\ 1 & a & a & x \end{vmatrix},$$ taking out $(x + 3a)$ common

$$= (x + 3a)\begin{vmatrix} 1 & a & a & a \\ 0 & x-a & 0 & 0 \\ 0 & 0 & x-a & 0 \\ 0 & 0 & 0 & x-a \end{vmatrix}$$, replacing R_2, R_3 and R_4 by $R_2 - R_1$, $R_3 - R_1$ and $R_4 - R_1$ respectively.

$$= (x + 3a)\begin{vmatrix} x-a & 0 & 0 \\ 0 & x-a & 0 \\ 0 & 0 & x-a \end{vmatrix}$$, expanding with respect to C_1

$$= (x + 3a)(x - a)\begin{vmatrix} x-a & 0 \\ 0 & x-a \end{vmatrix}$$, expanding with respect to C_1

$$= (x + 3a)\,(x - a)\,(x - a)\,(x - a) = (x + 3a)\,(x - a)^2.$$ **Ans.**

Example 24(b):

Evaluate $\begin{vmatrix} 1 & x & 1 & y \\ x & 1 & y & 1 \\ 1 & y & 1 & x \\ y & 1 & x & 1 \end{vmatrix}$

Solution:

We have

$$= \begin{vmatrix} x+y+2 & x+y+2 & x+y+2 & x+y+2 \\ x & 1 & y & 1 \\ 1 & y & 1 & x \\ y & 1 & x & 1 \end{vmatrix}$$, replacing R_1 by $R_1 + R_2 + R_3 + R_4$

$$= (x+y+2)\begin{vmatrix} 1 & 1 & 1 & 1 \\ x & 1 & y & 1 \\ y & 1 & x & 1 \\ y & 1 & x & 1 \end{vmatrix}$$, taking out $(x + y + 2)$ common from R_1

$$= (x+y+2)\begin{vmatrix} 1 & 0 & 0 & 0 \\ x & 1-x & y-x & 1-x \\ 1 & y-1 & 0 & x-1 \\ y & 1-y & x-y & 1-y \end{vmatrix},$$ replacing C_2, C_3 and C_4 by C_2-C_1, C_3-C_1 and C_4-C_1 respectively.

$$= (x+y+2)\begin{vmatrix} 1-x & y-x & 1-x \\ y-1 & 0 & x-1 \\ 1-y & x-y & 1-y \end{vmatrix},$$ expanding with respect to R_1

$$= (x+y+2)\begin{vmatrix} 0 & y-x & 1-x \\ y-x & 0 & x-1 \\ 0 & x-y & 1-y \end{vmatrix},$$ replacing C_1 by $C_1 - C_2$

$$= -(x+y+2)(y-x)\begin{vmatrix} y-x & 1-x \\ x-y & 1-y \end{vmatrix},$$ expanding with respect to C_1

$$= -(x+y+2)(y-x)(y-x)\begin{vmatrix} -1 & 1-x \\ -1 & 1-y \end{vmatrix},$$ taking out $(y - x)$ common

$$= -(x+y+2)(y-x)^2[(1-y)-(-1)(1-x)]$$
$$= -(x+y+2)(x-y)^2(2-x-y)$$
$$= (x+y+2)(x-y)^2(x+y-2).$$

Example 25:

Prove that $$\begin{vmatrix} x+a & b & c & d \\ a & x+b & c & d \\ a & b & x+c & d \\ a & b & c & x+d \end{vmatrix} = x^3(x+a+b+c+d).$$

Solution:

We have

$$= \begin{vmatrix} x+a+b+c+d & b & c & d \\ x+a+b+c+d & x+b & c & d \\ x+a+b+c+d & b & x+c & d \\ x+a+b+c+d & b & c & x+d \end{vmatrix},$$ replacing C_1 by $C_1+C_2+C_3+C_4$

$$=(x+a+b+c+d)\begin{vmatrix} 1 & b & c & d \\ 1 & x+b & c & d \\ 1 & b & x+c & d \\ 1 & b & c & x+d \end{vmatrix}, \text{ taking out } (x+a+b+c+d) \text{ common from } C_1$$

$$=(x+a+b+c+d)\begin{vmatrix} 1 & b & c & d \\ 0 & x & 0 & 0 \\ 0 & 0 & x & 0 \\ 0 & 0 & 0 & x \end{vmatrix}, \text{ replacing } R_2, R_3, R_4 \text{ by } R_2-R_1, R_3-R_1 \text{ and } R_4-R_1 \text{ respectively.}$$

$$=(x+a+b+c+d)\begin{vmatrix} x & 0 & 0 \\ 0 & x & 0 \\ 0 & 0 & x \end{vmatrix}, \text{ expanding with respect to } C_1$$

$$=(x+a+b+c+d)\begin{vmatrix} x & 0 \\ 0 & x \end{vmatrix}, \text{ expanding with respect to } C_1$$

$= x\,(x + a + b + c + d)\,x^2$

$= x^3\,(x + a + b + c + d).$ **Hence proved.**

Example 26(a):

Evaluate $\begin{vmatrix} x & a & b & c & 1 \\ d & x & f & h & 1 \\ d & e & x & k & 1 \\ d & e & g & x & 1 \\ d & e & g & m & 1 \end{vmatrix}$

Solution:

The given determinant

$$=\begin{vmatrix} x-d & a-x & b-f & c-h & 0 \\ 0 & x-e & f-x & h-k & 0 \\ 0 & 0 & x-g & k-x & 0 \\ 0 & 0 & 0 & x-m & 0 \\ d & e & g & m & 1 \end{vmatrix},$$

replacing R_1, R_2, R_3 and R_4 by $R_1 - R_2$, $R_2 - R_3$, $R_3 - R_4$ and $R_{4-R}5$ respectively.

$$= (x - m) \begin{vmatrix} x - d & a - x & b - f & 0 \\ 0 & x - e & f - x & 0 \\ 0 & 0 & x - g & 0 \\ d & e & g & 1 \end{vmatrix},$$

expanding with respect to R_4 **(Note)**

$$= (x - m) \begin{vmatrix} x - d & a - x & b - f \\ 0 & x - e & f - x \\ 0 & 0 & x - g \end{vmatrix},$$ expanding with respect to C_4

$$= (x - m)(x - d) \begin{vmatrix} x - c & f - x \\ 0 & x - g \end{vmatrix},$$ expanding with respect to C_1

$$= (x - m)(x - d)(x - c)(x - g).$$ **Ans.**

Example 26(b):

Evaluate $\begin{vmatrix} b + c & a - c & a - b \\ b - c & c + a & b - a \\ c - b & c - a & a + b \end{vmatrix}$

Solution:

$$= \begin{vmatrix} 2b & 2a & 0 \\ b - c & c + a & b - a \\ 0 & 2c & 2b \end{vmatrix},$$ replacing R_1 and R_2 by $R_1 + R_2$ and $R_3 + R_2$ respectively.

$$= 4 \begin{vmatrix} b & a & 0 \\ b - c & c + a & b - a \\ 0 & c & b \end{vmatrix},$$ taking out 2 common from R_1 and R_2

$$= 4 \begin{vmatrix} b & a & 0 \\ -c & c & b - a \\ 0 & c & b \end{vmatrix},$$ replacing R_2 by $R_2 - R_1$

$$= 4\left[b \begin{vmatrix} c & b - a \\ c & b \end{vmatrix} + c \begin{vmatrix} a & 0 \\ c & b \end{vmatrix}\right],$$ expanding with respect to C_1

$$= 4\left[b\begin{vmatrix} c & b-a \\ 0 & a \end{vmatrix} + c\begin{vmatrix} a & 0 \\ c & b \end{vmatrix}\right],$$ replacing R_2 by $R_2 - R_1$ in first det er min ant.

$= 4\ [b\ (ca) + c\ (ab)] = 4\ [2abc] = 8abc.$ **Ans.**

Example 26(c):

Evaluate $$\begin{vmatrix} b+c & a & a \\ b & c+a & b \\ c & c & a+b \end{vmatrix}$$

Solution:

We have

$$= \begin{vmatrix} b+c & 0 & 0 \\ b & c+a-b & b \\ c & c-a-b & a+b \end{vmatrix},$$ replacing C_2 by $C_2 - C_3$

$$= (b+c)\begin{vmatrix} c+a-b & b \\ c-a-b & a+b \end{vmatrix} + a\begin{vmatrix} b & c+a-b \\ c & c-a-b \end{vmatrix},$$

expanding with respect to R_1

$$= (b+c)\begin{vmatrix} c+a-b & b \\ -2a & a \end{vmatrix} + a\begin{vmatrix} b & c+a-b \\ c-b & -2a \end{vmatrix},$$

Replacing R_2 by $R_2 - R_1$ in each determinant.

$= (b + c)\ [a\ (c + a - b) - b\ (-2a)] + a\ [b - (-2a) - (c - b)\ (c + a - b)]$

$= (b + c)\ (ac + a^2 + ab) + a\ (-2ab - c^2 - ca + bc + bc + ab - b^2)$

$= abc + a^2b + ab^2 + ac^2 + a^2c + abc - 2a^2b - ac^2 - ca^2$

$+ 2abc + a^2b - ab^2 = 4\ 4abc.$ **Ans.**

Example 27:

Evaluate $$\begin{vmatrix} 1 & 1 & 1 \\ a & b & c \\ a^3 & b^3 & c^3 \end{vmatrix}$$

Solution:

The given determinant

$$= \begin{vmatrix} 1 & 0 & 0 \\ a & b-a & c-a \\ a^3 & b^3-a^3 & c^3-a^3 \end{vmatrix},$$ replacing C_2, C_3 by $C_3 - C_1$ and $C_3 - C_1$ respectively.

$$= \begin{vmatrix} b-a & c-a \\ b^3-a^3 & c^3-a^3 \end{vmatrix},$$ expanding with respect to R_1

$$= (b-a)(c-a)\begin{vmatrix} 1 & 1 \\ b^2+ab+a^2 & c^2+ac+a^2 \end{vmatrix},$$

taking common factors out from C_1 and C_2

$$= (b-a)(c-a)[(c^2+ac+a^2)-(b^2+ab+a^2)]$$

$$= (b-a)(c-a)[(c^2-b^2)+a(c-b)]$$

$$= (b-a)(c-a)(c-b)[(c+b)+a]$$

$$= (a-b)(b-c)(c-a)(a+b+c).$$ **Ans.**

Example 28:

If a, b, c are all different and

$$\begin{vmatrix} a & a^2 & 1+a^3 \\ b & b^2 & 1+b^3 \\ c & c^2 & 1+c^3 \end{vmatrix} = 0,$$ *prove that* $1 + abc = 0$.

Solution:

The given determinant

$$= \begin{vmatrix} a & a^2 & 1 \\ b & b^2 & 1 \\ c & c^2 & 1 \end{vmatrix} + \begin{vmatrix} a & a^2 & a^3 \\ b & b^2 & b^3 \\ c & c^2 & c^3 \end{vmatrix},$$ breaking into two determinants. **(Note)**

$$= \begin{vmatrix} a & b & c \\ a^2 & b^2 & c^2 \\ 1 & 1 & 1 \end{vmatrix} + \begin{vmatrix} a & b & c \\ a^2 & b^2 & c^2 \\ a^3 & b^3 & c^3 \end{vmatrix},$$ interchanging rows and columns in each determinant. **(Note)**

The value of given determinant = (1 + abc) [(a – b) (b – c) (c – a)]

Example 29(a):

Prove that $\begin{vmatrix} a^3 & 3a^2 & 3a & 1 \\ a^2 & a^2+2a & 2a+1 & 1 \\ a & 2a+1 & a+2 & 1 \\ 1 & 3 & 3 & 1 \end{vmatrix} = (a-1)^6$

Solution:

The given determinant

$$= \begin{vmatrix} a^3 - 3a^2 + 3a - 1 & 3a^2 & 3a & 1 \\ 0 & a^2+2a & 2a+1 & 1 \\ 0 & 2a+1 & a+2 & 1 \\ 0 & 3 & 3 & 1 \end{vmatrix},$$

replacing C_1 by $C_1 - C_2 + C_3 - C_4$ **(Note)**

$$= (a^3 - 3a^2 + 3a - 1) \begin{vmatrix} a^2+2a & 2a+1 & 1 \\ 2a+1 & a+2 & 1 \\ 3 & 3 & 1 \end{vmatrix},$$ expanding with respect to C_1

$$= (a-1)^3 \begin{vmatrix} (a^2+2a) - (2a+1) + 1 & 2a+1 & 1 \\ (2a+1) - (a+2) + 1 & a+2 & 1 \\ 3 - 3 + 1 & 3 & 1 \end{vmatrix},$$

replacing C_1 by $C_1 - C_2 + C_3$

$$= (a-1)^3 \begin{vmatrix} a^2 & 2a+1 & 1 \\ a & a+2 & 1 \\ 1 & 3 & 1 \end{vmatrix},$$ on simplifying

$$= (a-1)^3 \begin{vmatrix} a^2-1 & 2a-2 & 0 \\ a-1 & a-1 & 0 \\ 1 & 3 & 1 \end{vmatrix},$$ replacing R_1 and R_2 by $R_1 - R_2$ and $R_2 - R_3$ respectively.

$$= (a-1)^3 \begin{vmatrix} a^2-1 & 2(a-1) \\ a-1 & a-1 \end{vmatrix},$$ expanding with respect to C_3

$$= (a-1)^3(a-1)(a-1) \begin{vmatrix} a+1 & 2 \\ 1 & 1 \end{vmatrix},$$ taking out common factors.

$$= (a-1)^3\,[(a+1)-2] = (a-1)^6.$$ **Hence proved.**

Example 29(b):

Show that

$$\begin{vmatrix} x & y & z \\ x^2 & y^2 & z^2 \\ yz & zx & xy \end{vmatrix} = \begin{vmatrix} 1 & 1 & 1 \\ x^2 & y^2 & z^2 \\ x^3 & y^3 & z^3 \end{vmatrix}$$

$= (y-x)\,(z-x)\,(x-y)\,(yz+zx+xy).$

Solution:

$$\begin{vmatrix} x & y & z \\ x^2 & y^2 & z^2 \\ yz & zx & xy \end{vmatrix} = \frac{1}{xyz} \begin{vmatrix} x^2 & y^2 & z^2 \\ x^3 & y^3 & z^3 \\ xyz & xyz & xyz \end{vmatrix},$$ multiplying C_1, C_2 C_3 by x, y, z respectively **(Note)**

$$= \frac{1}{xyz}.xyz \begin{vmatrix} x^2 & y^2 & z^2 \\ x^3 & y^3 & z^3 \\ 1 & 1 & 1 \end{vmatrix},$$ taking out xyz common from R_2

$$= \begin{vmatrix} 1 & 1 & 1 \\ x^2 & y^2 & z^2 \\ x^3 & y^3 & z^3 \end{vmatrix},$$ interchanging R_2 and R_3 and then R_1 and R_2

(Note)

For the 2nd part do your self.

Example 30:

Evaluate $\begin{vmatrix} 1+a & 1 & 1 & 1 \\ 1 & 1+b & 1 & 1 \\ 1 & 1 & 1+c & 1 \\ 1 & 1 & 1 & 1+d \end{vmatrix}$

Solution:

We have

$$= abcd \begin{vmatrix} \frac{1}{a}+1 & \frac{1}{a} & \frac{1}{a} & \frac{1}{a} \\ \frac{1}{b} & \frac{1}{b}+1 & \frac{1}{b} & \frac{1}{b} \\ \frac{1}{c} & \frac{1}{c} & \frac{1}{c}+1 & \frac{1}{c} \\ \frac{1}{d} & \frac{1}{d} & \frac{1}{d} & \frac{1}{d}+1 \end{vmatrix},$$

taking a, b, c, d common from R_1, R_2, R_3 and R_4 respectively.

$$= abcd\left(\frac{1}{a}+\frac{1}{b}+\frac{1}{c}+\frac{1}{d}+1\right) \begin{vmatrix} 1 & 1 & 1 & 1 \\ \frac{1}{b} & \frac{1}{b}+1 & \frac{1}{b} & \frac{1}{b} \\ \frac{1}{c} & \frac{1}{c} & \frac{1}{c}+1 & \frac{1}{c} \\ \frac{1}{d} & \frac{1}{d} & \frac{1}{d} & \frac{1}{d}+1 \end{vmatrix},$$

replacing R_1 by $R_1 + R_2 + R_3 + R_4$ and taking $\left(\frac{1}{a}+\frac{1}{b}+\frac{1}{c}+\frac{1}{d}+1\right)$ common from R_1

$$= abcd\left(\frac{1}{a}+\frac{1}{b}+\frac{1}{c}+\frac{1}{d}+1\right) \begin{vmatrix} 1 & 0 & 0 & 0 \\ \frac{1}{b} & 1 & 0 & 0 \\ \frac{1}{c} & 0 & 1 & 0 \\ \frac{1}{d} & 0 & 0 & 1 \end{vmatrix},$$

replacing C_2, C_3 and C_4 by $C_2 - C_1$, $C_3 - C_1$ and $C_4 - C_1$ respectively.

$$= abcd\left(\frac{1}{a}+\frac{1}{b}+\frac{1}{c}+\frac{1}{d}+1\right)\begin{vmatrix} 1 & 0 & 0 \\ 0 & 1 & 0 \\ 0 & 0 & 0 \end{vmatrix}, \text{ expanding with respect to } R_1$$

$$= abcd\left(\frac{1}{a}+\frac{1}{b}+\frac{1}{c}+\frac{1}{d}+1\right)\begin{vmatrix} 1 & 0 \\ 0 & 1 \end{vmatrix}, \text{ expanding with respect to } R_1$$

$$= abcd\left(\frac{1}{a}+\frac{1}{b}+\frac{1}{c}+\frac{1}{d}+1\right)$$

Example 31:

In the determinant $|A| = \begin{vmatrix} a_1 & b_1 & c_1 \\ a_2 & b_2 & c_2 \\ a_3 & b_3 & c_3 \end{vmatrix}$ *prove that*

$a_1A_2 + b_1B_2 + c_1C_2 = 0$, $b_1C_1 + b_2C_2 + b_3C_3 = 0$ *and*

$c_1B_1 + c_2B_2 + c_3B_3 = 0$, *where capital letters denote the cofactors of the corresponding small letters.*

Also prove that

$$a_1A_1 + b_1B_1 + c_1C_1 = |A| = a_2A_2 + b_2B_2 + c_2C_2 = a_3A_3 + b_3B_3 + c_3C_3$$

Solution:

In the det $|A| = \begin{vmatrix} a_1 & b_1 & c_1 \\ a_2 & b_2 & c_2 \\ a_3 & b_3 & c_3 \end{vmatrix}$, we have

$$A_1 = \begin{vmatrix} b_2 & c_2 \\ b_3 & c_3 \end{vmatrix};\ A_2 = -\begin{vmatrix} b_1 & c_1 \\ b_3 & c_3 \end{vmatrix};\ B_1 = -\begin{vmatrix} a_2 & c_2 \\ a_3 & c_3 \end{vmatrix};\ B_2 = \begin{vmatrix} a_1 & c_1 \\ a_3 & c_3 \end{vmatrix};$$

$$B_3 = -\begin{vmatrix} a_1 & c_1 \\ a_2 & c_2 \end{vmatrix};\ C_1 = \begin{vmatrix} a_2 & b_2 \\ a_3 & b_3 \end{vmatrix};\ C_2 = -\begin{vmatrix} a_1 & b_1 \\ a_3 & b_3 \end{vmatrix};\ A_3 = \begin{vmatrix} b_1 & c_1 \\ b_2 & c_2 \end{vmatrix};$$

$$C_3 = \begin{vmatrix} a_1 & b_1 \\ a_2 & b_2 \end{vmatrix}$$

$\therefore\ a_1A_2 + b_1B_2 + c_1C_2$

$$= -a_1 \begin{vmatrix} b_1 & c_1 \\ b_3 & c_3 \end{vmatrix} + b_1 \begin{vmatrix} a_1 & c_1 \\ a_3 & c_3 \end{vmatrix} - c_1 \begin{vmatrix} a_1 & b_1 \\ a_3 & b_3 \end{vmatrix}$$

$$= -a_1(b_1c_3 - b_3c_1) + b_1(a_1c_3 - a_3c_1) - c_1(a_1b_3 - a_3b_1)$$

$= 0$, on simplifying

$$b_1C_1 + b_2C_2 + b_3C_3 = b_1 \begin{vmatrix} a_2 & b_2 \\ a_3 & b_3 \end{vmatrix} - b_2 \begin{vmatrix} a_1 & b_1 \\ a_3 & b_3 \end{vmatrix} + b_3 \begin{vmatrix} a_1 & b_1 \\ a_2 & b_2 \end{vmatrix}$$

$$= b_1(a_2b_3 - a_3b_2) - b_2(a_1b_3 - a_3b_1) + b_3(a_1b_2 - a_2b_1)$$

$= 0$, on simplifying.

In a similar way the remaining part can also be proved.

$$\text{Also } |A| = \begin{vmatrix} a_1 & b_1 & c_1 \\ a_2 & b_2 & c_2 \\ a_3 & b_3 & c_3 \end{vmatrix}$$

$$= a_1 \begin{vmatrix} b_2 & c_2 \\ b_3 & c_3 \end{vmatrix} - b_2 \begin{vmatrix} a_2 & c_2 \\ a_3 & c_3 \end{vmatrix} + c_1 \begin{vmatrix} a_2 & b_2 \\ a_3 & b_3 \end{vmatrix} \quad \text{...(i)}$$

expanding with respect to first row.

Again $a_1A_1 + b_1B_1 + c_1C_1$

$$= a_1 \begin{vmatrix} b_2 & c_2 \\ b_3 & c_3 \end{vmatrix} + b_2 \left\{ - \begin{vmatrix} a_2 & c_2 \\ a_3 & c_3 \end{vmatrix} \right\} + c_1 \begin{vmatrix} a_2 & b_2 \\ a_3 & b_3 \end{vmatrix}$$

$$= a_1 \begin{vmatrix} b_2 & c_2 \\ b_3 & c_3 \end{vmatrix} - b_2 \begin{vmatrix} a_2 & c_2 \\ a_3 & c_3 \end{vmatrix} + c_1 \begin{vmatrix} a_2 & b_2 \\ a_3 & b_3 \end{vmatrix}$$

$= |A|$, from (i)

Similarly, $a_2A_2 + b_2B_2 + c_2C_2$

$$= a_2 \left\{ - \begin{vmatrix} b_1 & c_1 \\ b_3 & c_3 \end{vmatrix} \right\} + b_2 \begin{vmatrix} a_1 & c_1 \\ a_3 & c_3 \end{vmatrix} + c_2 \left\{ - \begin{vmatrix} a_1 & b_1 \\ a_3 & b_3 \end{vmatrix} \right\}$$

$$= -a_2 \begin{vmatrix} b_1 & c_1 \\ b_3 & c_3 \end{vmatrix} + b_2 \begin{vmatrix} a_1 & c_1 \\ a_3 & c_3 \end{vmatrix} - c_2 \begin{vmatrix} a_1 & b_1 \\ a_3 & b_3 \end{vmatrix} \quad \text{...(ii)}$$

$$\text{Also } |A| = \begin{vmatrix} a_1 & b_1 & c_1 \\ a_2 & b_2 & c_2 \\ a_3 & b_3 & c_3 \end{vmatrix}$$

$$= -a_2 \begin{vmatrix} b_1 & c_1 \\ b_3 & c_3 \end{vmatrix} + b_2 \begin{vmatrix} a_1 & c_1 \\ a_3 & c_3 \end{vmatrix} - c_2 \begin{vmatrix} a_1 & b_1 \\ a_3 & b_3 \end{vmatrix}, \quad \ldots\text{(ii)}$$

expanding with respect to second row.

∴ From (ii) and (iii), we get

$$a_2A_2 + b_2B_2 + c_2C_2 = |\ A\ |.$$

In a similar way we can prove that $a_2A_2 + b_2B_2 + c_2C_2 = |\ A\ |$, by expanding $|\ A\ |$ with respect to third row.

Example 32:

Expand the determinant $\begin{vmatrix} a & h & g \\ h & b & f \\ g & f & c \end{vmatrix}$ *by the elements of 1st row.*

Solution:

Elements of firs row are a, h, g.

Let A, H, and G denote the cofactors of a, h, g.

Then $A = \begin{vmatrix} b & f \\ f & c \end{vmatrix}$; $H = -\begin{vmatrix} h & f \\ g & c \end{vmatrix}$ and $G = \begin{vmatrix} h & b \\ g & f \end{vmatrix}$

Hence $\begin{vmatrix} a & h & g \\ h & b & f \\ g & f & c \end{vmatrix} = aA + hH + gG$

$$= a\begin{vmatrix} b & f \\ f & c \end{vmatrix} - h\begin{vmatrix} h & f \\ g & c \end{vmatrix} + g\begin{vmatrix} h & b \\ g & f \end{vmatrix}$$

$$= a\,(bc - f^2) - h\,(ch - fg) + g\,(hf - bg)$$

$$= abc + 2fgh - af^2 - bg^2 - ch^2.$$ **Ans.**

Example 33:

Find the value of $\begin{vmatrix} a & h & g \\ h & b & f \\ g & f & c \end{vmatrix}$

Solution:

$$\begin{vmatrix} a & h & g \\ h & b & f \\ g & f & c \end{vmatrix} = a\ \{b'c - f'f\} - h\ \{hc - f'g\} + g\ \{hf - bg\}$$

$$= abc - af^2 - ch^2 + fgh + fgh - bg^2$$

$$= abc + 2fgh - af^2 - bg^2 - ch^2.$$ **Ans.**

Example 34:

Evaluate $\begin{vmatrix} 1 & 2 & 3 \\ 4 & 5 & 6 \\ 7 & 8 & 9 \end{vmatrix}$

Solution:

$$\begin{vmatrix} 1 & 2 & 3 \\ 4 & 5 & 6 \\ 7 & 8 & 9 \end{vmatrix} = 1\ \{5.9 - 6.8\} - 2\ \{4.9 - 6.7\} + 3\ \{4.8 - 5.7\},$$

$$= 1\ \{45 - 48\} - 2\{36 - 42\} + 3\ \{32 - 35\}$$

$$= -3 + 12 - 9 = 0.$$ **Ans.**

Example 35:

Evaluate $\begin{vmatrix} 1 & 1 & 1 & 1 \\ 1 & 1+x & 1 & 1 \\ 1 & 1 & 1+y & 1 \\ 1 & 1 & 1 & 1+z \end{vmatrix}$

Solution:

We have

$$= \begin{vmatrix} 1 & 1 & 1 & 1 \\ 0 & x & 0 & 0 \\ 0 & 0 & y & 0 \\ 0 & 0 & 0 & z \end{vmatrix},$$ replacing R_2, R_3 and R_4 by $R_4 - R_1$, $R_3 - R_1$ and $R_4 - R_1$ respectively.

$$= \begin{vmatrix} x & 0 & 0 \\ 0 & y & 0 \\ 0 & 0 & z \end{vmatrix}, \text{ expanding with respect to } C_1$$

$$= x \begin{vmatrix} y & 0 \\ 0 & z \end{vmatrix}, \text{ expanding with respect to } R_1$$

$= xyz.$ **Ans.**

Example 36(a):

Show that $\begin{vmatrix} (b+c)^2 & a^2 & a^2 \\ b^2 & (c+a)^2 & b^2 \\ c^2 & c^2 & (a+b)^2 \end{vmatrix} = 2abc\,(a+b+c)^3$

Solution:

We have

$$= \begin{vmatrix} (b+c)^2 - a^2 & 0 & a^2 \\ 0 & (c+a)^2 - b^2 & b^2 \\ c^2 - (a+b)^2 & c^2 - (a+b)^2 & (a+b)^2 \end{vmatrix},$$

replacing C_1 and C_2 by $C_1 - C_2$ and $C_2 - C_3$ respectively.

$$= \begin{vmatrix} (b+c+a)(b+c-a) & 0 & a^2 \\ 0 & (c+a+b)(c+a-b) & b^2 \\ (c+a+b)(c-a-b) & (c+a+b)(c-a-b) & (a+b)^2 \end{vmatrix},$$

$$= (a+b+c)^2 \begin{vmatrix} b+c-a & 0 & a^2 \\ 0 & c+a-b & b^2 \\ c-a-b & c-a-b & (a+b)^2 \end{vmatrix},$$ taking out the common factors from C_1 and C_2

$$= (a+b+c)^2 \begin{vmatrix} b+c-a & 0 & a^2 \\ 0 & c+a-b & b^2 \\ -2b & -2a & 2ab \end{vmatrix},$$ replacing R_2 by $R_3 - R_1 - R_2$

$$= (a+b+c)^2 \begin{vmatrix} b+c & a^2/b & a^2 \\ b^2/a & c+a & b^2 \\ 0 & 0 & 2ab \end{vmatrix},$$ replacing C_1 and C_2 by $C_1 + \frac{1}{a}C_2$ and $C_2 + \frac{1}{b}C_2$ respectively. **(Note)**

$$= 2ab\,(a+b+c)^2 \begin{vmatrix} b+c & a^2/b \\ b^2/a & c+a \end{vmatrix},$$ expanding with respect to R_2

$= 2ab\ (a + b + c)^2\ [(b + c)\ (c + a) - (b^2/a)\ (a^2/b)]$

$= 2ab\ (a + b + c)^2\ [bc + ba + c^2 + ca - ab]$

$= 2ab\ (a + b + c)^2\ [b + a + c] = 2abc\ (a + b + c)^2.$ **Ans.**

Example 36(b):

Prove that $\begin{vmatrix} a & b & b & b \\ a & b & a & a \\ a & a & b & a \\ b & b & b & a \end{vmatrix} = -(b-a)^4$

Solution:

We have

$$= \begin{vmatrix} a & b-a & b-a & b-a \\ a & b-a & 0 & 0 \\ a & 0 & b-a & 0 \\ b & 0 & 0 & a-b \end{vmatrix},$$ replacing C_2, C_3 and C_4 by $C_2 - C_1$, $C_3 - C_1$ and $C_4 - C_1$ respectively.

$$= -(b-a) \begin{vmatrix} a & b-a & 0 \\ a & 0 & b-a \\ b & 0 & 0 \end{vmatrix} + (a-b) \begin{vmatrix} a & b-a & b-a \\ a & b-a & 0 \\ a & 0 & b-a \end{vmatrix},$$

expanding with respect to C_1

$$= -(b-a)\,b\begin{vmatrix} b-a & 0 \\ 0 & b-a \end{vmatrix} + (a-b)\begin{vmatrix} 0 & 0 & b-a \\ a & b-a & 0 \\ a & 0 & b-a \end{vmatrix},$$

expanding first determinant with respect to R_2 and in the second determinant replacing R_1 by $R_1 - R_2$.

$$= -b\,(b-a)^2 + (a-b)\,(b-a)\begin{vmatrix} a & b-a \\ a & 0 \end{vmatrix},$$

expanding the second det. with respect to R_1

$$= -b\,(b-a)^3 + (a-b)\,(b-a)\,[(0 - a\,(b-a)]$$

$$= -b\,(b-a)^3 + a(b-a)^2 = -(b-a)^3\,(b-a)$$

$$= -(b-a)^4.$$

Hence proved.

Example 36(c):

Evaluate $\begin{vmatrix} 1 & bc+ad & b^2c^2+a^2d^2 \\ 1 & ca+bd & c^2a^2+b^2d^2 \\ 1 & ab+cd & a^2b^2+c^2d^2 \end{vmatrix}$

Solution:

We have

$$= \begin{vmatrix} 1 & bc+ad & b^2c^2+a^2d^2 \\ 0 & ca+bd-bc-ad & c^2a^2+b^2d^2-b^2c^2-a^2d^2 \\ 0 & ab+cd-bc-ad & a^2b^2+c^2d^2-b^2c^2-c^2d^2 \end{vmatrix},$$

replacing R_2 and R_3 by $R_2 - R_1$ and $R_3 - R_1$ respectively.

$$= \begin{vmatrix} ca-bc+bd-ad & c^2a^2-b^2c^2+b^2d^2-a^2d^2 \\ ab-bc+cd-ad & a^2b^2-b^2c^2+c^2d^2-a^2d^2 \end{vmatrix},$$

expanding with respect to C_1

$$= \begin{vmatrix} (c-d)\,(a-b) & (c^2-d^2)\,(a^2-b^2) \\ (b-d)\,(a-c) & (b^2-d^2)\,(a^2-c^2) \end{vmatrix},$$ factorising the elements

$$= (c - d)(a - b)(b - d)(a - c)\begin{vmatrix} 1 & (c + d)(a + b) \\ 1 & (b + d)(a + c) \end{vmatrix},$$

taking out the common factors

$$= (c - d)(a - b)(b - d)(a - c)\begin{vmatrix} 1 & ca + bc + da + db \\ 0 & ba + dc - ca - db \end{vmatrix},$$

replacing R_2 by $R_3 - R_1$

$$= (c - d)(a - b)(b - d)(a - c)(ba + dc - ca - db)$$

$$= (c - d)(a - b)(b - d)(a - c)(a - d)(b - c)$$ **Ans.**

Example 37:

Show that
$$\begin{vmatrix} 1 & 1 & 1 \\ bc(b + c) & ca(c + a) & ab(a + b) \\ c^2c^2 & c^2a^2 & a^2b^2 \end{vmatrix}$$

$$= abc(a - b)(b - c)(c - a)(a + b + c).$$

Solution:

We have

$$= \begin{vmatrix} 1 & 1 & 1 \\ b^2c + bc^2 & c^2a + ca^2 & a^2b + ab^2 \\ b^2c^2 & c^2a^2 & a^2b^2 \end{vmatrix}$$

$$= \begin{vmatrix} 1 & 0 & 0 \\ b^2c + bc^2 & c(a - b)(a + b + c) & b(a - c)(a + b + c) \\ c^2c^2 & c^2(a - b)(a + b) & b^2(a - c)(a + c) \end{vmatrix},$$

replacing C_2, C_3 by $C_2 - C_1$, $C_3 - C_1$

$$= \begin{vmatrix} c(a - b)(a + b + c) & b(a - c)(a + b + c) \\ c^2(a - b)(a + b) & b^2(a - c)(a + c) \end{vmatrix},$$

expanding with respect to R_1

$$= c\,(a-b)\,b(a-c)\begin{vmatrix} a+b+c & a+b+c \\ c(a+b) & b(b+c) \end{vmatrix},$$

$$= bc\,(a-b)\,(a-c)\,(a+b+c)\begin{vmatrix} 1 & 1 \\ ca+cb & ba+bc \end{vmatrix},$$

taking out a + b + c common from R_1

$$= bc\,(a-b)\,(a-c)\,(a+b+c)\begin{vmatrix} 1 & 0 \\ ca+cb & ba-bc \end{vmatrix},$$

replacing C_2 by $C_2 - C_1$

$$= bc\,(a-b)\,(a-c)\,(a+b+c)\,(ab-ca)$$

$$= -abc\,(a-b)\,(b-c)\,(c-a)\,(a+b+c)$$ **Ans.**

Example 38:

Evaluate $\begin{vmatrix} a & b & ax+by \\ b & c & bx+cy \\ ax+by & bx+cy & 0 \end{vmatrix}$

Solution:

We have

$$= \begin{vmatrix} a & b & 0 \\ b & c & 0 \\ ax+by & bx+cy & -x(ax+by) \\ & & -y(bx+cy) \end{vmatrix},$$ replacing C_2 by $C_3 - x\,C_1 - y.C_2$ **(Note)**

$$= -(ax^2 + 2bxy + cy^2)\begin{vmatrix} a & b \\ b & c \end{vmatrix},$$ expanding with respect to C_2

$$= -\,(ax^2 + 2bxy + cy^2)\,(ac - b^2).$$ **Ans.**

Example 39:

Prove that $\begin{vmatrix} 0 & -c & b & -l \\ c & 0 & -a & -m \\ -b & a & 0 & -n \\ x & y & z & 0 \end{vmatrix}, = (al + bm + cn)\,(ax + by + cz)$

Solution:

We have

$$= \frac{1}{a}\begin{vmatrix} 0 & -ac & ab & -al \\ c & 0 & -a & -m \\ -b & a & 0 & -n \\ x & y & z & 0 \end{vmatrix}, \text{ taking 1/a common from } R_1 \qquad \textbf{(Note)}$$

$$= \frac{1}{a}\begin{vmatrix} 0 & 0 & 0 & -al - bm - cn \\ c & 0 & -a & -m \\ -b & a & 0 & -n \\ x & y & z & 0 \end{vmatrix}, \text{ replacing } R_1 \text{ by } R_1 + b\,R_2 + c\,R_3$$

$$= \frac{(al + bm + cn)}{a}\begin{vmatrix} c & 0 & -a \\ -b & a & 0 \\ x & y & z \end{vmatrix}, \text{ expanding with respect to } R_1$$

$$= \frac{(al + bm + cn)}{a^2}\begin{vmatrix} ac & 0 & -a \\ -ab & a & 0 \\ ax & y & z \end{vmatrix}, \text{ taking 1/a common from } C_1. \qquad \textbf{(Note)}$$

$$= \frac{(al + bm + cn)}{a^2}\begin{vmatrix} 0 & 0 & -a \\ 0 & a & 0 \\ ax + by + cz & y & z \end{vmatrix}, \text{ replacing } C_1 \text{ by } C_1 + b\,C_2 + c\,C_2$$

$$= \frac{(al + bm + cn)(ax + by + cz)}{a^2}\begin{vmatrix} 0 & -a \\ a & 0 \end{vmatrix},$$

$= (al + bm + cn)\ (ax + by + cz).$ **Hence proved.**

Example 40:

Evaluate $\begin{vmatrix} a^2 & a^2 - (b-c)^2 & bc \\ b^2 & b^2 - (c-a)^2 & ca \\ c^2 & c^2 - (a-b)^2 & ab \end{vmatrix}$

Solution:

We, have

$$= \begin{vmatrix} a^2 & -(b-c)^2 & bc \\ b^2 & -(c-a)^2 & ca \\ c^2 & -(a-b)^2 & ab \end{vmatrix}, \text{ replacing } C_2 \text{ by } C_2 - C_1$$

$$= -\begin{vmatrix} a^2 & (b^2+c^2)-2bc & bc \\ b^2 & (c^2-a^2)-2ca & ca \\ c^2 & (a^2+b^2)-2ab & ab \end{vmatrix} \quad \textbf{(Note)}$$

$$= -\begin{vmatrix} a^2 & b^2+c^2 & bc \\ b^2 & c^2+a^2 & ca \\ c^2 & a^2+b^2 & ab \end{vmatrix}, \text{ replacing } C_2 \text{ by } C_2 + 2C_3$$

$$= -\begin{vmatrix} a^2 & b^2+c^2+a^2 & bc \\ b^2 & c^2+a^2+b^2 & ca \\ c^2 & a^2+b^2+c^2 & ab \end{vmatrix}, \text{ replacing } C_2 \text{ by } C_2 + C_1$$

$$= -(a^2+b^2+c^2)\begin{vmatrix} a^2 & 1 & bc \\ b^2 & 1 & ca \\ c^2 & 1 & ab \end{vmatrix}, \text{ taking out the common factor from } C_2$$

$$= -(a^2+b^2+c^2)\begin{vmatrix} a^2 & 1 & bc \\ b^2-a^2 & 0 & ca-bc \\ c^2-a^2 & 0 & ab-bc \end{vmatrix}, \text{ replacing } R_2 \text{ and } R_3 \text{ by } R_2 - R_1 \text{ and } R_3 - R_1 \text{ respectively.}$$

$$= (a^2+b^2+c^2)\begin{vmatrix} b^2-a^2 & c(a-b) \\ c^2-a^2 & b(a-c) \end{vmatrix}, \text{ expanding with respect to } C_2$$

$$= (a-b)(a-c)(a^2+b^2+c^2)\begin{vmatrix} -(b+a) & c \\ -(c+a) & b \end{vmatrix},$$

taking out the common factors

$= (a - b)(a - c)(a^2 + b^2 + c^2)[-b(b + a) + (c + a)]$

$= (a - b)(a - c)(a^2 + b^2 + c^2)[-b^2 - ab + c^2 + ac]$

$= (a - b)(a - c)(a^2 + b^2 + c^2)[(c^2 - b^2) + a(c - b)]$

$= (a - b)(a - c)(a^2 + b^2 + c^2)(c - b)(a + b + c)$

$= (a - b)(a - c)(a + b + c)(a^2 + b^2 + c^2).$ **Ans.**

Example 41:

Prove that

$$\begin{vmatrix} a^2 + \lambda & ab & ac & ad \\ bd & b^2 + \lambda & bc & bd \\ ca & cb & c^2 + \lambda & cd \\ da & db & dc & d^2 + \lambda \end{vmatrix} = \lambda^3 (a^2 + b^2 + c^2 + d^2 + \lambda)$$

Solution:

We have

$$= abcd \begin{vmatrix} a + \frac{\lambda}{a} & a & a & a \\ b & b + \frac{\lambda}{b} & b & b \\ c & c & c + \frac{\lambda}{c} & c \\ d & d & d & d + \frac{\lambda}{d} \end{vmatrix},$$ taking out a, b, c, d common from C_1, C_2, C_3 and C_4 respectively.

$$= abcd \begin{vmatrix} a + \frac{\lambda}{a} & -\frac{\lambda}{a} & -\frac{\lambda}{a} & -\frac{\lambda}{a} \\ b & \frac{\lambda}{b} & 0 & 0 \\ c & 0 & \frac{\lambda}{c} & 0 \\ d & 0 & 0 & \frac{\lambda}{d} \end{vmatrix},$$ replacing C_2 , C_3 and C_4 by $C_2 - C_1$, $C_3 - C_1$ and $C_4 - C_1$ respectively.

$$= abcd \begin{vmatrix} a & 0 & 0 & -\frac{\lambda}{a} \\ b & \frac{\lambda}{b} & 0 & 0 \\ c & 0 & \frac{\lambda}{c} & 0 \\ d + \frac{\lambda}{} & -\frac{\lambda}{} & -\frac{\lambda}{} & \frac{\lambda}{} \end{vmatrix},$$ replacing C_1, C_2 and C_3 by $C_1 + C_4$, $C_2 - C_4$ and $C_3 - C_4$ respectively.

$$= a^2bcd \begin{vmatrix} \frac{\lambda}{b} & 0 & 0 \\ 0 & \frac{\lambda}{c} & 0 \\ -\frac{\lambda}{d} & -\frac{\lambda}{d} & \frac{\lambda}{d} \end{vmatrix} + \lambda\, bcd \begin{vmatrix} b & \frac{\lambda}{b} & 0 \\ 0 & 0 & \frac{\lambda}{c} \\ d+\frac{\lambda}{d} & -\frac{\lambda}{d} & -\frac{\lambda}{d} \end{vmatrix},$$

expanding with respect to R_1

$$= a^2bcd.\frac{\lambda}{b} \begin{vmatrix} \frac{\lambda}{c} & 0 \\ -\frac{\lambda}{d} & \frac{\lambda}{d} \end{vmatrix} + \lambda\, bcd.b \begin{vmatrix} 0 & \frac{\lambda}{c} \\ -\frac{\lambda}{d} & -\frac{\lambda}{d} \end{vmatrix}$$

$$- \lambda bcd.\frac{\lambda}{b} \begin{vmatrix} c & \frac{\lambda}{c} \\ d+\frac{\lambda}{d} & -\frac{\lambda}{d} \end{vmatrix},$$

expanding each determinant with respect to R_1

$$= \lambda\, a^2\, cd \left(\frac{\lambda^2}{cd}\right) + \lambda b^2\, cd \left(\frac{\lambda^2}{cd}\right) - \lambda^2\, cd \left(-\frac{\lambda c}{d} - \frac{\lambda d}{c} - \frac{\lambda^2}{cd}\right)$$

$$= \lambda^2\, a^2 + \lambda^2\, b^2 + \lambda^2\, cd \left(\frac{\lambda c^2 + \lambda d^2 + \lambda^2}{cd}\right)$$

$$= \lambda^2\, a^2 + \lambda^2\, b^2 + \lambda^2\, (c^2 + d^2 + \lambda) = \lambda^2\, (a^2 + b^2 + c^2 + d^2 + \lambda)$$

Hence proved.

Example 42:

Prove that

$$\begin{vmatrix} 1 & 0 & x & 0 & x \\ 0 & 1 & 0 & x & 0 \\ x & 0 & x+1 & 0 & x \\ 0 & x & 0 & 1 & 0 \\ x & 0 & x & 0 & 1 \end{vmatrix} = (x-1)^2\, (x+1)\, (1+2x-x^2)$$

Solution:

The given determinant

$$= \begin{vmatrix} 1 & 0 & 0 & 0 & 0 \\ 0 & 1 & 0 & x & 0 \\ x & 0 & x+1-x^2 & 0 & -1 \\ 0 & x & 0 & 1 & 0 \\ x & 0 & x-x^2 & 0 & 1-x \end{vmatrix}$$, replacing C_3 and C_5 by $C_3 - x\,C_1$ and $C_5 - C_3$ respectively.

$$= \begin{vmatrix} 1 & 0 & x & 0 \\ 0 & x+1-x^2 & 0 & -1 \\ x & 0 & 1 & 0 \\ 0 & x-x^2 & 0 & 1-x \end{vmatrix}$$, expanding with respect to R_1

$$= \begin{vmatrix} 1 & 0 & 0 & 0 \\ b & x+1-x^2 & 0 & -1 \\ x & 0 & 1-x^2 & 0 \\ 0 & x-x^2 & 0 & 1-x \end{vmatrix}$$, replacing C_3 by $C_2 - x\,C_1$

$$= \begin{vmatrix} x+1-x^2 & 0 & -1 \\ 0 & 1-x^2 & 0 \\ x(1-x) & 0 & 1-x \end{vmatrix}$$, expanding with respect to R_1

$$= -(1-x) \begin{vmatrix} 0 & 1-x^2 & 0 \\ x+1-x^2 & 0 & -1 \\ x & 0 & 1 \end{vmatrix}$$, interchanging R_1 and R_2 and taking out $(1-x)$ common from R_2

$$= (1-x)(1-x^2) \begin{vmatrix} x+1-x^2 & -1 \\ x & 1 \end{vmatrix}$$, expanding with respect to R_1

$$= (1-x)^2 (1+x) [(x+1-x^2)\,1 - (-1).x]$$

$$= (x-1)^2 (1+x) (2x+1-x^2).$$

Hence proved.

Example 43:

Evaluate $\begin{vmatrix} 13 & 16 & 19 \\ 14 & 17 & 20 \\ 15 & 18 & 21 \end{vmatrix}$

Solution:

We have

$= \begin{vmatrix} 13 & 16 & 3 \\ 14 & 17 & 3 \\ 15 & 18 & 3 \end{vmatrix}$, replacing C_2 by $C_3 - C_2$

$= \begin{vmatrix} 13 & 3 & 3 \\ 14 & 3 & 3 \\ 15 & 3 & 3 \end{vmatrix}$, replacing C_2 by $C_3 - C_1$

= 0, since two columns are identical. **Ans.**

Example 44:

Evaluate $\begin{vmatrix} 1 & 2 & 5 \\ 2 & 3 & 1 \\ -1 & 1 & 1 \end{vmatrix}$

Solution:

The given determinant

$= \begin{vmatrix} 1 & 3 & 6 \\ 2 & 5 & 3 \\ -1 & 0 & 0 \end{vmatrix}$, replacing C_2 and C_3 by $C_2 + C_1$ and $C_3 + C_1$ respctively.

$= \begin{vmatrix} 0 & 3 & 6 \\ 0 & 5 & 3 \\ -1 & 0 & 0 \end{vmatrix}$, replacing R_1 by $R_1 + R_3$ and R_2 by $R_2 + 2R_3$

$= -1 \times \begin{vmatrix} 3 & 6 \\ 5 & 3 \end{vmatrix}$, expanding with respect to first column

$= -[3 (3) - 5 (6)] = 21.$ **Ans.**

Example 45:

Evaluate $\begin{vmatrix} 1 & a & b+c \\ 1 & b & c+a \\ 1 & c & a+b \end{vmatrix}$

Solution:

We have

$$= \begin{vmatrix} 1 & a & a+b+c \\ 1 & b & b+c+a \\ 1 & c & c+a+b \end{vmatrix}, \text{ replacing } C_3 \text{ by } C_3 + C_2$$

$$= (a+b+c)\begin{vmatrix} 1 & a & 1 \\ 1 & b & 1 \\ 1 & c & 1 \end{vmatrix}, \text{ taking out } (a+b+c) \text{ common from } C_3$$

= 0, since two columns are identical. **Ans.**

PRODUCT OF DETERMINANTS

Theorem:

If A is an n × n matrix and E is an elementary matrix obtained from the identity matrix I_n, then

$$|EA| = |AE| = |E|.|A| = |A|\,|E|.$$

Proof:

$\because I_n$ is the n × n identity matrix.

$\therefore |I_n| = 1.$ ...(i)

Let E_a, E_b, E_c be three elementary matrices.

Then $|E_a| = -|I_n|$

$\therefore$ From (i) we get $|E_a| = -1$ ...(ii)

Again $|E_b| = c\,|I_n|$

$\therefore$ From (i) we get $|E_b| = c$ ...(iii)

Similarly $|E_c| = |I_n|$

$\therefore$ From (i) we get $|E_c| = 1$ ...(iv)

Now, as the matrix E_a A can be obtained by interchanging two rows of the matrix A, so we have from (ii)

$|E_a| = -|A| = |E_a|.|A|$ ...(v)

Similarly, the product E_bA can be obtained by applying second elementary row operation on the matrix A, so we have from (iii)

$|E_bA| = c\,|A| = |E_b|.|A|$ ...(vi)

Again the product E_c A can be obtained by applying third elementary row operation on the matrix A, so we have from (iv)

$$|E_c A| = |A| = |E_c| \cdot |A| \qquad \text{...(vii)}$$

Hence from (v). (vi) and (vii) we conclude that

$$|EA| = |E| \cdot |A|,$$

where E is any one of the elementary matrices.

In similar way we can prove that

$$|AE| = |A| \cdot |E|,$$

LAPLACE'S EXPANSION OF A DETERMINANT BY THE MINORS OF FIRST R COLUMNS

If $|B_i|$ is r × r minor of an n × n matrix A formed by the elements of the first r columns of A and $|B'|$ is the complementary minor of $|B_i|$, then

$$|A| = \Sigma \pm |B_i|, |B'_i|,$$

where the summation is extended over all the possible r × r minor of A which can be formed from the elements of the first r columns and + or – sign is taken according as an even or odd number of interchanges of adjacent rows of A is required to bring the submatrix B_i into the first r rows of A.

The following solved examples will explain the above theorem.

CANONICAL FORM (OR NORMAL FORM OF A MATRIX)

Every non-zero m × n matrix A can be reduced by means of elementary transformation (*i.e.*, elementary row and column operations) to the form

$$\begin{bmatrix} I & 0 \\ 0 & 0 \end{bmatrix}'$$

where I_r is the r × r identity matrix and the remaining sub-matrices are zero matrices.

The above form is called the *canonical form or orthogonal form or normal form* of the matrix A.

COMPLEMENTARY MINOR OF A DETERMINANT

Definition: *If B is r × r sub-matrix of an n × n matrix A, then the determinant E' of A formed by removing the rows and columns of A containing the elements of B is called the complementary minor of B.*

In the matrix $\begin{bmatrix} a_1 & b_1 & c_1 & d_1 \\ a_2 & b_2 & c_2 & d_2 \\ a_3 & b_3 & c_3 & d_3 \\ a_4 & b_4 & c_4 & d_4 \end{bmatrix}$

the complementary minor of the det. $\begin{vmatrix} a_1 & b_1 \\ a_2 & b_2 \end{vmatrix}$ is $\begin{vmatrix} c_3 & d_3 \\ c_4 & d_4 \end{vmatrix}$;

the complementary minor of $\begin{vmatrix} b_1 & c_1 & d_1 \\ b_2 & c_2 & d_2 \\ b_3 & c_3 & d_3 \end{vmatrix}$ is a_4

and the complementary minor of $\begin{vmatrix} b_2 & c_2 \\ b_3 & c_3 \end{vmatrix}$ is $\begin{vmatrix} a_1 & d_1 \\ a_4 & d_4 \end{vmatrix}$

MISCELLANEOUS EXAMPLES

Example 1:

Show that

$$\Delta = \begin{vmatrix} b^2+c^2 & ab & ac \\ ab & c^2+a^2 & bc \\ ca & cb & a^2+b^2 \end{vmatrix} = 4a^2b^2c^2$$

Solution:

Multiplying C_1 by a, C_2 by b and C_3 by c, we get

$$\Delta = \frac{1}{abc}\begin{vmatrix} a(b^2+c^2) & ab^2 & ac^2 \\ a^2b & b(c^2+a^2) & bc^2 \\ a^2c & cb^2 & (b^2+a^2)c \end{vmatrix}$$

Taking a, b and c common from R_1, R_2 and R_3 respectively, we get

$$\Delta = \frac{abc}{abc}\begin{vmatrix} b^2+c^2 & b^2 & c^2 \\ a^2 & c^2+a^2 & c^2 \\ a^2 & b^2 & b^2+a^2 \end{vmatrix}$$

Applying $C_1 \to C_1 - C_2 - C_3$, we get

$$\Delta = \begin{vmatrix} 0 & b^2 & c^2 \\ -2c^2 & c^2+a^2 & c^2 \\ -2b^2 & b^2 & b^2+a^2 \end{vmatrix}$$

Taking 2 common from C_1, and applying $C_2 \to C_2 + C_1$ and $C_3 \to C_3 + C_1$, we get

$$\Delta = 2\begin{vmatrix} 0 & b^2 & c^2 \\ -c^2 & a^2 & 0 \\ -b^2 & 0 & a^2 \end{vmatrix}$$

Evaluating along C_1, we get

$\Delta = 2[a^2 (b^2 a^2 - 0) - b^2 (0 - a^2c^2)] = 4a^2b^2c^2$.

Example 2:

If x, y and z are unequal and

$$\begin{vmatrix} x^3 & (x+a)^3 & (x-a)^3 \\ y^3 & (y+a)^3 & (y-a)^3 \\ z^3 & (z+a)^3 & (z-a)^3 \end{vmatrix} = 0$$

prove that $a^2 (x + y + z) = 3xyz$.

Solution:

Applying $C_2 \to C_2 - C_3$, we get

$$\Delta = \begin{vmatrix} x^3 & 6x^2a+2a^3 & (x-a)^3 \\ y^3 & 6y^2a+2a^3 & (y-a)^3 \\ z^3 & 6z^2a+2a^3 & (z-a)^3 \end{vmatrix}$$

Taking 2 common from C_2 and applying $C_3 \to C_3 - C_1 + C_2$, we get

$$\Delta = -2\begin{vmatrix} x^3 & 3x^2a+a^3 & 3xa^2 \\ y^3 & 3y^2a+a^3 & 3ya^2 \\ z^3 & 3z^2a+a^3 & 3za^2 \end{vmatrix} = -18a^2\Delta_1 - 6a^5\Delta_2$$

$$\text{where } \Delta_1 = \begin{vmatrix} x^3 & x^2 & x \\ y^3 & y^2 & y \\ z^3 & z^2 & z \end{vmatrix} = xyz \begin{vmatrix} x^2 & x & 1 \\ y^2 & y & 1 \\ z^2 & z & 1 \end{vmatrix}$$

$$\text{and} \quad \Delta_2 = \begin{vmatrix} x^3 & 1 & x \\ y^3 & 1 & y \\ z^3 & 1 & z \end{vmatrix}$$

Applying $R_2 \to R_2 - R_1$ and $R_3 \to R_3 - R_1$ in Δ_1 and Δ_2 we obtain

$$\Delta_1 = xyz \begin{vmatrix} x^2 & x & 1 \\ y^2 - x^2 & y - x & 0 \\ z^2 - x^2 & z - x & 0 \end{vmatrix}$$

$$= xyz(y-x)(z-x) \begin{vmatrix} y+x & 1 \\ z+x & 1 \end{vmatrix}$$

$$= xyz\ (y - x)\ (z - x)\ (y - z)$$

$$\text{and } \Delta_2 = \begin{vmatrix} x^3 & 1 & x \\ y^3 - x^3 & 0 & y - x \\ z^3 - x^3 & 0 & z - x \end{vmatrix}$$

$$= -(y-x)(z-x) \begin{vmatrix} y^2 + x^2 + yx & 1 \\ z^2 + x^2 + zx & 1 \end{vmatrix}$$

$$= (y - x)\ (z - x)\ (z - y)\ (x + y + z)$$

Thus

$$\Delta = 6a^3\ (y - x)\ (z - x)\ (y - z)$$

$$[a^2\ (x + y + z) - 3xyz]$$

As $\Delta = 0$ and x, y and z are unequal we get

$$a^2\ (x + y + z) - 3xyz = 0$$

$$\Rightarrow \quad a^2\ (x + y + z) = 3yz$$

Example 3:

If a, b and c are distinct, solve the equation

$$\Delta = \begin{vmatrix} x^2 - a^2 & x^2 - b^2 & x^2 - c^2 \\ (x-a)^3 & (x-b)^3 & (x-c)^3 \\ (x+a)^3 & (x+b)^3 & (x+c)^3 \end{vmatrix} = 0$$

Solution:

Applying $R_3 \to R_3 - R_2$, we obtain

$$\Delta = \begin{vmatrix} x^2 - a^2 & x^2 - b^2 & x^2 - c^2 \\ (x-a)^3 & (x-b)^3 & (x-c)^3 \\ 6x^2a + 2a^3 & 6x^2b + 2b^3 & 6x^2c + 2c^3 \end{vmatrix}$$

Taking 2 common from R_3 and applying $R_2 \to R_2 + R_3$, we get

$$\Delta = 2\begin{vmatrix} x^2 - a^2 & x^2 - b^2 & x^2 - c^2 \\ x^3 + 3xa^2 & x^3 + 3xb^2 & x^3 + 3xc^2 \\ 3x^2a + a^3 & 3x^2b + b^3 & 3x^2c + c^3 \end{vmatrix}$$

Taking x common from R_2 and applying $C_2 \to C_2 - C_1$ and $C_3 \to C_3 - C_1$, we get

$$\Delta = 2x\begin{vmatrix} x^2 - a^2 & x^2 - b^2 & x^2 - c^2 \\ x^2 + 3a^2 & 3(b^2 - a^2) & 3(c^2 - a^2) \\ 3x^2a + a^3 & 3x^2(b-a) + b^3 - a^3 & 3x^2(c-a) + c^3 - a^3 \end{vmatrix}$$

$$= 2x\ (b - a)\ (c - a)\ \Delta_1$$

$$\text{where } \Delta_1 = \begin{vmatrix} x^2 - a^2 & -(a+b) & -(a+c) \\ x^2 + 3a^2 & 3(b+a) & 3(a+c) \\ 3x^2a + a^3 & 3x^2 + b^2 + a^2 + ba & 3x^2 + c^2 + a^2 + ac \end{vmatrix}$$

Applying $R_2 \to R_2 + 3R_1$, we get $\Delta_1 =$

$$\begin{vmatrix} x^2 - a^2 & -(a+b) & -(a+c) \\ 4x^2 & 0 & 0 \\ 3x^2a + a^3 & 3x^2 + b^2 + a^2 + ba & 3x^2 + c^2 + a^2 + ac \end{vmatrix}$$

Expanding along R_2, we get

$$\Delta = 8x^3 (b - a)(c - a)\begin{vmatrix} a+b & a+c \\ 3x^2+b^2+a^2+ba & 3x^2+c^2+a^2+ac \end{vmatrix}$$

Applying $C_2 \to C_2 - C_1$, we get

$$\Delta = 8x^3 (b - a)(c - a)\begin{vmatrix} a+b & c-b \\ 3x^2+b^2+a^2+ba & c^2-b^2+ac-ab \end{vmatrix}$$

$= 8x^3 (b - a)(c - a)(c - b)[(a + b)(a + b + c) - 3x^2 - b^2 - a^2 - ba]$

$= 8x^2 (b - a)(c - a)(c - b)[(ab + bc + ca) - 3x^2]$

$\therefore\ \Delta = 0 \Rightarrow 8(b - a)(c - a)(c - b)x^3 [(ab + bc + ca) - 3x^2] = 0$

As a, b and c are distinct, we get

$$x = 0, 0, 0, \pm \sqrt{\frac{1}{3}(bc+ca+ab)}$$

Example 4:

Without expanding at any stage, show that

$$\Delta = \begin{vmatrix} x^2+x & x+1 & x-2 \\ 2x^3+3x-1 & 3x & 3x-3 \\ x^2+2x+3 & 2x-1 & 2x-1 \end{vmatrix} = xA + B$$

where A and B are determinants of order 3 not involving x.

Solution:

Applying $R_2 \to R_2 - R_1 - R_3$, we obtain

$$\Delta = \begin{vmatrix} x^2+x & x+1 & x-2 \\ -4 & 0 & 0 \\ x^2+2x+3 & 2x-1 & 2x-1 \end{vmatrix}$$

Applying $R_1 \to R_1 + \frac{1}{4}x^3 R_2$ and $R_3 \to R_3 + \frac{1}{4}x^3 R_2$, we get

$$\Delta = \begin{vmatrix} x & x+1 & x-2 \\ -4 & 0 & 0 \end{vmatrix}$$

Applying $R_3 \to R_3 - 2R_1$ we get

$$\Delta = \begin{vmatrix} x & x+1 & x-2 \\ -4 & 0 & 0 \\ 3 & -3 & 3 \end{vmatrix} = xA + B$$

where $A = \begin{vmatrix} 1 & 1 & 1 \\ -4 & 0 & 0 \\ 3 & -3 & 3 \end{vmatrix}$ and $B \begin{vmatrix} 0 & 1 & -2 \\ -4 & 0 & 0 \\ 3 & -3 & 3 \end{vmatrix}$

Example 5:

Let α be a repeated root of the quadratic equation $f(x) = 0$ and $A(x)$, $B(x)$ and $C(x)$ polynomials of degrees 3, 4 and 5 respectively. Show that

$$\begin{vmatrix} A(x) & B(x) & C(x) \\ A(\alpha) & B(\alpha) & C(\alpha) \\ A'(\alpha) & B'(\alpha) & C'(\alpha) \end{vmatrix}$$

is divisible by $f(x)$, where prime denotes the derivative.

Solution:

Since α is a repeated root of the quadratic equation $f(x) = 0$, we have $f(x) = a(x - \alpha)^2$, where α is some constant, and $a \neq 0$.

Let $$\phi(x) = \begin{vmatrix} A(x) & B(x) & C(x) \\ A(\alpha) & B(\alpha) & C(\alpha) \\ A'(\alpha) & B'(\alpha) & C'(\alpha) \end{vmatrix}$$

Note that $\phi(x)$ is a polynomial in x of degree at most 5. We have

$$\phi(\alpha) = \begin{vmatrix} A(\alpha) & B(\alpha) & C(\alpha) \\ A(\alpha) & B(\alpha) & C(\alpha) \\ A'(\alpha) & B'(\alpha) & C'(\alpha) \end{vmatrix}$$

$= 0$ [$\because$ R_1 and R_2 are identical]

Also $$\phi'(x) = \begin{vmatrix} A'(x) & B'(x) & C'(x) \\ A(\alpha) & B(\alpha) & C(\alpha) \\ A'(\alpha) & B'(\alpha) & C'(\alpha) \end{vmatrix} \Rightarrow \phi'(\alpha) = 0$$

Thus, α is a repeated root of $\phi(x) = 0$. Therefore, $\phi(x) = (x - \alpha)^2 \psi(x)$, where $\psi(x)$ is a polynomial in x of degree at most 3. This shows that $f(x) | \phi(x)$.

Example 6(a):

Show that

$$\Delta = \begin{vmatrix} a_1 & x & x \\ x & a_2 & x \\ x & x & a_3 \end{vmatrix} = xf'(x) - f(x)$$

where $f(x) = (x - a_1)(x - a_2)(x - a_3)$

Solution:

Applying $C_3 \to C_3 - C_2$ and $C_2 \to C_2 - C_1$ we get

$$\Delta = \begin{vmatrix} a_1 & x-a_1 & 0 \\ x & a_2-x & x-a_2 \\ x & x & a_3-x \end{vmatrix}$$

Expanding along R_1 we get

$$\Delta = \begin{vmatrix} a_2-x & x-a_2 \\ 0 & a_3-x \end{vmatrix} - (x-a_1)\begin{vmatrix} x & x-a_2 \\ x & a_3-x \end{vmatrix}$$

$= a_1[(a_2 - x)(a_3 - x)] - (x - a_1) x \{(a_3 - x) - (x - a_2)\}$

$= x[(x - a_1)(x - a_2) + (x - a_1)(x - a_2)] + a_1 (x - a_2)(x - a_3)$

$= x[x - a_2)(x - a_3) + (x - a_3)(x - a_1) + (x - a_1)(x - a_2)] + a_1$

$(x - a_2)(x - a_2) - x(x - a_2)(x - a_3)$

$= xf'(x) - (x - a_1)(x - a_2)(x - a_3) = xf'(x) - f(x).$

Example 6(b):

Show that $\begin{vmatrix} bc-a^2 & ca-b^2 & ab-c^2 \\ ca-b^2 & ab-c^2 & bc-a^2 \\ ab-c^2 & bc-a^2 & ca-b^2 \end{vmatrix}$

$$= \begin{vmatrix} a^2 & c^2 & 2ac-b^2 \\ 2ab-c^2 & b^2 & a^2 \\ b^2 & 2bc-a^2 & c^2 \end{vmatrix}$$

Solution:

$$\text{Let}\quad \Delta = \begin{vmatrix} a & b & c \\ b & c & a \\ c & a & b \end{vmatrix}$$

Replacing each element of Δ by its cofactor we obtain

$$\Delta_1 = \begin{vmatrix} bc-a^2 & ca-b^2 & ab-c^2 \\ ca-b^2 & ab-c^2 & bc-a^2 \\ ab-c^2 & bc-a^2 & ca-b^2 \end{vmatrix} = \Delta^2 \qquad ...(1)$$

$$\text{Also}\quad \Delta^2 = \begin{vmatrix} a & b & c \\ b & c & a \\ c & a & b \end{vmatrix}^2 = \begin{vmatrix} a & b & c \\ b & c & a \\ c & a & b \end{vmatrix}\begin{vmatrix} a & -c & b \\ b & -a & c \\ c & -b & a \end{vmatrix}$$

$$= \begin{vmatrix} a^2 & c^2 & 2ac-b^2 \\ 2ab-c^2 & b^2 & a^2 \\ b^2 & 2bc-a^2 & c^2 \end{vmatrix} \qquad ...(2)$$

From (1) and (2) we get

$$\begin{vmatrix} bc-a^2 & ca-b^2 & ab-c^2 \\ ca-b^2 & ab-c^2 & bc-a^2 \\ ab-c^2 & bc-a^2 & ca-b^2 \end{vmatrix} = \begin{vmatrix} a^2 & c^2 & 2ac-b^2 \\ 2ab-c^2 & b^2 & a^2 \\ b^2 & 2bc-a^2 & c^2 \end{vmatrix}$$

Example 6(c):

Evaluate the determinant

$$\Delta = \begin{vmatrix} \sqrt{p}+\sqrt{q} & 2\sqrt{r} & \sqrt{r} \\ \sqrt{qr}+\sqrt{2p} & r & \sqrt{2r} \\ q+\sqrt{pr} & \sqrt{qr} & r \end{vmatrix}$$

where p, q and r are positive real numbers.

Solution:

Taking $\sqrt{r}$ common from C_2 and C_3 we get

$$\Delta = r\begin{vmatrix} \sqrt{p}+\sqrt{q} & 2 & 1 \\ \sqrt{qr}+\sqrt{2p} & \sqrt{r} & \sqrt{2} \\ q+\sqrt{pr} & \sqrt{q} & \sqrt{r} \end{vmatrix}$$

Applying $C_1 \to C_1 - \sqrt{q}\,C_2 - \sqrt{p}\,C_3$ we get

$$\Delta = r\begin{vmatrix} -\sqrt{p} & 2 & 1 \\ 0 & \sqrt{r} & \sqrt{2} \\ 0 & \sqrt{q} & \sqrt{r} \end{vmatrix} = -\,r\sqrt{q}\,(r-\sqrt{2q})$$

$$= r\,(2\sqrt{2q} - r\sqrt{q})$$

Example 7:

If $a \neq p$, $b \neq q$, $c \neq r$ and

$$\Delta = \begin{vmatrix} p & b & c \\ a & q & c \\ a & b & r \end{vmatrix} = 0, \text{ then find the value of}$$

$$\frac{p}{p-a} + \frac{q}{q-b} + \frac{r}{r-c}$$

Solution:

Applying $R_2 \to R_2 - R_1$, $R_3 \to R_3 - R_1$, we get

$$\Delta = \begin{vmatrix} p & b & c \\ a-p & q-b & 0 \\ a-p & 0 & r-c \end{vmatrix}$$

$$= c\begin{vmatrix} a-p & q-b \\ a-p & 0 \end{vmatrix} + (r-c)\begin{vmatrix} p & b \\ a-p & q-b \end{vmatrix}$$

$= -c(a - p)\,(q - b) + (r - c)\,(q - b)\,p - (r - c)\,(a - p) = 0$

As $\Delta = 0$, we get

$-c(a - p)\,(q - b) + (r - c)\,(q - b)\,p - (r - c)\,(a - p)\,b = 0$

Dividing by $(p - a)\,((q - b)\,(r - c)$ we get

$$\frac{c}{r-c} + \frac{p}{p-a} + \frac{b}{q-b} = 0 \qquad ...(1)$$

Let $\quad k = \dfrac{p}{p-a} + \dfrac{q}{q-b} + \dfrac{r}{r-c} \qquad ...(2)$

Subtracting (1) from (2) we get

$$k - 0 = \frac{q-b}{q-b} + \frac{r-c}{r-c} = 1 + 1 = 2 \Rightarrow k = 2.$$

Example 8:

For a fixed positive integer n, if

$$D = \begin{vmatrix} n! & (n+1)! & (n+2)! \\ (n+1)! & (n+2)! & (n+3)! \\ (n+2)! & (n+3)! & (n+4)! \end{vmatrix}$$

then show that $\left(\dfrac{D}{(n!)^3} - 3\right)$ *is divisible by n.*

Solution:

Taking n! ((n + 1) !) {(n + 2) !} common from C_1 (C_2) (C_3) we get

D = n! (n + 1)! (n + 2)!

$$\begin{vmatrix} 1 & 1 & 1 \\ n+1 & n+2 & n+3 \\ (n+1)(n+2) & (n+2)(n+3) & (n+3)(n+4) \end{vmatrix}$$

Applying $C_3 \to C_3 - C_2$ and $C_2 \to C_2 - C_1$ we get

D = n! (n + 1)! (n + 2)!

$$\begin{vmatrix} 1 & 0 & 0 \\ n+1 & 1 & 1 \\ (n+1)(n+2) & 2n+4 & 2n+6 \end{vmatrix}$$

$= n!\ (n + 1)!\ (n + 2)!\ (2n + 6 - 2n - 4)$

$D = 2(n!)\ (n + 1)!\ (n + 2)!$

$$\Rightarrow \qquad \frac{D}{(n!)^3} - 4 = 2(n + 1)\ (n + 1)\ (n + 2) - 4$$

$$= 2n(n^2 + 4n + 5)$$

$$\Rightarrow \quad \left(\frac{D}{(n!)^3} - 3\right) \text{ is divisible by n.}$$

Example 9:

For all values of A, B, C and P, Q, R show that

$$\Delta = \begin{vmatrix} \cos(A-P) & \cos(A-Q) & \cos(A-R) \\ \cos(B-P) & \cos(B-Q) & \cos(B-R) \\ \cos(C-P) & \cos(C-Q) & \cos(C-R) \end{vmatrix} = 0$$

Solution:

The given determinant can be written as product of two determinants as follows:

$$\Delta = \begin{vmatrix} \cos(A-P) & \cos(A-Q) & \cos(A-R) \\ \cos(B-P) & \cos(B-Q) & \cos(B-R) \\ \cos(C-P) & \cos(C-Q) & \cos(C-R) \end{vmatrix}$$

$$= \begin{vmatrix} \cos A & \sin A & 0 \\ \cos B & \sin B & 0 \\ \cos C & \sin C & 0 \end{vmatrix} \begin{vmatrix} \cos P & \sin P & 0 \\ \cos Q & \sin Q & 0 \\ \cos R & \sin R & 0 \end{vmatrix}$$

$$= (0)\ (0) = 0.$$

Alternative Solution

Using $\cos(\alpha - \beta) = \cos\alpha \cos\beta + \sin\alpha \sin\beta$, we can write Δ as follows:

$$\Delta = \cos P \begin{vmatrix} \cos A & \cos(A-Q) & \cos(A-R) \\ \cos B & \cos(B-Q) & \cos(B-R) \\ \cos C & \cos(C-Q) & \cos(C-R) \end{vmatrix}$$

$$+ \sin P \begin{vmatrix} \sin A & \cos(A-Q) & \cos(A-R) \\ \sin B & \cos(B-Q) & \cos(B-R) \\ \sin C & \cos(C-Q) & \cos(C-R) \end{vmatrix}$$

Let us denote the first determinant on the right side by Δ_1 and the second by Δ_2. In Δ_1, we apply

$C_2 \to C_2 - C_1 \cos Q$ and $C_3 \to C_3 - C_1 \cos R$

$$\text{so that } \Delta_1 = \begin{vmatrix} \cos A & \sin A \sin Q & \sin A \sin R \\ \cos B & \sin B \sin Q & \sin B \sin R \\ \cos C & \sin C \sin Q & \sin C \sin R \end{vmatrix}$$

$= 0$ [$\because$ C_2 and C_3 are proportional]

In Δ_2, we apply, $C_2 \to C_2 - C_1 \sin Q$ and $C_3 \to C_3 - C_1 \sin R$ so that

$$\Delta_2 = \begin{vmatrix} \sin A & \cos A \cos Q & \cos A \cos R \\ \sin B & \cos B \cos Q & \cos B \cos R \\ \sin C & \cos C \cos Q & \cos C \cos R \end{vmatrix}$$

$= 0$ [$\because$ C_2 and C_3 are proportional]

Thus, $\Delta = (\cos p)(0) + (\sin p)(0) = 0$.

Example 10:

Let $a > 0$, $d > 0$. Find the value of the determinant

$$\begin{vmatrix} \frac{1}{a} & \frac{1}{a(a+d)} & \frac{1}{(a+d)(a+2d)} \\ \frac{1}{(a+d)} & \frac{1}{(a+d)(a+2d)} & \frac{1}{(a+2d)(a+3d)} \\ \frac{1}{(a+2d)} & \frac{1}{(a+2d)(a+3d)} & \frac{1}{(a+3d)(a+4d)} \end{vmatrix}$$

Solution:

Let us denote the given determinant by Δ. Taking

$\frac{1}{a(a+d)(a+2d)}$ common from R_1,

$\frac{1}{(a+d)(a+2d)(a+3d)}$ from R_2 and

$\frac{1}{(a+2d)(a+3d)(a+4d)}$ from R_3, we get

$$\Delta = \frac{1}{a(a+d)^2(a+2d)^3(a+3d)^2(a+4d)}\Delta_1$$

$$\text{where } \Delta_1 = \begin{vmatrix} (a+d)(a+2d) & a+2d & a \\ (a+2d)(a+3d) & a+3d & a+d \\ (a+3d)(a+4d) & a+4d & a+2d \end{vmatrix}$$

Applying $R_3 \to R_3 - R_2$ and $R_2 \to R_2 - R_1$, we get

$$\Delta_1 = \begin{vmatrix} (a+d)(a+2d) & d & a \\ (a+2d)(2d) & d & d \\ (a+3d)(2d) & d & d \end{vmatrix}$$

Applying $R_3 \to R_3 - R_2$, we get

$$\Delta_1 = \begin{vmatrix} (a+d)(a+2d) & a+2d & a \\ (a+2d)(2d) & d & d \\ 2d^2 & 0 & 0 \end{vmatrix}$$

Expanding along R_3, we get

$$\Delta_1 = (2d)^2 \begin{vmatrix} a+2d & a \\ d & d \end{vmatrix}.$$

$$= (2d^2)\,(d)\,(a + 2d - a) = 4d^2$$

$$\text{Thus } \Delta = \frac{4d^4}{a(a+d)^2(a+2d)^3(a+3d)^2(a+4d)}$$

Example 11:

Show that

$$\Delta = \begin{vmatrix} \cos(\alpha-\beta) & \cos(\beta-\gamma) & \cos(\gamma-\alpha) \\ \cos(\alpha+\beta) & \cos(\beta+\gamma) & \cos(\gamma+\alpha) \\ \sin(\alpha+\beta) & \sin(\beta+\gamma) & \sin(\gamma+\alpha) \end{vmatrix}$$

$$= -2\sin(\beta-\gamma)\sin(\gamma-\alpha)\sin(\alpha-\beta)$$

Solution:

Applying $R_1 \to R_1 - R_2$, we get

$$\Delta = \begin{vmatrix} 2\sin\alpha\sin\beta & 2\sin\beta\sin\gamma & 2\sin\gamma\sin\alpha \\ \cos(\alpha+\beta) & \cos(\beta+\gamma) & \cos(\gamma+\alpha) \\ \sin(\alpha+\beta) & \sin(\beta+\gamma) & \sin(\gamma+\alpha) \end{vmatrix}$$

Taking 2 common from R_1 and applying $R_2 \to R_2 + R_1$, we get

$$\Delta = 2\begin{vmatrix} \sin\alpha\sin\beta & \sin\beta\sin\gamma & \sin\gamma\sin\alpha \\ \cos\alpha\cos\beta & \cos\beta\cos\gamma & \cos\gamma\cos\alpha \\ \sin(\alpha+\beta) & \sin(\beta+\gamma) & \sin(\gamma+\alpha) \end{vmatrix}$$

Taking sin α sin β common from C_1, sin β sin γ from C_2 and sin γ sin α from C_3 and using the fact that sin (x + y)/sin x sin y = cot x + cot y, we get

$$\Delta = 2\sin^2\alpha\sin^2\beta\sin^2\gamma\begin{vmatrix} 1 & 1 & 1 \\ \cot\alpha\cot\beta & \cot\beta\cot\gamma & \cot\gamma\cot\alpha \\ \cot\alpha+\cot\beta & \cot\beta+\cot\gamma & \cot\gamma+\cot\alpha \end{vmatrix}$$

Applying $C_2 \to C_2 - C_1$ and $C_3 \to C_3 - C_1$, we get

$\Delta = 2\sin^2\alpha\sin^2\beta\sin^2\gamma\Delta_1$

$$\text{where } \Delta_1 = \begin{vmatrix} 1 & 0 & 1 \\ \cot\alpha\cot\beta & \cot\beta(\cot\gamma-\cot\alpha) & \cot\alpha(\cot\gamma-\cot\beta) \\ \cot\alpha+\cot\beta & \cot\gamma-\cot\alpha & \cot\gamma-\cot\beta \end{vmatrix}$$

Expanding along R_1, we get

$$\Delta = 2\sin^2\alpha\sin^2\beta\sin^2\gamma(\cot\gamma-\cot\alpha)\ (\cot\gamma-\cot\beta)\begin{vmatrix} \cot\beta & \cot\alpha \\ 1 & 1 \end{vmatrix}$$

$$= 2\sin^2\alpha\sin^2\beta\sin^2\gamma(\cot\gamma-\cot\alpha)\ (\cot\gamma-\cot\beta)\ (\cot\beta-\cot\alpha)$$

$$= 2\sin^2\alpha\sin^2\beta\sin^2\gamma\ \frac{\sin(\alpha-\gamma)\sin(\beta-\gamma)\sin(\alpha-\beta)}{\sin^2\alpha\sin^2\beta\sin^2\gamma}$$

$$= -2\sin(\beta-\gamma)\sin(\gamma-\alpha)\sin(\alpha-\beta).$$

Example 12(a):

Show that

$$\Delta = \begin{vmatrix} \sin 3\alpha & \sin 2\alpha & \sin\alpha \\ \sin 3\beta & \sin 2\beta & \sin\beta \\ \sin 3\gamma & \sin 2\gamma & \sin\gamma \end{vmatrix}$$

$$= 2^6 \sin\alpha\sin\beta\sin\gamma\sin\frac{\alpha+\beta}{2}\sin\frac{\beta+\gamma}{2}$$

Solution:

We know that $\sin 3\theta = 3\sin\theta - 4\sin^3\theta$ and $\sin 2\theta = 2\sin\theta\cos\theta$. Applying $C_1 \to C_1 - 3C_3$ we get

$$\Delta = \begin{vmatrix} -4\sin^3\alpha & 2\sin\alpha\cos\alpha & \sin\alpha \\ -4\sin^3\beta & 2\sin\beta\cos\beta & \sin\beta \\ -4\sin^3\gamma & 2\sin\gamma\cos\gamma & \sin\gamma \end{vmatrix}$$

$$= 8\sin\alpha\sin\beta\sin\gamma \begin{vmatrix} -\sin^2\alpha & \cos\alpha & 1 \\ -\sin^2\beta & \cos\beta & 1 \\ -\sin^2\gamma & \cos\gamma & 1 \end{vmatrix}$$

Applying $R_2 \to R_2 - R_1$ and $R_3 \to R_3 - R_1$ we get

$$\Delta = 8\sin\alpha\sin\beta\sin\gamma \begin{vmatrix} -\sin^2\alpha & \cos\alpha & 1 \\ \sin^2\alpha - \sin^2\beta & \cos\beta - \cos\alpha & 0 \\ \sin^2\alpha - \sin^2\gamma & \cos\gamma - \cos\alpha & 1 \end{vmatrix}$$

$$= 8\sin\alpha\sin\beta\sin\gamma \cdot \begin{vmatrix} \cos^2\beta - \cos^2\alpha & \cos\beta - \cos\alpha \\ \cos^2\gamma - \cos^2\alpha & \cos\gamma - \cos\alpha \end{vmatrix}$$

$$= 8\sin\alpha\sin\beta\sin\gamma(\cos\beta - \cos\alpha)(\cos\gamma - \cos\alpha) \begin{vmatrix} \cos\beta + \cos\alpha & 1 \\ \cos\gamma + \cos\alpha & 1 \end{vmatrix}$$

$$= 8\sin\alpha\sin\beta\sin\gamma(\cos\beta - \cos\alpha)(\cos\gamma - \cos\alpha)(\cos\beta - \cos\gamma)$$

$$= 2^6 \sin\alpha\sin\beta\sin\gamma \sin\frac{\alpha-\beta}{2}\sin\frac{\alpha+\beta}{2}\sin\frac{\alpha-\gamma}{2} \times \sin\frac{\alpha+\gamma}{2}\sin\frac{\gamma-\beta}{2}\sin\frac{\beta+\gamma}{2}$$

$$= 2^6 \sin\alpha\sin\beta\sin\gamma \sin\frac{\beta+\gamma}{2}\sin\frac{\gamma+\alpha}{2}\sin\frac{\gamma+\beta}{2} \times \sin\frac{\beta-\alpha}{2}\sin\frac{\gamma-\alpha}{2}\sin\frac{\gamma-\beta}{2}$$

Example 12(b):

By differentiation or otherwise, show that the determinant

$$\begin{vmatrix} \sin(x+\alpha) & \cos(x+\alpha) & a+x\sin\alpha \\ \sin(x+\beta) & \cos(x+\beta) & b+x\sin\beta \\ \cos(x+\gamma) & \cos(x+\beta) & c+x\sin\gamma \end{vmatrix}$$ *is independent of x.*

Solution:

Let us denote the given determinant as $\Delta(x)$. Then

$$\Delta'(x) = \begin{vmatrix} \cos(x+\alpha) & \cos(x+\alpha) & a+x\sin\alpha \\ \cos(x+\beta) & \cos(x+\beta) & b+x\sin\beta \\ \cos(x+\gamma) & \cos(x+\beta) & c+x\sin\gamma \end{vmatrix}$$

$$+\begin{vmatrix} \sin(x+\alpha) & -\sin(x+\alpha) & a+x\sin\alpha \\ \sin(x+\beta) & -\sin(x+\beta) & b+x\sin\beta \\ \sin(x+\gamma) & -\sin(x+\beta) & c+x\sin\gamma \end{vmatrix} + \begin{vmatrix} \sin(x+\alpha) & \cos(x+\alpha) & \sin\alpha \\ \sin(x+\beta) & \cos(x+\beta) & \sin\beta \\ \sin(x+\gamma) & \cos(x+\beta) & \sin\gamma \end{vmatrix}$$

The first two determinants are zero. In the third, apply $C_1 \rightarrow C_1 - (\cos x) C_2$ and $C_2 \rightarrow C_2 + (\sin x) C_3$.

$$\Delta'(x) = \begin{vmatrix} \sin x\cos\alpha & \cos x\cos\alpha & \sin\alpha \\ \sin x\cos\beta & \cos x\cos\beta & \sin\beta \\ \sin x\cos\gamma & \cos x\cos\beta & \sin\gamma \end{vmatrix}$$

$= 0$ [$\because$ C_1 and C_2 are proportional]

This show that D(x) is a constant and hence independent of x.

Example 13:

If $\Delta(x) = \begin{vmatrix} a_1+x & b_1+x & c_1+x \\ a_2+x & b_2+x & c_2+x \\ a_3+x & b_3+x & c_3+x \end{vmatrix}$

show that $\Delta''(x) = 0$ *and that* $\Delta(x) = \Delta(0) + Sx$*, where S denotes the sum of all the cofactors of the elements in* $\Delta(0)$*.*

Solution:

We have

$$\Delta'(x) = \begin{vmatrix} 1 & b_1+x & c_1+x \\ 1 & b_2+x & c_2+x \\ 1 & b_3+x & c_3+x \end{vmatrix} + \begin{vmatrix} a_1+x & 1 & c_1+x \\ a_2+x & 1 & c_2+x \\ a_3+x & 1 & c_3+x \end{vmatrix} + \begin{vmatrix} a_1+x & b_1+x & 1 \\ a_2+x & b_2+x & 1 \\ a_3+x & b_3+x & 1 \end{vmatrix}$$

Applying $C_2 \rightarrow C_{2-x}\, C_1$ and $C_3 \rightarrow C_{3-x}\, C_1$ in the first; $C_1 \rightarrow C_{1-x}\, C_2$ and $C_3 - C_{3-x}\, C_2$ in the second; and $C_1 \rightarrow C_{1-x}\, C_3$ and $C_2 \rightarrow C_{2-x}\, C_3$ in the third determinant of $\Delta'(x)$, we get

$$\Delta'(x) = \begin{vmatrix} 1 & b_1 & c_1 \\ 1 & b_2 & c_2 \\ 1 & b_3 & c_3 \end{vmatrix} + \begin{vmatrix} a_1 & 1 & c_1 \\ a_2 & 1 & c_2 \\ a_3 & 1 & c_3 \end{vmatrix} + \begin{vmatrix} a_1 & b_1 & 1 \\ a_2 & b_2 & 1 \\ a_3 & b_3 & 1 \end{vmatrix} = 5$$

where S denotes the sum of all the cofactors of all the elements in $\Delta(0)$, so that $\Delta''(x) = 0$. Since $\Delta'(x) = S$, we get $\Delta(x) = xS + k$. Also, $k = \Delta(0)$. Therefore, $\Delta(x) = \Delta(0) + Sx$.

Example 14:

Prove that

$$\Delta = \begin{vmatrix} a^2 & (s-a)^2 & (s-a)^2 \\ (s-b)^2 & b^2 & (s-b)^2 \\ (s-c)^2 & (s-c)^2 & c^2 \end{vmatrix} = 2s^3\,(s-a)\,(s-b)\,(s-c)$$

where $2s = a + b + c$.

Solution:

Let $s - a = \alpha$, $s - b = \beta$ and $s - c = \gamma$, so that $\alpha + \beta + \gamma = 3s - (a + b + c) = s$. Also, $\beta + \gamma = 2s - (b + c) = a$, $\gamma + \alpha = b$ and $\alpha + \beta = c$. Now Δ can be written is

$$\Delta = \begin{vmatrix} (\beta+\gamma)^2 & \alpha^2 & \alpha^2 \\ \beta^2 & (\gamma+\alpha)^2 & \beta^2 \\ \gamma^2 & \gamma^2 & (\alpha+\beta)^2 \end{vmatrix}$$

$$= 2(\alpha + \beta + \gamma)^2\, \alpha\beta\gamma = 2s^3\,(s - a)(s - b)(s - c).$$

Example 15:

Show that

$$\Delta = \begin{vmatrix} -bc & b^2+bc & c^2+bc \\ a^2+ac & -ac & c^2+ac \\ a^2+ab & b^2+ab & -ab \end{vmatrix} = (bc+ca+ab)^2$$

Solution:

If any of a, b, c, is zero, the result is trivial. Applying $R_1 \to aR_1$, $R_2 \to bR_2$ and $R_3 \to cR_3$, we get

$$\Delta = \frac{1}{abc}\begin{vmatrix} -abc & ab^2+abc & ac^2+abc \\ a^2b+abc & -abc & bc^2+abc \\ a^2c+abc & b^2c+abc & -abc \end{vmatrix}$$

Applying $C_1 \to (1/a)C_1$, $C_2 \to (1/b)C_2$ and $C_3 \to (1/c)C_3$, we get

$$\Delta = \frac{abc}{abc}\begin{vmatrix} -bc & ab+ac & ac+ab \\ ab+bc & -ac & bc+ab \\ ac+bc & bc+ac & -ab \end{vmatrix}$$

Applying $R_1 \to R_1 + R_2 + R_3$ and taking ab + bc + ca common from R_1, we get

$$\Delta = (bc+ca+ab)\begin{vmatrix} 1 & 1 & 1 \\ ab+bc & -ac & bc+ab \\ ac+bc & bc+ac & -ab \end{vmatrix}$$

Applying $C_2 \to C_2 - C_1$ and $C_3 \to C_3 - C_1$, we get

Δ = (ab + bc + ca)

$$\begin{vmatrix} 1 & 0 & 0 \\ ab+bc & -(ab+bc+ac) & 0 \\ ac+bc & 0 & -(ab+bc+ca) \end{vmatrix}$$

Expanding along R_1, we get $\Delta = (bc + ca + ab)^3$.

Example 16:

Reduce $A = \begin{bmatrix} 13 & 16 & 19 \\ 14 & 17 & 20 \\ 15 & 18 & 21 \end{bmatrix}$ *to the canonical form.*

Solution:

We have $\begin{bmatrix} 13 & 16 & 19 \\ 14 & 17 & 20 \\ 15 & 18 & 21 \end{bmatrix}$

$A \sim \begin{bmatrix} 13 & 16 & 19 \\ 1 & 1 & 1 \\ 1 & 1 & 1 \end{bmatrix}$, replacing R_2 and R_3 by $R_2 - R_1$ and $R_3 - R_2$ respectively.

$\sim \begin{bmatrix} 13 & 3 & 3 \\ 1 & 0 & 0 \\ 1 & 0 & 0 \end{bmatrix}$, replacing C_2 and C_3 by $C_2 - C_1$, $C_3 - C_2$ respectively.

$\sim \begin{bmatrix} 13 & 3 & 0 \\ 1 & 0 & 0 \\ 1 & 0 & 0 \end{bmatrix}$, replacing C_3 by $C_3 - C_2$.

$\sim \begin{bmatrix} 13 & 3 & 0 \\ 1 & 0 & 0 \\ 1 & 0 & 0 \end{bmatrix}$, replacing R_3 by $R_3 - R_2$.

$\sim \begin{bmatrix} 3 & 13 & 0 \\ 0 & 1 & 0 \\ 0 & 0 & 0 \end{bmatrix}$, int erchanging C_1 and C_2.

$\sim \begin{bmatrix} 3 & 0 & 0 \\ 0 & 1 & 0 \\ 0 & 0 & 0 \end{bmatrix}$, replacing C_2 by $C_3 - (13/3)C_1$.

$\sim \begin{bmatrix} 1 & 0 & 0 \\ 0 & 1 & 0 \\ 0 & 0 & 0 \end{bmatrix}$, replacing R_1 by $\frac{1}{3} R_1$.

$$\sim \begin{bmatrix} I_2 & 0 \\ 0 & 0 \end{bmatrix}, \text{ which is the required canonical form}$$

Example 17:

Reduce $A = \begin{bmatrix} 1 & 2 & -1 & 4 \\ 2 & 4 & 3 & 5 \\ 1 & 2 & 3 & 5 \\ -1 & -2 & 6 & -7 \end{bmatrix}$ *to the canonical form.*

Solution:

$$A \sim \begin{bmatrix} 1 & 2 & -1 & 4 \\ 0 & 0 & 5 & -3 \\ 0 & 0 & 4 & 0 \\ 0 & 0 & 5 & -3 \end{bmatrix},$$ replacing R_2, R_3 and R_4 by $R_2 - 2R_1$, $R_3 - R_1$ and $R_4 + R_1$ respectively.

$$\sim \begin{bmatrix} 1 & 0 & 0 & 0 \\ 0 & 0 & 5 & -3 \\ 0 & 0 & 4 & 0 \\ 0 & 0 & 5 & -3 \end{bmatrix},$$ replacing C_2, C_3 and C_4 by $C_2 - 2C_1$, $C_3 + C_1$ and $C_4 - 4C_1$ respectively.

$$\sim \begin{bmatrix} 1 & 0 & 0 & 0 \\ 0 & 0 & 5 & -3 \\ 0 & 0 & 1 & 0 \\ 0 & 0 & 0 & 0 \end{bmatrix},$$ replacing R_3 and R_4 by $\frac{1}{4}R_3$ and $R_4 - R_3$ respectively.

$$\sim \begin{bmatrix} 1 & 0 & 0 & 0 \\ 0 & -3 & 5 & 0 \\ 0 & 0 & 1 & 0 \\ 0 & 0 & 0 & 0 \end{bmatrix},$$ interchanging C_2 and C_4

$$\sim \begin{bmatrix} 1 & 0 & 0 & 0 \\ 0 & -3 & 0 & 0 \\ 0 & 0 & 1 & 0 \\ 0 & 0 & 0 & 4 \end{bmatrix}, \text{ replacing } R_2 \text{ by } R_2 - 5R_3.$$

or $$\sim \begin{bmatrix} 1 & 0 & 0 & 0 \\ 0 & 1 & 0 & 0 \\ 0 & 0 & 1 & 0 \\ 0 & 0 & 0 & 0 \end{bmatrix}, \text{ replacing } R_2 \text{ by } -\frac{1}{3}R_2$$

$$\sim \begin{bmatrix} I_3 & 0 \\ 0 & 0 \end{bmatrix}, \text{ which is the required canonical form}$$

Definition: *If A and B be two $m \times n$ matrices, then $B \sim A$ if and only if $B = SAT$, where S is an $m \times m$ non-singular matrix and T is an $n \times n$ non-singular matrix.*

The following two properties of the above relation are fundamental.

1. **Symmetry:** If $A \sim B$, then $B \sim A$ for if $A = PBQ$

   then $B \sim P^{-1} AQ^{-1}$, where P^{-1} and Q^{-1} are non-singular matrices.

2. **Reflexivity:** Every matrix A is equivalent to itself since we can write $A = IAI$, so that $P = I = Q$.

Theorem 1:

An $n \times n$ matrix A is non-singular (or invertible) if and only if the determinant $|A| \neq 0$.

Proof:

If C be the canonical form of the matrix A, then

$$C \sim A$$

Therefore $C = S\,A\,T$

where S and T are non-singular.

Hence $A = S^{-1}\,C\,T^{-1}$

$\Rightarrow$ $A = E_r \ldots E_2\,E_1\,C\,D_1\,D_2 \ldots D_s$,

where E_1 and D_1 are elementary matrices.

By the successive application, we have

$$| A | = | E_r | \dots | E_2 | | E_1 | | C | | D_1 1 | | D_2 \dots | D_s |$$

$\therefore$ If $| A | = 0$, then $| C | = 0$, as $| E_i | \neq 0$ and $| D_i | \neq 0$.

If $| C | = 0$, then it has at least one row of zero.

$\therefore$ The rank of matrix A is less than n

i.e., the matrix A is singular.

If the matrix A is non-singular, then

$C = I_n$, where I_n is the n × n identity matrix.

i.e., $| C | = | I_n | = 1$

$\therefore$ From (i) above, we have $| A | \neq 0$. **Hence the theorem.**

Th :orem 2:

$| A_1 A_2 | = | A_1 | . | A_2 |$, *where* $| A_2 |$ *and* $| A_3 |$ *are two determinants.*

Proof:

Let C_1 and C_2 be the canonical form of the matrices A_1 and A_2 *i.e.*, $A_1 \sim C_1$ and $A_2 \sim C_2$.

If $A_1 \sim C_1$, we have

$A_1 \sim S\ C_1\ T$, where S and T are non-singular matrices.

$\Rightarrow \quad A_1 = E_r \dots E_2\ E_1\ C_1\ D_1\ D_2 \dots D_s$,

where E_i and D_i are elementary matrices.

Similarly $A_2 = F_i \dots F_2\ F_1\ C_2\ K_1\ K_2 \dots K_s$,

were F_i and K_i are elementary matrices.

$\therefore A_1\ A_2 = E_r \dots E_2\ E_1\ C_1\ D_1\ D_2 \dots D_s\ F_t \dots F_2\ F_1\ C_2\ K_2 \dots K_s$

We have

$$| A_1 A_2 | = | E_r \dots E_2\ E_1 | . | C_1\ D_1\ D_2 \dots D_s\ F_t \dots F_2\ F_1\ C_2 | . | K_1 \dots K_s | \quad \dots(i)$$

Now the following cases arise.

Case 1: Let A_1 be a singular matric.

Then C_1 has at least one row of zero.

$\therefore | C_1\ D_1\ D_2 \dots D_s\ F_t \dots F_2\ F_1\ C_2 | = 0$

since the matrix $C_1\ D_1\ D_2 \dots D_s\ F_t \dots F_2\ F_1\ C_2$ has a row of zero.

$\therefore$ From (i) above we have $| A_1\ A_2 | = 0$.

Case 2: If A_2 is a singular matrix. Then C_2 has at least one column of zero, hence as in Case I above

$$| C_1 D_1 D_2 \; D_s F_t \ldots F_1 C_2 | = 0$$

$\therefore$ From (i) above we have $| A_1 A_2 | = 0$.

Case 3: If either A_1 or A_2 is singular, then

$$| A_1 | \cdot | A_2 | = 0$$

Hence $| A_1 A_2 | = 0 = | A_1 | \cdot | A_2 |$.

Case 4: If A_1 and A_2 are non-singular matrices. Then C_1 and C_2 are identity matrices. We have

$| A_1 A_2 | = | E_r \ldots E_2 E_1 C_1 D_1 D_2 \ldots D_s | \cdot | F_t \ldots F_2 F_1 C_2$ $K_1 \ldots K_s|$

$$= | A_1 | \cdot | A_2 |$$

$\therefore$ From all the above cases it is clear that

$$| A_1 A_2 | = | A_1 | \cdot | A_2 |.$$

Cor.

$| A_1 A'_2 | = | A_1 | \cdot | A_2 |$, where A'_2 is the transpose of A_2.

Proof:

$| A'_2 | = | A_2 |$, $\quad \because A'_2$ is the transpose of A_2.

$\therefore | A_1 A'_2 | = | A_1 | | A'_2 |$,

$= | A_1 | | A_2 |$, $\quad \because | A'_2 | = | A_2 |$.

The corollary leads to the row by row rule of multiplication of the determinants as given in Examples below:

Example 18:

Express $\begin{vmatrix} (a-x)^2 & (b-x)^2 & (c-x)^2 \\ (a-y)^2 & (b-y)^2 & (c-y)^2 \\ (a-z)^2 & (b-z)^2 & (c-z)^2 \end{vmatrix}$

as the product of two determinants.

Solution:

We have

$$= \begin{vmatrix} a^2 - 2ax + x^2 & b^2 - 2bx + x^2 & c^2 - 2cx + x^2 \\ a^2 - 2ay + y^2 & b^2 - 2by + y^2 & c^2 - 2cy + y^2 \\ a^2 - 2az + z^2 & b^2 - 2bz + z^2 & c^2 - 2cz + z^2 \end{vmatrix}$$

The element in the first row and first column is $a^2 - 2ax + x^2$, which can be written as $1\ (a^2) + (-2x)\ (a) + x^2\ (1)$. (Note)

This suggests that the first row of the required determinants are $1, -2x, x^2$ and $a^2, a, 1$.

Example 19:

If $u = ax + by + cz$, $v = ay + bz + cx$, $w = az + bx + cy$, prove that

$$\begin{vmatrix} a & b & c \\ b & c & a \\ c & a & b \end{vmatrix} \times \begin{vmatrix} x & y & z \\ y & z & x \\ z & x & y \end{vmatrix} = u^3 + v^3 + w^3 - 3uvw.$$

Solution:

By row-by-row multiplication, the product of the given determinants

$$= \begin{vmatrix} ax + by + cz & ay + bz + cx & az + bx + cy \\ bx + cy + az & by + cz + ax & bz + cx + ay \\ cx + ay + bz & cy + az + bx & cz + ax + by \end{vmatrix}$$

$$= \begin{vmatrix} u & v & w \\ w & u & v \\ v & w & u \end{vmatrix}, \text{ since } ax + by + cz = u \text{ etc. (given)}$$

$$= u \begin{vmatrix} u & v \\ w & u \end{vmatrix} - v \begin{vmatrix} w & v \\ v & u \end{vmatrix} + w \begin{vmatrix} w & u \\ v & w \end{vmatrix}$$

$= u(u^2 - vw) - v\ (uw - v^2) + w\ (w^2 - uv)$

$= u^2 + v^2 + w^2 - 3uvw.$ **Hence proved.**

Example 20:

Evaluate $\begin{vmatrix} 0 & \cos x & -\sin x \\ \sin x & 0 & \cos x \\ \cos x & \sin x & 0 \end{vmatrix}^2$

Solution:

We have

$$= \begin{vmatrix} 0 & \cos x & -\sin x \\ \sin x & 0 & \cos x \\ \cos x & \sin x & 0 \end{vmatrix} \times \begin{vmatrix} 0 & \cos x & -\sin x \\ \sin x & 0 & \cos x \\ \cos x & \sin x & 0 \end{vmatrix}$$

$$= \begin{vmatrix} 0.0 + \cos x.\cos x + \sin x \sin x & 0.\sin x + \cos.0 - \sin x \cos x & 0.\cos x + \cos x \sin x - \sin x.0 \\ \sin x.0 + 0.\cos x - \cos x \sin x & \sin x \sin x + 0.0 + \cos x \cos x & \sin x.\cos x + 0 \sin x + \cos x.0 \\ \cos x.0 + \sin x \cos x + 0(-\sin x) & \cos x \sin x + \sin x.0 + 0.\cos x & \cos x.\cos x + \sin x.\sin x + 0.0 \end{vmatrix}$$

$$= \begin{vmatrix} \cos^2 x + \sin^2 x & -\sin x \cos x & \cos x \sin x \\ -\cos x \sin x & \sin^2 + \cos^2 x & \sin x \cos x \\ \sin x \cos x & \cos x \sin x & \cos^2 x + \sin^2 x \end{vmatrix}$$

$$= \begin{vmatrix} 1 & -\lambda & \lambda \\ -\lambda & 1 & \lambda \\ \lambda & \lambda & 1 \end{vmatrix}, \text{ where } \lambda \sin x \cos x$$

Example 21:

Prove that $\begin{vmatrix} a & b & c \\ b & c & a \\ c & a & b \end{vmatrix}^2 = \begin{vmatrix} 2bc - a^2 & c^2 & b^2 \\ c^2 & 2ac - b^2 & a^2 \\ b^2 & a^2 & 2ab - c^2 \end{vmatrix}$

$= (a^3 + b^3 + c^3 - 3abc)^2.$

Solution:

$$\begin{vmatrix} a & b & c \\ b & c & a \\ c & a & b \end{vmatrix}^2 = \begin{vmatrix} a & b & c \\ b & c & a \\ c & a & b \end{vmatrix} \times \begin{vmatrix} a & b & c \\ b & c & a \\ c & a & b \end{vmatrix}$$

$$= -\begin{vmatrix} a & b & c \\ b & c & a \\ c & a & b \end{vmatrix} \times \begin{vmatrix} a & c & b \\ b & a & c \\ c & b & a \end{vmatrix}$$, interchanging C_2, C_3 of the second determinant. **(Note)**

$$= \begin{vmatrix} -a & b & c \\ -b & c & a \\ -c & a & b \end{vmatrix} \times \begin{vmatrix} a & c & b \\ b & a & c \\ c & b & a \end{vmatrix}$$, multiplying C_1 of first determinant by -1. **(Note)**

$$= \begin{vmatrix} -aa + bc + cb & -ab + ba + cc & -ac + bb + ca \\ -ba + cc + ab & -bb + ca + ac & -bc + cb + aa \\ -ca + ac + bb & -cb + aa + bc & -cc + ab + a \end{vmatrix}$$

$$= \begin{vmatrix} 2bc - a^2 & c^2 & b^2 \\ c^2 & 2ac - b^2 & a^2 \\ b^2 & a^2 & 2ab - c^2 \end{vmatrix}$$ **Hence proved.**

Also $$\begin{vmatrix} a & b & c \\ b & c & a \\ c & a & b \end{vmatrix} = a\begin{vmatrix} c & a \\ a & b \end{vmatrix} - b\begin{vmatrix} b & a \\ c & b \end{vmatrix} + c\begin{vmatrix} b & c \\ c & a \end{vmatrix}$$

$$= a(cb - a^2) - b(b^2 - ac) + c(ab - c^2)$$

$$= abc - a^3 - b^3 + abc - c^3$$

$$= -(a^3 + b^3 + c^3 - 3abc)$$

$$\therefore \begin{vmatrix} a & b & c \\ b & c & a \\ c & a & b \end{vmatrix}^2 = (a^3 + b^3 + c^3 - 3abc)^2$$ **Hence proved.**

Example 22:

Evaluate $$\begin{vmatrix} a_1 & b_1 & c_1 \\ a_2 & b_2 & c_2 \\ a_3 & b_3 & c_3 \end{vmatrix} \times \begin{vmatrix} x_1 & y_1 & z_1 \\ x_2 & y_2 & z_2 \\ x_3 & y_3 & z_3 \end{vmatrix}$$

Solution:

The required product

$$= \begin{vmatrix} a_1x_1 + b_1y_1 + c_1z_1 & a_1x_2 + b_1y_2 + c_1z_2 & a_1x_3 + b_1y_3 + c_1z_3 \\ a_2x_1 + b_2y_1 + c_2z_1 & a_2x_2 + b_2y_2 + c_2z_2 & a_2x_3 + b_2y_3 + c_2z_3 \\ a_3x_1 + b_3y_1 + c_3z_1 & a_3x_2 + b_3y_2 + c_3z_2 & a_3x_3 + b_3y_3 + c_3z_3 \end{vmatrix}$$

Example 23:

Evaluate $\begin{vmatrix} 0 & c & b \\ c & 0 & a \\ b & a & 0 \end{vmatrix}^2$

Solution:

$$\begin{vmatrix} 0 & c & b \\ c & 0 & a \\ b & a & 0 \end{vmatrix}^2 = \begin{vmatrix} 0 & c & b \\ c & 0 & a \\ b & a & 0 \end{vmatrix} \times \begin{vmatrix} 0 & c & b \\ c & 0 & a \\ b & a & 0 \end{vmatrix}$$

$$= \begin{vmatrix} 0.0 + c c + b.b & 0.c + c.0 + b.a & 0.b + c.a + b.0 \\ c.0 + 0c + a.b & c.c + 00 + a.a & c.b + 0.a + a0 \\ b.0 + a.c + 0.b & b.c + a.0 + 0.a & b.b + a.a + 0.0 \end{vmatrix}$$

$$= \begin{vmatrix} c^2 + b^2 & ba & ca \\ ab & c^2 + a^2 & bc \\ ac & bc & b^2 + a^2 \end{vmatrix}$$ **Ans.**

Example 24:

Express $\begin{vmatrix} b^2 + c^2 & ab & ca \\ ab & c^2 + a^2 & bc \\ ca & bc & a^2 + b^2 \end{vmatrix}$

as the square of a determinant.

Hence evaluate.

Solution:

The element in first row and first column is $b^2 + c^2$ which can be written as $0.0 + c.c + b.b$.

So by trial and error method, we get the given determinant

$$= \begin{vmatrix} 0 & c & b \\ c & 0 & a \\ b & a & 0 \end{vmatrix} \times \begin{vmatrix} 0 & c & b \\ c & 0 & a \\ b & a & 0 \end{vmatrix}$$

$$= \begin{vmatrix} 0 & c & b \\ c & 0 & a \\ b & a & 0 \end{vmatrix}^2$$

Now $\begin{vmatrix} 0 & c & b \\ c & 0 & a \\ b & a & 0 \end{vmatrix}$

$= -c \begin{vmatrix} c & a \\ b & 0 \end{vmatrix} + b \begin{vmatrix} c & 0 \\ b & a \end{vmatrix}$, expanding with respect to R_1.

$= -c[c.0 - a.b] + b[c.a - b.0] = 2abc.$

$\therefore$ The given determinant

$$= \begin{vmatrix} 0 & c & b \\ c & 0 & a \\ b & a & 0 \end{vmatrix}^2 = (2abc)^2 = 4a^2b^2c^2.$$

Ans.

Example 25:

Express $\begin{vmatrix} 2bc - a^2 & c^2 & b^2 \\ c^2 & 2ca - b^2 & a^2 \\ b^2 & a^2 & 2ab - c^2 \end{vmatrix}$

as the product of two determinants.

Solution:

The element in the first row and first column is $2bc - a^2$ which can be written as

$$a(-a) + b(c) + c(b) \qquad ...(i)$$

The element in the first row and second column ins c^2 which can be written as

$$a(-b) + b(a) + c(c) \qquad ...(ii)$$

The element in the first row and third column is b^2 which can be written as

$$a(-c) + b(b) + c(a) \qquad ...(iii)$$

(i), (ii) and (iii) suggest that the given determinant can be written tentatively as

$$\begin{vmatrix} a & b & c \\ b & c & a \\ c & a & b \end{vmatrix} \times \begin{vmatrix} -a & c & b \\ -b & a & c \\ -c & b & a \end{vmatrix}$$

But actually multiplying these two determinants we get the given determinant. Hence these determinants are the required ones.

Example 26:

Find the product of two determinants of different orders or evaluate

$$\begin{vmatrix} \alpha_1 & \beta_1 & \gamma_1 \\ \alpha_2 & \beta_2 & \gamma_2 \\ \alpha_3 & \beta_3 & \gamma_3 \end{vmatrix} \times \begin{vmatrix} a_1 & b_1 \\ a_2 & b_2 \end{vmatrix}$$

Solution:

Here the two given determinants are of different orders, so we adopt the following method :

$$\begin{vmatrix} \alpha_1 & \beta_1 & \gamma_1 \\ \alpha_2 & \beta_2 & \gamma_2 \\ \alpha_3 & \beta_3 & \gamma_3 \end{vmatrix} \times \begin{vmatrix} a_1 & b_1 \\ a_2 & b_2 \end{vmatrix}$$

$$= \begin{vmatrix} \alpha_1 & \beta_1 & \gamma_1 \\ \alpha_2 & \beta_2 & \gamma_2 \\ \alpha_3 & \beta_3 & \gamma_3 \end{vmatrix} \times \begin{vmatrix} a_1 & b_1 & 0 \\ a_2 & b_2 & 0 \\ 0 & 0 & 1 \end{vmatrix}, \qquad \textbf{(Note)}$$

making the two determinants of the same order

$$= \begin{vmatrix} \alpha_1 a_1 + \beta_1 b_1 + \gamma_1.0 & \alpha_1 a_2 + \beta_1 b_2 + \gamma_1.0 & \alpha_1.0 + \beta_1.0 + \gamma_1.1 \\ \alpha_2 a_1 + \beta_2 b_1 + \gamma_2.0 & \alpha_2 a_2 + \beta_2 b_2 + \gamma_2.0 & \alpha_2.0 + \beta_2.0 + \gamma_2.1 \\ \alpha_3 a_1 + \beta_3 b_1 + \gamma_3.0 & \alpha_3 a_2 + \beta_3 b_2 + \gamma_3.0 & \alpha_3.0 + \beta_3.0 + \gamma_3.1 \end{vmatrix}$$

$$= \begin{vmatrix} a_1\alpha_1 + b_1\beta_1 & a_2\alpha_1 + b_2\beta_1 & \gamma_1 \\ a_1\alpha_2 + b_1\beta_2 & a_2\alpha_2 + b_2\beta_2 & \gamma_2 \\ a_1\alpha_3 + b_1\beta_3 & a_2\alpha_3 + b_2\beta_3 & \gamma_3 \end{vmatrix}$$

Ans.

Example 27:

Express $\begin{vmatrix} (1+ax)^2 & (1+ay)^2 & (1+az)^2 \\ (1+bx)^2 & (1+by)^2 & (1+bz)^2 \\ (1+cx)^2 & (1+cy)^2 & (1+cz)^2 \end{vmatrix}$

as the product of two determinants.

Solution:

The given determinant

$$= \begin{vmatrix} 1 + 2ax + a^2x^2 & 1 + 2ay + a^2y^2 & 1 + 2az + a^2z^2 \\ 1 + 2bx + b^2x^2 & 1 + 2by + b^2y^2 & 1 + 2bz + b^2z^2 \\ 1 + 2cx + c^2x^2 & 1 + 2cy + c^2y^2 & 1 + 2cz + c^2z^2 \end{vmatrix}$$

The element in the first row and first column is $1 + 2ax + a^2x^2$ which can be written as

$$(1)\,(1) + (2a)\,.\,(x) + (a^2)\,.\,(x^2)$$

This suggests that the first rows of the two required determinants are 1, 2a, a^2 and 1, x, x^2.

Hence the given determinant may be written as

$$\begin{vmatrix} 1 & 2a & a^2 \\ 1 & 2b & b^2 \\ 1 & 2c & c^2 \end{vmatrix} \times \begin{vmatrix} 1 & x & x^2 \\ 1 & y & y^2 \\ 1 & z & z^2 \end{vmatrix}$$ **Ans.**

Theorem:

If C_{ij} be the cofactor of a_{ij} in the $n \times n$, matrix $A = [a_{ij}]$ then $| C_{ij} | = \{ | a_{jk} | \}^{n-1}$

Proof:

If $A = [a_{ij}]$, then

$A' = [a'_{ki}]$, where A' is the transpose of A and $a'_{ki} = a_{ik}$.

Now $A' . [C_{ij}] = [a'_{ki}] [C_{ij}] = [b_{kj}]$, say. ...(i)

where $b_{kj} = \sum_{i=1}^{n} a'_{ki} C_{ij} = \sum_{i=1}^{n} a_{ij} C_{ij}$, $\quad \because a'_{ki} = a_{ik}$

We know that

$$b_{kj} = \sum_{i=1}^{n} a_{ik} C_{ij} = 0, \text{ if } j \neq k$$

$$= | A |, \text{ if } j = k.$$

$\therefore$ From (i) we conclude that for the product $A', [C_{ij}]$ *i.e.*, $[b_{ki}]$ all the diagonal terms (for which $j = k$) are $| A |$, whereas the nondiagonal terms (for which $j \neq k$) are zero.

i.e.,, $$A'.[C_{ij}] = \begin{bmatrix} |A| & 0 & \dots & 0 \\ 0 & |A| & \dots & 0 \\ \dots & \dots & \dots & \dots \\ \dots & \dots & \dots & \dots \\ 0 & 0 & \dots & |A| \end{bmatrix}$$

Hence $$| A' . [C_{ij}] | = \begin{bmatrix} |A| & 0 & \dots & 0 \\ 0 & |A| & \dots & 0 \\ \dots & \dots & \dots & \dots \\ \dots & \dots & \dots & \dots \\ 0 & 0 & \dots & |A| \end{bmatrix}$$

$\Rightarrow |A'| \cdot |C_{ij}| = \{|A|\}^n \qquad \because \qquad |A_1 \cdot A_2| = |A_1| \cdot |A_2|$

$\Rightarrow |A| \cdot |C_{ij}| = \{|A|\}^n, \qquad \because \qquad |A'| = |A|$

$\Rightarrow |C_{ij}| = \{|A|\}^{n-1}$

Example 28(a):

Expand

$$\begin{vmatrix} 3 & 2 & 1 & 4 \\ 5 & 29 & 2 & 14 \\ 16 & 19 & 3 & 17 \\ 33 & 39 & 8 & 38 \end{vmatrix}$$

by Laplace's expansion by the minors of the first two columns.

Solution:

All the possible minors of the first two columns and their complementary minors and given by :

$$|B_1| = \begin{vmatrix} 3 & 2 \\ 15 & 29 \end{vmatrix}, \; |B'_1| = \begin{vmatrix} 3 & 17 \\ 8 & 38 \end{vmatrix}$$

$$|B_2| = \begin{vmatrix} 3 & 2 \\ 16 & 19 \end{vmatrix}, \; |B'_2| = \begin{vmatrix} 2 & 14 \\ 8 & 38 \end{vmatrix}$$

$$|B_3| = \begin{vmatrix} 3 & 2 \\ 33 & 39 \end{vmatrix}, \; |B'_3| = \begin{vmatrix} 2 & 14 \\ 3 & 17 \end{vmatrix}$$

$$|B_4| = \begin{vmatrix} 15 & 29 \\ 16 & 19 \end{vmatrix}, \; |B'_4| = \begin{vmatrix} 1 & 4 \\ 8 & 38 \end{vmatrix}$$

$$|B_5| = \begin{vmatrix} 15 & 29 \\ 33 & 39 \end{vmatrix}, \; |B'_5| = \begin{vmatrix} 1 & 4 \\ 3 & 17 \end{vmatrix}$$

$$|B_6| = \begin{vmatrix} 16 & 19 \\ 33 & 39 \end{vmatrix}, \; |B'_6| = \begin{vmatrix} 1 & 4 \\ 2 & 14 \end{vmatrix}$$

$\therefore$ The given determinant

$$= \begin{vmatrix} 13 & 2 \\ 15 & 29 \end{vmatrix} . \begin{vmatrix} 3 & 17 \\ 8 & 38 \end{vmatrix} - \begin{vmatrix} 3 & 2 \\ 16 & 19 \end{vmatrix} . \begin{vmatrix} 2 & 14 \\ 8 & 38 \end{vmatrix}$$

$$+ \begin{vmatrix} 3 & 2 \\ 33 & 39 \end{vmatrix} . \begin{vmatrix} 2 & 14 \\ 2 & 17 \end{vmatrix} + \begin{vmatrix} 15 & 29 \\ 16 & 19 \end{vmatrix} . \begin{vmatrix} 1 & 4 \\ 8 & 38 \end{vmatrix}$$

$$- \begin{vmatrix} 15 & 29 \\ 33 & 39 \end{vmatrix} . \begin{vmatrix} 1 & 14 \\ 3 & 17 \end{vmatrix} + \begin{vmatrix} 16 & 19 \\ 33 & 39 \end{vmatrix} . \begin{vmatrix} 1 & 4 \\ 2 & 14 \end{vmatrix}$$

Example 28(b):

Expand $\begin{vmatrix} a & x & y & a \\ x & 0 & 0 & y \\ y & 0 & 0 & x \\ a & y & x & a \end{vmatrix}$

by Laplace's expansion by the minors of the first two columns. Hence evaluate it.

Solution:

All the possible minors of the first two columns and their complementary minors are given by

$$| B_1 | = \begin{vmatrix} a & x \\ x & 0 \end{vmatrix} ; | B'_1 | = \begin{vmatrix} 0 & x \\ x & a \end{vmatrix} ;$$

$$| B_2 | = \begin{vmatrix} a & x \\ y & 0 \end{vmatrix} ; | B'_2 | = \begin{vmatrix} 0 & y \\ x & a \end{vmatrix} ;$$

$$| B_3 | = \begin{vmatrix} a & x \\ y & a \end{vmatrix} ; | B'_2 | = \begin{vmatrix} 0 & y \\ 0 & x \end{vmatrix} ;$$

$$| B_4 | = \begin{vmatrix} x & 0 \\ y & 0 \end{vmatrix} ; | B'_4 | = \begin{vmatrix} y & a \\ x & a \end{vmatrix} ;$$

$$|B_5| = \begin{vmatrix} x & 0 \\ a & y \end{vmatrix}; \ |B'_5| = \begin{vmatrix} y & 0 \\ 0 & x \end{vmatrix};$$

and $$|B_6| = \begin{vmatrix} y & 0 \\ a & y \end{vmatrix}; \ |B'_6| = \begin{vmatrix} y & a \\ 0 & y \end{vmatrix};$$

Therefore the given determinant

$$= \begin{vmatrix} a & x \\ x & 0 \end{vmatrix} . \begin{vmatrix} 0 & x \\ x & a \end{vmatrix} - \begin{vmatrix} a & x \\ y & 0 \end{vmatrix} . \begin{vmatrix} 0 & y \\ x & a \end{vmatrix} + \begin{vmatrix} a & x \\ a & y \end{vmatrix} . \begin{vmatrix} 0 & y \\ 0 & x \end{vmatrix}$$

$$+ \begin{vmatrix} x & 0 \\ y & 0 \end{vmatrix} . \begin{vmatrix} y & a \\ x & a \end{vmatrix} - \begin{vmatrix} x & 0 \\ a & y \end{vmatrix} . \begin{vmatrix} y & a \\ 0 & x \end{vmatrix} + \begin{vmatrix} y & 0 \\ a & y \end{vmatrix} . \begin{vmatrix} y & a \\ 0 & y \end{vmatrix} \quad ...(i)$$

The submatrix B_2 requires one interchange of rows viz. of second and third rows to bring it into the first two rows therefore –sign is put before the product $|B_2||B'_2|$. Again the submatrix B_2 requires two interchanges of rows to bring fourth row to the position of second row *i.e.*, to bring B_2 into first two rows, therefore +sign is put before the product $|B_3| |B'_3|$.

Similarly, B_4 requires two interchanges, B_5 requires three interchanges and B_3 requires four interchanges, hence +, – and + signs are put before $|B_4||B'_4|$, $|B'_5| |B'_5|$ and $|B_6| |B'_6|$ respectively.

Hence from (i) we have (expanding the determinants) the given determinant

$= (-x^2)(-x^2) - (-xy) + (ay - ax)(0) + 0; (ay - ax) - (xy),(xy) + (y^2)(y^2)$

$= x^4 - 2x^2y^2 + y^4 = (x^2 - y^2)^2$. **Ans.**

Example 28(c):

Expand $\begin{vmatrix} a & b & c & d \\ e & f & g & h \\ 0 & 0 & i & k \\ 0 & 0 & l & m \end{vmatrix}$ *by Laplace's expansion by the minors of the first two columns.*

Solution:

All the possible minors of the first two columns and their complementary minors are given by:

$$|B_1| = \begin{vmatrix} a & b \\ e & f \end{vmatrix}; \quad |B'_1| = \begin{vmatrix} j & k \\ l & m \end{vmatrix};$$

$$|B_2| = \begin{vmatrix} a & b \\ 0 & 0 \end{vmatrix} = 0, \text{ hence } |B'_2| \text{ need not be calculated}$$

$$\text{Similarly } |B_3| = \begin{vmatrix} a & b \\ 0 & 0 \end{vmatrix} = 0; \quad |B_4| = \begin{vmatrix} e & f \\ 0 & 0 \end{vmatrix} = 0;$$

$$|B_5| = \begin{vmatrix} e & f \\ 0 & 0 \end{vmatrix} = 0 \text{ and } |B_6| = \begin{vmatrix} 0 & 0 \\ 0 & 0 \end{vmatrix} = 0 \text{ and therefore their}$$

complementary minors need not be calculated.

Then the given determinant by Laplace's Expansion method

$$= \begin{vmatrix} a & b \\ e & f \end{vmatrix} . \begin{vmatrix} j & k \\ l & m \end{vmatrix}$$ **Ans.**

Example 28(d):

Expand $\begin{vmatrix} a & 1 & 0 & 0 & 0 \\ b & a & 1 & 0 & 0 \\ 0 & b & a & 1 & 0 \\ 0 & 0 & b & a & 1 \\ 0 & 0 & 0 & b & a \end{vmatrix}$ *by Laplace's expansion by the minors of the first two columns. Hence evaluate it.*

Solution:

All the possible minors of the first two columns and their complementary minors are given by:

$$|B_1| = \begin{vmatrix} a & 1 \\ b & a \end{vmatrix}, \quad |B'_1| = \begin{vmatrix} a & 1 & 0 \\ b & a & 1 \\ 0 & b & a \end{vmatrix}$$

$$|B_2| = \begin{vmatrix} a & 1 \\ 0 & b \end{vmatrix}, \quad |B'_2| = \begin{vmatrix} 1 & 0 & 0 \\ b & a & 1 \\ 0 & b & a \end{vmatrix}$$

$$|B_3| = \begin{vmatrix} b & a \\ 0 & b \end{vmatrix}, \quad |B'_3| = \begin{vmatrix} 0 & 0 & 0 \\ b & a & 1 \\ 0 & b & a \end{vmatrix} = 0$$

All other minors of the first two columns are equal to zero as they have at least one row of zero.

Hence the given determinant

$$= \begin{vmatrix} a & 1 \\ 0 & b \end{vmatrix} \cdot \begin{vmatrix} a & 1 & 0 \\ b & a & 1 \\ 0 & b & a \end{vmatrix} - \begin{vmatrix} a & 1 \\ 0 & b \end{vmatrix} \cdot \begin{vmatrix} 1 & 0 & 0 \\ b & a & 1 \\ 0 & b & a \end{vmatrix} \quad \text{...(i)}$$

Now $\begin{vmatrix} a & 1 & 0 \\ b & a & 1 \\ 0 & b & a \end{vmatrix} = \begin{vmatrix} a & 1 \\ b & a \end{vmatrix} .a - \begin{vmatrix} b & 1 \\ 0 & a \end{vmatrix} .1 + \begin{vmatrix} b & a \\ 0 & b \end{vmatrix} .0,$

expanding by the minors of first two columns.

$= (a^2 - b)\, a - (ab) = a^3 - 2ab$

$\therefore$ From (i), the given determinant

$$= \begin{vmatrix} a & 1 \\ b & a \end{vmatrix} .(a^3 - 2ab) - \begin{vmatrix} a & 1 \\ 0 & b \end{vmatrix} . \begin{vmatrix} a & 1 \\ b & a \end{vmatrix},$$

expanding the last determinant with respect to R_1.

$= (a^2 - b)\,(a^2 - 2ab) - (ab)\,(a^2 - b) = (a^2 - b)\,[a^2 + 3ab]$

$= a\,(a^2 - b)\,(a^2 - 3b)$ **Ans.**

Example 29:

Show that

$$\Delta = \begin{vmatrix} (b+c)^2 & a^2 & bc \\ (c+a)^2 & b^2 & ca \\ (a+b)^2 & c^2 & ab \end{vmatrix}$$

$= (a^2 + b^2 + c^2)\,(a + b + c)\,(a - b)\,(b - c)(c - a)$

Solution:

Applying $C_1 \to C_1 + C_2 - 2C_3$, we get

$$\Delta = \begin{vmatrix} a^2+b^2+c^2 & a^2 & bc \\ a^2+b^2+c^2 & b^2 & ca \\ a^2+b^2+c^2 & c^2 & ab \end{vmatrix}$$

Taking $(a^2 + b^2 + c^2)$ common from C_1 and applying $R_2 \to R_2 - R_1$ and $R_3 \to R_3 - R_1$, we get

$$\Delta = (a^2 + b^2 + c^2) \begin{vmatrix} 1 & a^2 & bc \\ 0 & b^2-a^2 & ca-bc \\ 0 & c^2-a^2 & ab-bc \end{vmatrix}$$

$$= (a^2 + b^2 + c^2)(b - a)(c - a) \begin{vmatrix} 1 & a^2 & bc \\ 0 & b+a & -c \\ 0 & c+a & -b \end{vmatrix}$$

Expanding along C_1, we get

$$\Delta = (a^2 + b^2 + c^2)(b - a)(c - a)[c(c + a) - b(b + a)]$$

$$= (a^2 + b^2 + c^2)(a + b + c)(a - b)(b - c)(c - a)$$

Example 30:

Show that

$$\Delta = \begin{vmatrix} a & b-c & c+b \\ a+c & b & c-a \\ a-b & b+a & c \end{vmatrix} = (a+b+c)(a^2+b^2+c^2)$$

Solution:

Multiplying C_1 and dividing Δ by a, we get

$$\Delta = \frac{1}{a}\begin{vmatrix} a & b-c & c+b \\ a+c & b & c-a \\ a-b & b+a & c \end{vmatrix}$$

Applying $C_1 \to C_1 + bC_2 + cC_3$, we get

$$\Delta = \frac{1}{a}\begin{vmatrix} a^2+b^2+c^2 & b-c & c+b \\ a^2+b^2+c^2 & b & c-a \\ a^2+b^2+c^2 & b+a & c \end{vmatrix}$$

Taking $a^2 + b^2 + c^2$ common from C_1 and applying $R_2 \to R_2 - R_1$ and $R_3 \to R_3 - R_1$, we get

$$\Delta = \frac{a^2+b^2+c^2}{a}\begin{vmatrix} 1 & b-c & c+b \\ 0 & c & -(a+b) \\ 0 & a+c & -b \end{vmatrix}$$

Expanding along C_1, we get

$$\Delta = \frac{a^2+b^2+c^2}{a}\,[-bc + (a+b)(a+c)]$$

$$= \frac{a^2+b^2+c^2}{a}\,[a^2 + ab + ac]$$

$$= (a + b + c)(a^2 + b^2 + c^2),$$

Example 31:

Show that

$$\Delta = \begin{vmatrix} (b+c)^2 & a^2 & a^2 \\ b^2 & (c+a)^2 & b^2 \\ c^2 & c^2 & (a+b)^2 \end{vmatrix} = 2abc(a+b+c)^3$$

Solution:

Applying $C_2 \to C_2 - C_1$ and $C_3 \to C_3 - C_1$, we get

$$\Delta = \begin{vmatrix} (b+c)^2 & a^2-(b+c)^2 & a^2-(b+c)^2 \\ b^2 & (c+a)^2-b^2 & 0 \\ c^2 & 0 & (a+b)^2-c^2 \end{vmatrix}$$

Taking $(a + b + c)$ common from C_2 and C_3, we obtain

$$\Delta = (a+b+c)^2\begin{vmatrix} (b+c)^2 & a-b-c & a^2-(b+c)^2 \\ b^2 & c+a-b & 0 \\ c^2 & 0 & a+b-c \end{vmatrix}$$

Applying $R_1 \to R_1 - R_2 - R_3$, we get

$$\Delta = (a+b+c)^2 \begin{vmatrix} 2bc & -2c & -2b \\ b^2 & c+a-b & 0 \\ c^2 & 0 & a+b-c \end{vmatrix}$$

Applying $C_2 \to C_2 + (1/b)C_1$ and $C_3 \to C_3 + (1/c)C_1$, we obtain

$$\Delta = (a+b+c)^2 \begin{vmatrix} 2bc & 0 & 0 \\ b^2 & c+a & b^2/c \\ c^2 & c^2/b & a+b \end{vmatrix}$$

Expanding along R_1, we get

$$\Delta = (a+b+c)^2\, 2bc\, [(c+a)(a+b) - bc]$$

$$= (a+b+c)^2\, 2bc\, (ab + ac + bc + a^2 - bc)$$

$$= 2bc\, (a+b+c)^3$$

Example 32:

Show that

$$\Delta = \begin{vmatrix} (a+b)^2 & ca & cb \\ ca & (b+c)^2 & ab \\ bc & ab & (c+a)^2 \end{vmatrix} = 2abc(a+b+c)^3$$

Solution:

Multiplying R_1 by c, R_2 by a and R_3 by b, we get

$$\Delta = \frac{1}{abc} \begin{vmatrix} (a+b)^2 c & c^2 a & c^2 b \\ ca^2 & (b+c)^2 a & a^2 b \\ b^2 c & ab^2 & (c+a)^2 b \end{vmatrix}$$

Taking c common from C_1, a from C_2 and b from C_3, we get

$$\Delta = \begin{vmatrix} (a+b)^2 c & c^2 & c^2 \\ a^2 & (b+c)^2 & a^2 \\ b^2 & b^2 & (c+a)^2 \end{vmatrix}$$

$$= 2abc\, (a+b+c)^3$$

Example 33:

Expand $\begin{vmatrix} 0 & 1 & x & y \\ 0 & 0 & y & x \\ z & w & 0 & 0 \\ w & z & 0 & 0 \end{vmatrix}$ *by Laplace's Expansion by the minors of the first two columns.*

Solution:

All the possible minors of the first two columns and their complementary minors are given below:

$$|B_1| = \begin{vmatrix} 0 & 1 \\ z & w \end{vmatrix}, \; |B'_1| = \begin{vmatrix} y & x \\ 0 & 0 \end{vmatrix} = 0$$

$$|B_2| = \begin{vmatrix} 0 & 1 \\ w & z \end{vmatrix}, \; |B'_2| = \begin{vmatrix} y & 0 \\ 0 & 0 \end{vmatrix} = 0$$

$$|B_3| = \begin{vmatrix} z & w \\ w & z \end{vmatrix}, \; |B'_3| = \begin{vmatrix} x & y \\ y & x \end{vmatrix}$$

$\therefore$ The given determinant $= \begin{vmatrix} z & w \\ w & z \end{vmatrix} \cdot \begin{vmatrix} x & y \\ y & x \end{vmatrix}$

the remaining minors or complementary minors are zero

$$= (z^2 - w^2)(x^2 - y^2)$$ **Ans.**

Example 34:

Write down as a determinant the product

$$\begin{vmatrix} a & b & c \\ c & a & b \\ b & c & a \end{vmatrix} \cdot \begin{vmatrix} x & y & z \\ z & x & y \\ y & z & x \end{vmatrix}$$

Solution:

Multiplying by the 'row-by-row' rule we get

$$\begin{vmatrix} ax + by + cz & az + bx + cy & ay + bz + cx \\ cx + ay + bz & cz + ax + by & cy + az + bx \\ bx + cy + az & bz + cx + ay & by + cz + ax \end{vmatrix}$$

$$= [(a + b + c)(x + y + z)] \begin{vmatrix} 1 & 1 & 1 \\ cx + ay + bz & cz + ax + by & cy + az + bx \\ bx + cy + az & bz + cx + ay & by + cz + ax \end{vmatrix}$$

replacing R_1 by $R_1 + R_2 + R_3$ and taking the common factors out.

Example 35:

Show that $\begin{vmatrix} -1 & 0 & 0 & a \\ 0 & -1 & 0 & b \\ 0 & 0 & -1 & c \\ x & y & z & -1 \end{vmatrix} = 1 - ax - by - cz$

Solution:

The given determinant

$$= \begin{vmatrix} -1 & 0 & 0 & -a + 0 + 0 + a \\ 0 & -1 & 0 & 0 - b + 0 + b \\ 0 & 0 & -1 & 0 + 0 - c + c \\ x & y & z & 0 + 0 - c + c \end{vmatrix}$$, expanding with respect to C_4

$$= \begin{vmatrix} -1 & 0 & 0 & 0 \\ 0 & -1 & 0 & 0 \\ 0 & 0 & -1 & 0 \\ x & y & z & ax + by + cz - 1 \end{vmatrix}$$

$$= (ax + by + cz - 1) \begin{vmatrix} -1 & 0 & 0 \\ 0 & -1 & 0 \\ 0 & 0 & -1 \end{vmatrix}$$, expanding with respect to C_4

$$= -(ax + by + cz - 1)\begin{vmatrix} -1 & 0 \\ 0 & -1 \end{vmatrix}$$, expanding with respect to C_1

$$= (1 - ax - by - cz)\,[(-1)(-1) - 0.0]$$

$$= (1 - ax - by - cz)\,[1] = (1 - ax - by - cz).$$ **Hence proved.**

Example 36(a):

Evaluate $$\begin{vmatrix} 265 & 240 & 219 \\ 240 & 225 & 198 \\ 219 & 198 & 181 \end{vmatrix}$$

Solution:

The given determinant

$$= \begin{vmatrix} 25 & 21 & 219 \\ 15 & 27 & 198 \\ 21 & 17 & 181 \end{vmatrix}$$, replacing C_1, C_2 by $C_1 - C_2$ and $C_2 - C_3$ respectively.

$$= \begin{vmatrix} 4 & 21 & 219 \\ -12 & 27 & 198 \\ 4 & 17 & 181 \end{vmatrix}$$, replacing C_1 by $C_1 - C_2$

$$= \begin{vmatrix} 0 & 4 & 38 \\ 0 & 78 & 741 \\ 4 & 17 & 181 \end{vmatrix}$$, replacing R_1, R_2 by $R_1 - R_3$ and $R_2 + 3R_3$ respectively

$$= 4\begin{vmatrix} 4 & 38 \\ 78 & 741 \end{vmatrix}$$, expanding with respect to C_1

$$= 4\,[(741 \times 4) - (78 \times 38)] = 4\,[2964 - 2964] = 0$$ **Ans.**

Example 36(b):

Prove that $$\begin{vmatrix} a^2 & bc & ac + c^2 \\ a^2 + ab & b^2 & ac \\ ab & b^2 + bc & c^2 \end{vmatrix} = 4a^2 b^2 c^2.$$

Solution:

The given determinant

$$= abc\begin{vmatrix} a & c & a+c \\ a+b & b & a \\ b & b+c & c \end{vmatrix}$$, taking out a, b, c common from C_1, C_2, C_3 respectively.

$$= abc\begin{vmatrix} a+c & c & a+c \\ a+2b & b & a \\ 2b+c & b+c & c \end{vmatrix}$$, replacing C_1 by $C_1 + C_2$

$$= abc\begin{vmatrix} 0 & c & a+c \\ 2b & b & a \\ 2b & b+c & c \end{vmatrix}$$, replacing C_1 by $C_1 - C_3$

$$= abc\begin{vmatrix} 0 & c & a+c \\ 2b & b & a \\ 0 & c & c-a \end{vmatrix}$$, replacing R_2 by $R_3 - R_2$

$$= -2b\,(abc)\begin{vmatrix} c & a+c \\ c & c-a \end{vmatrix}$$, expanding with respect to C_1

$$= -2ab^2c\begin{vmatrix} 0 & 2a \\ c & c-a \end{vmatrix}$$, replacing R_1 by $R_1 - R_2$

$= -2ab^2c\,(-2ac)$, expanding the det.

$= 4a^2b^2c^2$. **Hence proved.**

Example 37:

Show that $(a + b + c)$ and $(a^2 + b^2 + c^2)$ are factors of determinant

$$\begin{vmatrix} a^2 & (b+c)^2 & bc \\ b^2 & (c+a)^2 & ca \\ c^2 & (a+b)^2 & ab \end{vmatrix}$$ *and find the remaining factors.*

Solution:

The given determinant

$$= \begin{vmatrix} a^2 & (b^2 + c^2 + 2bc) + a^2 & bc \\ b^2 & (c^2 + a^2 + 2ca) + b^2 & ca \\ c^2 & (a^2 + b^2 + 2ab) + c^2 & ab \end{vmatrix}, \text{ replacing } C_2 \text{ by } C_2 + C_1$$

$$= \begin{vmatrix} a^2 & b^2 + c^2 + a^2 & bc \\ b^2 & c^2 + a^2 + b^2 & ca \\ c^2 & a^2 + b^2 + c^2 & ab \end{vmatrix}, \text{ replacing } C_2 \text{ by } C_2 - 2C_3$$

$$= (a^2 + b^2 + c^2) \begin{vmatrix} a^2 & 1 & bc \\ b^2 & 1 & ca \\ c^2 & 1 & ab \end{vmatrix}, \text{ taking out } (a^2 + b^2 + c^2) \text{ common from } C_2$$

$$= \frac{(a^2 + b^2 + c^2)}{abc} \begin{vmatrix} a^2 & a & abc \\ b^2 & b & bca \\ c^2 & c & cab \end{vmatrix},$$

taking 1/a, 1/b, 1/c common from R_1, R_2, and R_3 respectively. (Note)

$$= \frac{(a^2 + b^2 + c^2)}{abc} \times abc \begin{vmatrix} a^3 & a & 1 \\ b^3 & b & 1 \\ c^3 & c & 1 \end{vmatrix}, \text{ taking out abc common from } C_2.$$

Example 38:

Prove that the value of the determinant

$$\begin{vmatrix} x + 1 & x + 2 & x + a \\ x + 2 & x + 3 & x + b \\ x + 3 & x + 4 & x + c \end{vmatrix}$$ *is independent of x.*

Solution:

The given determinant

$$= \begin{vmatrix} x+1 & 1 & a-1 \\ x+2 & 1 & b-2 \\ a+3 & 1 & c-3 \end{vmatrix}, \text{ replacing } C_2, C_3 \text{ by } C_2 - C_1 \text{ and } C_3 - C_1 \text{ respectively}$$

$$= \begin{vmatrix} x+1 & 1 & a-1 \\ 1 & 0 & b-a-1 \\ 2 & 0 & c-a-2 \end{vmatrix}, \text{ replacing } R_2, R_3 \text{ by } R_2 - R_1, R_3 - R_1 \text{ respectively}$$

$$= -\begin{vmatrix} 1 & x+1 & a-1 \\ 0 & 1 & b-a-1 \\ 0 & 2 & c-a-2 \end{vmatrix}, \text{ interchanging } C_1 \text{ and } C_2$$

$$= -\begin{vmatrix} 1 & b-a-1 \\ 2 & c-a-2 \end{vmatrix} = -[(c-a-2) - 2(b-a-1)]$$

$$= -[c - a - 2 - 2b + 2a + 2] = -a + 2b - c.$$

which is independent of x. **Hence proved.**

Example 39:

Solve the equation $\begin{vmatrix} 3x-8 & 3 & 3 \\ 3 & 3x-8 & 3 \\ 3 & 3 & 3x-8 \end{vmatrix} = 0$

Solution:

Given that $\begin{vmatrix} 3x-8 & 3 & 3 \\ 3 & 3x-8 & 3 \\ 3 & 3 & 3x-8 \end{vmatrix} = 0$

$$\Rightarrow \begin{vmatrix} 3x-2 & 3 & 3 \\ 3x-2 & 3x-8 & 3 \\ 3x-2 & 3 & 3x-8 \end{vmatrix} = 0, \text{ replacing } C_1 \text{ by } C_1 + C_2 + C_3$$

$$\Rightarrow (3x-2)\begin{vmatrix} 1 & 3 & 3 \\ 1 & 3x-8 & 3 \\ 1 & 3 & 3x-8 \end{vmatrix} = 0, \text{ taking out } (3x-2) \text{ common}$$

$$\Rightarrow (3x-2)\begin{vmatrix} 1 & 3 & 3 \\ 0 & 3x-11 & 0 \\ 0 & 0 & 3x-11 \end{vmatrix} = 0, \text{ replacing } R_2, R_3 \text{ by } R_2 - R_1, R_3 - R_1 \text{ respectively.}$$

$$\Rightarrow (3x-2)\begin{vmatrix} 3x-11 & 0 \\ 0 & 3x-11 \end{vmatrix} = 0, \text{ expanding with respect to } C_1$$

$\Rightarrow$ $(3x-2)(3x-11)^2 = 0$

$\Rightarrow$ $x = 2/3 \Rightarrow 11/3$. **Ans.**

Example 40:

Give correct answer to the following:

The values of the determinant $\begin{vmatrix} -3 & 1 & 1 & 1 \\ 1 & -3 & 1 & 1 \\ 1 & 1 & -3 & 1 \\ 1 & 1 & 1 & -3 \end{vmatrix}$ *is (A) – 1, (B) 1, (C) 0, (D) 4.*

Solution:

The correct answer is (C) *i.e.,* 0, since replacing C_1 by $C_1 + C_2 + C_3 + C_4$ we find that all the elements of C_1 are zero. Hence the value of the given determinant is zero.

Example 41:

Evaluate $\begin{vmatrix} a & -a & -a & -a \\ b & b & -b & -b \\ c & c & c & -c \\ d & d & d & d \end{vmatrix}$

Solution:

The given determinant

$$= \begin{vmatrix} a & 0 & 0 & 0 \\ b & 2b & 0 & 0 \\ c & 2c & 2c & 0 \\ d & 2d & 2d & 2d \end{vmatrix},$$ replacing C_2, C_3 and C_4 by $C_2 + C_1$, $C_3 + C_1$ and $C_4 + C_1$ respectively.

$$= a \begin{vmatrix} 2b & 0 & 0 \\ 2c & 2c & 0 \\ 2d & 2d & 2d \end{vmatrix},$$ expanding with respect to R_1

$$= 2ab \begin{vmatrix} 2c & 0 \\ 2d & 2d \end{vmatrix},$$ expanding with respect to R_1

$$= 2ab\,[(2c \times 2d) - (2d) \times 0] = 8abcd.$$ **Ans.**

Example 42(a):

Prove that $\begin{vmatrix} 1 & \omega^3 & \omega^2 \\ \omega^3 & 1 & \omega \\ \omega^2 & \omega & 1 \end{vmatrix} = 3$, *where* ω *is one of the imaginary cube roots of unity.*

Solution:

If ω be one of the imaginary cube roots of unity, the

$\omega^2 = 1$ and $1 + \omega + \omega^2 = 0$ ...(i)

Now the given determinate

$$= \begin{vmatrix} 1 & 1 & \omega^2 \\ 1 & 1 & \omega \\ \omega^2 & \omega & 1 \end{vmatrix},$$ from (i) using $\omega^2 = 1$

$$= \begin{vmatrix} 0 & 1 & \omega^2 \\ 0 & 1 & \omega \\ \omega^2 - \omega & \omega & 1 \end{vmatrix},$$ replacing C_1 by $C_1 - C_2$

$$= (\omega^2 - \omega) \begin{vmatrix} 1 & \omega^2 \\ 1 & \omega \end{vmatrix},$$ expanding with respect to C_1

$= (\omega^2 - \omega)(\omega - \omega^2) = \omega^3 = \omega^4 - \omega^2 = 1 - \omega - \omega^2 + 1 \quad \because \omega^2 = 1$

$= 2 - (\omega + \omega^2) = 2 - (-1), \quad \because \; 1 + \omega + \omega^2 = 0$

$\Rightarrow \omega + \omega^2 = -1$

$= 2 + 1 = 3.$ **Hence proved.**

Example 42(b):

Prove that the determinant

$$\begin{vmatrix} 1 & \cos(\beta - \alpha) & \cos(\gamma - \alpha) \\ \cos(\alpha - \beta) & 1 & \cos(\gamma - \beta) \\ \cos(\alpha - \gamma) & \cos(\beta - \gamma) & 1 \end{vmatrix}$$

is a perfect square (of a determinant) and find its value.

Solution:

The given determinant

$$= \begin{vmatrix} \cos\alpha\cos\alpha + \sin\alpha\sin\alpha & \cos\alpha\cos\beta + \sin\alpha\sin\beta & \cos\alpha\cos\gamma + \sin\alpha\sin\gamma \\ \cos\beta\cos\alpha + \sin\beta\sin\alpha & \cos\beta\cos\beta + \sin\beta\sin\beta & \cos\beta\cos\gamma + \sin\beta\sin\gamma \\ \cos\gamma\cos\alpha + \sin\gamma\sin\alpha & \cos\gamma\cos\beta + \sin\gamma\sin\beta & \cos\gamma\cos\gamma + \sin\gamma\sin\gamma \end{vmatrix} \quad \text{(Note)}$$

The element in the first row and first column in cos α cos α + sin α sin α, which can be written as

$(\cos\alpha)(\cos\alpha) + (\sin\alpha)(\sin\alpha) + 0.0$ (Note)

Similarly the element in the first row and second column is cos α cos β + sin α sin β, which can be written as

$(\cos\alpha)(\cos\beta) + (\sin\alpha)(\sin\beta) + 0.0$ (Note)

Proceeding in this way we may write the given determinant

$$= \begin{vmatrix} \cos\alpha & \sin\alpha & 0 \\ \cos\beta & \sin\beta & 0 \\ \cos\gamma & \sin\gamma & 0 \end{vmatrix} \times \begin{vmatrix} \cos\alpha & \sin\alpha & 0 \\ \cos\beta & \sin\beta & 0 \\ \cos\gamma & \sin\gamma & 0 \end{vmatrix}$$

$$= \begin{vmatrix} \cos\alpha & \sin\alpha & 0 \\ \cos\beta & \sin\beta & 0 \\ \cos\gamma & \sin\gamma & 0 \end{vmatrix}^2$$ hence a perfect square of a determinant

= 0, since the value of this determinant is zero as all the elements of one of its columns are zero.

Example 42(c):

Solve the equation $\begin{vmatrix} x-2 & 2x-3 & 3x-4 \\ x-4 & 2x-9 & 3x-16 \\ x-8 & 2x-27 & 3x-64 \end{vmatrix} = 0$

Solution:

The given equation is

$$\begin{vmatrix} x-2 & 2x-3 & 3x-4 \\ -2 & -6 & -12 \\ -6 & -24 & -60 \end{vmatrix} = 0,$$ replacing R_2, R_3 by $R_3 - R_1$ and $R_3 - R_1$ respectively.

$$\Rightarrow \begin{vmatrix} x-2 & 2x-3 & 3x-4 \\ 1 & 3 & 6 \\ 1 & 4 & 10 \end{vmatrix} = 0,$$ replacing R_2, R_3 by $-\frac{1}{2}R_2$ and $\frac{1}{6}R_3$ respectively.

$$\Rightarrow \begin{vmatrix} x-2 & 1 & 2 \\ 1 & 1 & 3 \\ 1 & 2 & 7 \end{vmatrix} = 0,$$ replacing C_2, C_3 by $C_2 - 2C_1$, $C_2 - 3C_1$ respectively.

$$\Rightarrow \begin{vmatrix} x-2 & 1 & 2 \\ 1 & 1 & 3 \\ 0 & 1 & 4 \end{vmatrix} = 0,$$ replacing R_2 by $R_3 - R_2$

$$\Rightarrow \begin{vmatrix} x-2 & 0 & -2 \\ 1 & 0 & -1 \\ 0 & 1 & 4 \end{vmatrix} = 0,$$ replacing R_1, R_2 by $R_1 - R_2$, $R_2 - R_3$ respectively

$$\Rightarrow \quad -\begin{vmatrix} 0 & x-2 & -2 \\ 1 & 0 & -1 \\ 0 & 1 & 4 \end{vmatrix} = 0, \text{ interchanging } C_1 \text{ and } C_2$$

$$\Rightarrow \quad -\begin{vmatrix} x-2 & -2 \\ 1 & -1 \end{vmatrix} = 0, \text{ expanding with respect to } C_1$$

$$\Rightarrow \quad -[-(x-2)+2] = 0 \text{ or } x = 4.$$ **Ans.**

Example 43:

$$\textit{Evaluate} \begin{vmatrix} 0 & \alpha & \beta & \gamma \\ l & 0 & c & -b \\ m & -c & 0 & a \\ n & b & -a & 0 \end{vmatrix}$$

Solution:

The given determinant

$$-\frac{1}{a}\begin{vmatrix} 0 & \alpha & \beta & \gamma \\ al & 0 & ac & -ab \\ m & -c & 0 & a \\ n & b & -a & 0 \end{vmatrix}, \text{taking } (1/a) \text{ common from } R_2 \ \textbf{(Note)}$$

$$-\frac{1}{a}\begin{vmatrix} 0 & \alpha & \beta & \gamma \\ al+bm+cn & 0 & 0 & 0 \\ m & -c & 0 & a \\ n & b & -a & 0 \end{vmatrix}, \text{ replacing } R_2 \text{ by } R_2 + bR_3 + cR_4$$

$$= -\frac{1}{a}(al+bm+cn)\begin{vmatrix} \alpha & \beta & \gamma \\ -c & 0 & a \\ b & -a & 0 \end{vmatrix}, \text{ expanding with respect to } R_2$$

$$= -\frac{1}{a}(al+bm+cn)\begin{vmatrix} a\alpha & \beta & \gamma \\ -ac & 0 & a \\ ab & -a & 0 \end{vmatrix}, \text{taking out } 1/a \text{ common from } C_1$$

$$= -\frac{1}{a}(al + bm + cn)\begin{vmatrix} a\alpha + b\beta + c\gamma & \beta & \gamma \\ 0 & 0 & a \\ 0 & -a & 0 \end{vmatrix},$$

replacing C_1 by $C_1 + bC_2 + cC_2$

$$= -\frac{1}{a}(al + bm + cn)\begin{vmatrix} 0 & a \\ -a & 0 \end{vmatrix}, \text{ expanding with respect to } C_1$$

$$= -(1/a^2)(al + bm + cn)(a\alpha + b\beta + c\gamma)\, a^2$$

$$= -(al + bm + cn)(a\alpha + b\beta + c\gamma).$$ **Ans.**

Example 44:

Prove that $\begin{vmatrix} 4 & 5 & 6 & x \\ 5 & 6 & 7 & y \\ 6 & 7 & 8 & z \\ x & y & z & 0 \end{vmatrix} = (x - 2y + z)^2$

Solution:

The given determinant

$$= \begin{vmatrix} 10 & 5 & 6 & x \\ 12 & 6 & 7 & y \\ 14 & 7 & 8 & z \\ x + z & y & z & 0 \end{vmatrix}, \text{ replacing } C_1 \text{ by } C_1 + C_3$$

$$= \begin{vmatrix} 0 & 5 & 6 & x \\ 0 & 6 & 7 & y \\ 0 & 7 & 8 & z \\ x - 2y + z & y & z & 0 \end{vmatrix}, \text{ replacing } C_1 \text{ by } C_1 - 2C_2$$

$$= -(x - 2y + z)\begin{vmatrix} 5 & 6 & x \\ 6 & 7 & y \\ 7 & 8 & z \end{vmatrix}, \text{ expanding with respect to } C_1$$

$$= -(x - 2y + z)\begin{vmatrix} 12 & 14 & x + z \\ 6 & 7 & y \\ 7 & 8 & z \end{vmatrix}, \text{ replacing } R_1 \text{ by } R_1 + R_2$$

$$= -(x - 2y + z) \begin{vmatrix} 0 & 0 & x + z - 2y \\ 6 & 7 & y \\ 7 & 8 & z \end{vmatrix}$$, replacing R_1 by $R_1 - 2R_2$

$$= -(x - 2y + z)^2 \begin{vmatrix} 6 & 7 \\ 7 & 8 \end{vmatrix}$$, expading with respect to R_1

$= -(x - 2y + z)^2 [48 - 49]$, expanding the det.

$= (x - 2y + z)^2$. **Hence proved.**

Example 45:

If f, g and h are differentiable functions of x and

$$\Delta = \begin{vmatrix} f & g & h \\ (xf)' & (xg)' & (xh)' \\ (x^2f)'' & (x^2g)'' & (x^2h)'' \end{vmatrix}$$

prove that
$$\Delta' = \begin{vmatrix} f & g & h \\ f' & g' & h' \\ (x^3f'')' & (x^3g'')' & (x^3h'')' \end{vmatrix}$$

Solution:

We have

$(x^2f)'' = [x(xf)]'' = [x(xf)' + 1(xf)]'$

$= 1(xf)' + x(xf)'' + (xf)' = 2(xf)' + x(xf)''$

Write each element of the third row in Δ in the above form and apply $R_3 \to R_3 - 2R_2$, to get

$$\Delta = \begin{vmatrix} f & g & h \\ (xf)' & (xg)' & (xh)' \\ x(xf)'' & x(xg)'' & x(xh)'' \end{vmatrix}$$

Taking x common from R_3 and multiplying R_1 by x, we get

$$\Delta = \begin{vmatrix} xf & xg & xh \\ (xf)' & (xg)' & (xh)' \\ (xf)'' & (xg)'' & (xh)'' \end{vmatrix}$$

Differentiating w.r.t. x, we get

$$\Delta' = \begin{vmatrix} (xf)' & (xg)' & (xh)' \\ (xf)' & (xg)' & (xh)' \\ (xf)'' & (xg)'' & (xh)'' \end{vmatrix} + \begin{vmatrix} xf & xg & xh \\ (xf)'' & (xg)'' & (xh)'' \\ (xf)'' & (xg)'' & (xh)'' \end{vmatrix}$$

$$+ \begin{vmatrix} xf & xg & xh \\ (xf)' & (xg)' & (xh)' \\ (xf)''' & (xg)''' & (xh)''' \end{vmatrix}$$

The first two determinants are equal to zero, because in each of them two rows are identical. Thus, we can write

$$\Delta' = \begin{vmatrix} xf & xg & xh \\ (xf)' & (xg)' & (xh)' \\ (xf)''' & (xg)''' & (xh)''' \end{vmatrix} = x \begin{vmatrix} f & g & h \\ (xf)' & (xg)' & (xh)' \\ (xf)''' & (xg)''' & (xh)''' \end{vmatrix}$$

We know that $(xf)' = xf' + f$. Applying $R_2 \to R_2 - R_1$, we get

$$\Delta' = x \begin{vmatrix} f & g & h \\ xf' & xg' & xh' \\ (xf)''' & (xg)''' & (xh)''' \end{vmatrix}$$

$$= x^2 \begin{vmatrix} f & g & h \\ f' & g' & h' \\ (xf)''' & (xg)''' & (xh)''' \end{vmatrix}$$

$$= \begin{vmatrix} f & g & h \\ f' & g' & h' \\ x^2(xf)''' & x^2(xg)''' & x^2(xh)''' \end{vmatrix}$$

We have $x^2(xf)''' = x^2(xf' + f)''$

$$= x^2(xf'' + 2f')$$

$$= x^2(xf''' + 3f'')$$

$$= x^3f''' + 3x^2f''$$

$$= (x^3f'')$$

$$\text{Hence}\quad \Delta' = \begin{vmatrix} f & g & h \\ f' & g' & h' \\ x^2(xf'')' & x^2(xg'')' & x^2(xh'')' \end{vmatrix}$$

Example 46:

Prove that

$$\Delta = \begin{vmatrix} \beta\gamma & \beta\gamma'+\beta'\gamma & \beta'\gamma' \\ \gamma\alpha & \gamma\alpha'+\gamma'\alpha & \gamma'\alpha' \\ \alpha\beta & \alpha\beta'+\alpha'\beta & \alpha'\beta' \end{vmatrix}$$

$$= (\alpha\beta' - \alpha'\beta)\,(\beta\gamma' - \beta'\gamma)\,(\gamma\alpha' - \gamma'\alpha)$$

Solution:

Taking $\beta'\gamma'$ common from R_1, $\gamma'\alpha'$ from R_2 and $\alpha'\beta'$ from R_3, we get

$$\Delta = (\alpha'\beta'\gamma')^2 \begin{vmatrix} bc & b+c & 1 \\ ca & c+a & 1 \\ ab & a+b & 1 \end{vmatrix}$$

where $a = \alpha/\alpha'$, $b = \beta/\beta'$ and $c = \gamma/\gamma'$. Applying $R_2 \to R_1$ and $R_3 \to R_3 - R_1$, we get

$$\Delta = (\alpha'\beta'\gamma')^2 \begin{vmatrix} bc & b+c & 1 \\ c(a-b) & a-b & 0 \\ b(a-c) & a-c & 0 \end{vmatrix}$$

$$= (\alpha'\beta'\gamma')^2 (a-b)(a-c) \begin{vmatrix} bc & b+c & 1 \\ c & 1 & 0 \\ b & 1 & 0 \end{vmatrix}$$

Expanding along C_3, we get

$$\Delta = (\alpha'\beta'\gamma')^2\,(a - b)\,(a - c)\,(c - b)$$

$$= (\alpha\beta' - \alpha'\beta)\,(\beta\gamma' - \beta'\gamma)\,(\gamma\alpha' - \gamma'\alpha).$$

Example 47:

Find the value of $f(\pi/6)$, where

$$f(\theta) = \begin{vmatrix} \cos^2\theta & \cos\theta\sin\theta & -\sin\theta \\ \cos\theta\sin\theta & \sin^2\theta & \cos\theta \\ \sin\theta & -\cos\theta & 0 \end{vmatrix}$$

Solution:

Applying $C_1 \to C_1 - \sin\theta\, C_3$ and $C_2 \to C_2 + \cos\theta\, C_3$, we get

$$f(\theta) = \begin{vmatrix} 1 & 0 & -\sin\theta \\ 0 & 1 & \cos\theta \\ \sin\theta & -\cos\theta & 0 \end{vmatrix}$$

Applying $R_3 \to R_3 - \sin\theta\, R_1 + \cos\theta\, R_2$, we get

$$f(\theta) = \begin{vmatrix} 1 & 0 & -\sin\theta \\ 0 & 1 & \cos\theta \\ 0 & 0 & \sin^2\theta + \cos^2\theta \end{vmatrix} = 1$$

Thus, $f(\pi/6) = 1$.

Example 48:

Show that

$$\Delta = \begin{vmatrix} a^2 & bc & ac+c^2 \\ a^2+bc & b^2 & ac \\ ab & b^2+bc & c^2 \end{vmatrix} = 4a^2b^2c^2$$

Solution:

Taking a common from C_1, b from C_2 and c from C_3, we get

$$\Delta = abc \begin{vmatrix} a & c & a+c \\ a+b & b & a \\ b & b+c & c \end{vmatrix}$$

Applying $C_1 \to C_1 + C_2 - C_3$, we get

$$\Delta = abc\begin{vmatrix} 0 & c & a+c \\ 2b & b & a \\ 2b & b+c & c \end{vmatrix}$$

Taking 2b common from C_1 and applying $R_3 \to R_3 - R_2$, we get

$$\Delta = 2ab^2c\begin{vmatrix} 0 & c & a+c \\ 1 & b & a \\ 0 & c & c-a \end{vmatrix}$$

Expanding along C_1, we get

$$\Delta = 2ab^2c(-1)\begin{vmatrix} c & a+c \\ c & c-a \end{vmatrix}$$

$$= 2\ ab^2c\ (-1)\ [c(c-a) - c(a+c)]$$

$$= 4a^2b^2c^2.$$

EXERCISES

1. *Prove that* $\begin{vmatrix} -2a & a+b & c+a \\ b+a & -2b & c+b \\ c+a & c+b & -2c \end{vmatrix} = 4(a+b)(b+c)(c+a)$.

2. *Show that* $\begin{vmatrix} a-b & b-c & c-a \\ b-c & c-a & a-b \\ c-a & a-b & b-c \end{vmatrix} = 0$

3. *Evalute* $\begin{vmatrix} x+a & x+2a & x+3a \\ x+2a & x+3a & x+4a \\ x+4a & x+5a & x+6a \end{vmatrix}$

4. *Calculate the value of the determinant* $\begin{vmatrix} 7 & 13 & 10 & 6 \\ 5 & 9 & 7 & 4 \\ 8 & 12 & 11 & 7 \\ 4 & 10 & 6 & 3 \end{vmatrix}$

5. *Show that* $\begin{vmatrix} 21 & 17 & 7 & 10 \\ 24 & 22 & 6 & 10 \\ 6 & 8 & 2 & 3 \\ 5 & 7 & 1 & 2 \end{vmatrix} = 0$

6. *Show that* $\begin{vmatrix} a-b & b-c & c-a \\ b-c & c-a & a-b \\ c-a & a-b & b-c \end{vmatrix} = 0$

7. *Evaluate* $\begin{vmatrix} 1 & a & b+c \\ 1 & b & c+a \\ 1 & c & a+c \end{vmatrix}$ **(Ans. 0)**

3

Permutations and Combinations

INTRODUCTION OF PERMUTATIONS

Let n be a positive integer and r, a positive integer less than or equal to n. The number of different arrangements of r things taken out of n dissimilar things is denoted by ^nP_r. *Each such arrange ment is called a permutation of n things taken r at a time.*

For example, all the arrangements of two letters chosen out of $\{a, b, c\}$ are given by

ab, ba, ac, ca, bc, cb.

Thus $^3P_2 = 6$

Again all the arrangements of three letters chosen out of $\{a, b, c, d\}$ are given by

abc, bac, acb, bca, cab, cba, abd, adb, bad, bda, dab, dba, acd, adc, cda, cad, dac, dca, bcd, bdc, cdb, cbd, dbc, dcb.

Thus $^4P_3 = 24$.

SUPPOSE THAT NUMBER OF WAYS OF DOING A WORK IS l AND ONCE THIS WORK IS COMPLETED, THE NUMBER OF WAYS OF DOING SECOND WORK IS m, THEN NUMBER OF PERFORMING BOTH THE WORKS TOGETHER IS $l \times m$

Let A and B be the two works. For each way of performing A; B can be completed in m ways. But A itself can be done in l ways, so that the total number of ways in which A and B can be performed together is $l \times m$.

Similarly if after completing first two works, the number of ways inwhich a third work can be performed is n, then the total number of different mannners in which all the three works can be completed is $l \times m \times n$.

The above result can be understood by the following example.

Let A, B, C be three cities and from A to B there are l different routes while from B to C the number of different routes is m. What is the number of different paths ane can traverse in going from A to C ? Suppose a person goes from A to B by route no. 1.

After reachig B, he has at his disposal m paths. Thus for every choice of path from A to B, he has m paths at his disposal. So total number of different routes he can undertake is $l \times m$.

TO FIND THE NUMBER OF PERMUTATIONS OF r THINGS CHOSEN OUT OF n DISSIMILAR THINGS

We consider n chairs in a row and r persons, clearly the number of pernutations of r things chosen out of n things is same as that the number of ways in which these r persons can occupy these chairs.

We start with on person. He has n chair to sit in. So, he can chose any one of them in n ways. Once a chair is occupied by him only $n - 1$ chairs remain unoccupied. So, for the second person, the choice is limited to $n - 1$ chaires only. Thus the number of ways second person can occupy any chair is $n - 1$. Hence total number os ways which first two person selected at randon from the group of given r persons to occupy chairs in $n(n - 1)$. Prpceeding in this way, we get that

$$^nP_r = n(n-1)(n-2)\ldots(n-\overline{r-1})$$

$$= n(n-1)(n-2)\ldots(n-r+1)$$

FACTORIAL NOTATION

The product of all consecutive integers starting from 1 to t is denoted by $r!$ or $\underline{|t}$ and read as t–factorial.

Thus $\quad t! = 1 \times 2 \times 3 \times \ldots \times t.$

In this way $! = 1$, $2! = 1 \times 2 = 2$, $3! = 1 \times 2 \times 3 = 6$

$4! = 1 \times 2 \times 3 \times 4 = 24$ etc.

Note that for $n > 1$, $n! = n(n-1)!$

Now $\quad ^nP_r = n(n-1)(n-2)\ldots n-r+1)$

$$= \frac{n(n-1)\ldots(n-r+1)(n-r)(n-r-1)\ldots3.2.1}{(n-r)(n-r-1)\ldots3.2.1}$$

$$= \frac{\underline{|n}}{\underline{|n-r}}$$

CONVENTION

As a convention we take 0 ! equal to 1.

Example 1:

Find the value of ${}^6P_4$

Solution:

$${}^nP_4 = \frac{n!}{(n-4)!} = 12\frac{n!}{(n-2)!}$$

Example 2:

If ${}^nP_4 = 12\ {}^nP_2$, *find n*

Solution:

$${}^nP_4 = \frac{n!}{(n-4)!} = 12\frac{n!}{(n-2)!}$$

By hypothesis $\frac{n!}{(n-4)!} = 12\frac{n!}{(n-2)!}$

$\Rightarrow 12\ (n-4)! = (n-2)!$

$\Rightarrow 12\ (n-4)! = (n-2)(n-3)(n-4)!$

$\Rightarrow 12 = n^2 - 5n + 6$

$\Rightarrow \quad n^2 - 5n - 6 = 0$

$\Rightarrow (n-6)(n+1) = 0$

$\Rightarrow n = 6$ or $n = -1$

Since n is positive integer, we reject the second value of n.Thus $n = 6$.

Example 3:

In how many ways 5 passengers can sit in a compartment having 16 vacant seats ?

Solution:

Required number of ways = ${}^{16}P_5$

$$= \frac{16!}{(16-5)!}$$

$$= \frac{16!}{11!}$$

$$= \frac{16.15.14.13.12\underline{|11}}{\underline{|11}}$$

= 16.15.14.13.12

= 524160.

TO FIND OUT THE NUMBER OF PERNUTATIONS OF r THINGS CHOSEN OUT OF n DISSIMILAR THINGS, WHEN A PARTICULAR THING IS TO BE ALWAYS INCLUDED IN EACH ARRANGEMENT

Here we have to consider the number of ways in which r places can be filled up when we have n d9fferent things such that in each arragement a particular thing is always to be present.

Suppose $\{a_1, a_2, \ldots, a_n\}$ are n different things. Let a_i be a particular thing to be included in each arrangement of r things out of n different things given. Now a_i can occupy any of 1st, 2nd, 3rd, . . . rth place. Suppose a_i is at the first place.Then remaining $r - 1$ places can be fiilled up by $n - 1$ things *viz.* $\{a_1, a_2, \ldots a_i - 1, a_{i+1}, \ldots a_n\}$ in ${}^{n-1}P_{r-1}$ ways. Similarly, if a_i is at the second place, then the remaining $r - 1$ places can be filled up by $n - 1$ things, again in ${}^{n-1}P_{r-1}$ ways. Thus a_i can occupy r places and each position of a_i gives rise to ${}^{n-1}P_{r-1}$ arrangements. Hence requires number of permutations is $r \cdot {}^{n-1}P_{r-1}$.

TO FIND THE NUMBER OF PERMUTATIONS OF r THINGS CHOSEN OUT OF n DIFFERENT THINGS, WHEN A PARTICULAR THING IS TO BE ALWAYS EXCLUDED IN EACH ARRANGEMENT

Here we have to consider the number of ways in which r places can be filled up when we have n different things such that in each arrangement, a particular thing is alwaysto be missing.

To do so, we exclude that particular thing. Then we are left with $n - 1$ things and r places canbe filled up with these $n - 1$ things in ${}^{n-1}P_{r-1}$ ways. Hence required number of permutations is ${}^{n-1}P_{r-1}$.

TO PROVE THAT ${}^{n}P_r = {}^{n-1}P_r + r\ {}^{n-1}P_{r-1}$.

Proof. ${}^{n}P_r$ means that the number of ways in which r places can be filled up by$n$ different things. This can be done as follows.

consider first those arrangements which contain a particular thing. The number of such arrangements $= r\ {}^{n-1}P_{r-1}$.

Now, consider these arrangementss which do not contain that particular thing. The number of such arrangements $= {}^{n-1}P_r$.

Each arrangement of r places, either contains a particular thing or does not. Hence total number of arrangements of r places by n different things = ${}^{n-1}P_r + r^{n-1}P_{r-1}$.

Alternatively. RHS = ${}^{n-1}P_r + r.\ {}^{n-1}P_{r-1}$.

$$= \frac{\underline{|n-1}}{\underline{|n-r-1}} + r\frac{\underline{|n-1}}{\underline{|n-r}}$$

$$= \frac{\underline{|n-1}}{\underline{|n-r-1}} + \frac{r\underline{|n-1}}{(n-r)\underline{|n-r-1}}$$

$$= \frac{\underline{|n-1}}{\underline{|n-r-1}}\left[1 + \frac{r}{n-r}\right]$$

$$= \frac{\underline{|n-1}}{\underline{|n-r-1}}\left[\frac{n}{n-r}\right]$$

$$= \frac{n\underline{|n-1}}{(n-r)\underline{|n-r-1}}$$

$$= \frac{\underline{|n}}{\underline{|n-r}} = {}^{n}p_r = \text{LHS.}$$

TO FIND THE NUMBER OF WAYS IN WHICH *n* THINGS MAYBE AARANGED AMONG THEMSELVES, TAKEN ALL AT A TIME, WHEN *p* THINGS ARE OF ONE KIND, *q* OF SECOND KIND, AND *r* OF THIRD KIND AND REST ALL DIFFERENT

Let x be the required of permutations. Take any one of these permutations. Replace p like things in this permutation by p unlike things. If we arrange these p unlike things among themselves, we get $\underline{|p}$ permutations (keeping all other tjings unchanged). So, in all we get $x\underline{|p}$ permutations in which now q things are of one kind, r things of second kind and rest of them different. Consider any one of these permutations. Replace q like things by q unlike things. These q unlike things can be arranged among themselves, in $\underline{|p}$ ways. But there are $x\underline{|p}$ such permutations. In this way, we get $x\ \underline{|q}$ permutations in which r things are or one kind and rest of them different. Consider any such permutation and replace r like things by r unlike things. These r unlike things can be arranged in $\underline{|r}$ ways. These are $x\underline{|p}\underline{|q}$ such pwemutations. This gives rise to $x\ \underline{|p}\ \underline{|q}\ \underline{|r}$ permutations in which all tjings are different. But, we know n different things can be arranged among themselves in $\underline{|n}$ ways.

Thus $\quad x\ \underline{|p}\ \underline{|q}\ \underline{|r} = \underline{|n}$

$$x = \frac{\underline{|n}}{\underline{|p}\underline{|q}\underline{|r}}$$

which is required number of premutations.

Example 4:

How many numbers lying between 10 and 100can be formed with the help of digits 3,0, 4, 5, 6?

Solution:

Each number between 10 and 100 consists of two digits. Number of permutations of two digits out of given five digits = ${}^5P_2$ = 20.

But these permutations inchlude numbers of type 03, 04, 05, 06 which do not lie between 10 and 100.

Therefore 16 different numbers can be formed with the help of digits 3, 4, 0, 5, 6 lying between 10 and 100.

Example 5:

How many words can be formed with the help of 3 consonants and 2 vowels, such that no two consonants are adjacent ?

Solution:

Let C_1, C_2, C_3 be three consonants and V_1, V_2 be two vowels. Consider any arrangement in which no two consonants are gogether as shown in the figure.

Now vowels have only tow positions to occupy namely, between C_1 and C_2, C_2 and C_3 So each such arrrangements of consonants, gives rise to two arrangements of vowels. But consonants can be arranged in 3 places in 3 ! ways so that total number of arrangements is 2 × 3 ! i. e. 12. Thus the number of different words formed is 12.

Example 6:

How many numbers consisting of 2 digits can be formed with the help of the digits 1, 2, 3, 4 such that

(*i*) *each numbe contains* 1

(*ii*) *no number contains* 1?

Solution:

(*i*) By number of 2–digit numbers containing 1 are $2 \times {}^3P_1 = 2 \times \frac{\underline{|3}}{\underline{|2}} = \underline{|3} = 6$.

(*ii*) By number of 2–digit numbers which do not contain 1 are $^3P_1$ = $\frac{\lfloor 3}{\lfloor 1} = 6$.

Example 7:

How many wony woras can be formed from the letters of VALEDICTORY provided the letters A, E, I, O, Y are not to be separated at all ?

Solution:

Put together A, E, I, O, Y and count them as one letter. Thus we have V, L, D, C, T, R, (AELOY). These seven, letters can be arranged in 7 ! ways, But the group (AELOY) consisting of five letters can itself be itself be arranged in 5 ! ways.

Hence the required number = 5 ! × 7 ! = 604800.

Example 8:

Out of the letters A, B. C, P, q, r how many arrangements cannbe made (*i*) beginning with a capital *(ii) beginning and ending with a capital.*

Solution:

(*i*) A capital letter out of given letters can be chosen in 3 ways. Remaining five letters can be arranged among themsclves in 5 ! ways. Therefore total number of arrangements beginning with a capital =5 ! × 3 = 5 × 4 × 3 × 2 × 3 = 360.

(*ii*) Two capitals out of the given letters can be chosen in $^3P_2$ = 6 ways. For each choice of these two letters, remainiug four can be arranged in 4 ! ways. Hence the required numberof arrrangements = 6 × 4 ! = 6 × 24 = 144.

Example 9:

In how many ways can 3 boys and 5 girls be arranged in a row so that all the 3 boys are together.

Solution:

We count the 3 boys as one boy so that number of persons involved is now 6. They can be arranged in 6 ! ways. But these 3 boys themselves can be arranged in 3 ! ways. Hence required number of arrangements is

6 ! × 3 ! = 720 × 6 = 4320.

Example 10:

The letters of the word ZENITH are written in all possible orders. How many words are possible ? If all these words are written as in dictionary, what is the rank of word ZENITH.

Solution:

Total number of letters is ZENITH is 6. So the total number of possible words is 6 ! = 720. Total number of words begining with *E* is same as the number of possible arrangements of *Z, N I, T, H.* But total number of arrangements of *Z, N I, T, H* is 5 ! = 120.

Similarly total number of leters beginning with *N* is 120; with *I*, is 120; with *T*, is 120 and with *H* is 120.

Thus there are 600 words which begin with *E*, or *N* or *I*, or *T*, or *H*. Hence the words beginiing with Z will have their rank between 601 to 720.

Out of these 120 words, the number of words in which *E* is in second place is equal to 4 ! = 24. (Number of arrangements of *N, I, T, H*).

Thus the rank of words beginning with *Z E* will be between 601 to 624.

Out of these 24 words.

Words beginning with *ZEH* are 6, their ranks lie between 601 to 606.

Words beginiing with *ZEI* are 6, ranks lying between 607 – 612.

Words beginning with *ZEN* are in 6, ranks lying between 613 – 618.

And words beginning with *ZET* are 6, ranks between 619 – 624. We consider words beginning with *ZEN.*

According to Dictionary order, the words beginning with *ZEN* and containing, *ITH* are

ZEN HIT with rank 613

ZEN HTI with rank 614

ZEN IHT with rank 615

ZEN ITH with rank 616

ZEN THI with rank 617

ZEN TIH with rank 618

Hence rank of *ZENITH* is 616.

Example 11:

Find the number of arrangement that can be made out of the letters of the word ASSASSINATION".

Solution:

There are 13 letters in ASSASSINATION out of which there are 3 A's, 4 S's, 2 I's and two N's.

So the total number of arrangements

$$= \frac{13!}{3!4!2!2!}$$

$$= \frac{13.12.11.10.9.8.7.6.5.4.3.2.1}{(3.2.1).(4.3.2.1)(2.1)(2.1)}$$

$= 10810800.$

Example 12:

Find the number of permutations of the word "ACCOUNTANTS."

Solution:

Total number of letters in ACCOUNTANTS is 11 out of which there are two C's, two A's, two N's and two T's. So the required numer of permutations

$$= \frac{11!}{2!2!2!2!}$$

$= 2494800.$

Example 13:

Six papers are set in an examination of which two are mathematical. In how many different orders can the papers be arranged so that (i) the two mathematical papers are together (ii) the two mathematical papers are not consecutive.

Solution:

Counting the two mathematical papers as one, total number of arrangements is 5 !

The two mathematical papers can be arranged whthin themselves in two ways. So required number of arrangements in which the mathematical papers are always together

$= 2 \times 5! = 2 \times 120 = 240.$

Again total number of arrangements in which Mathematical papers are not consecutive is

$= 720 - 240 = 480.$

Example 14:

How many different words can be formed with the letters of 'HARYANA'? In how many of these H and N are together? How many of these words begin with H and end with N?

Solution:

Total number of letters in HARYANA is 7 with 3 *A*'s. So total number of different words

$$= \frac{.7\,!}{3\,!} = 840.$$

Counting *HN* as onё letter, we are left with 6 letters having 3 A's. *H* and *N* can be put together in two different manners namely *HN* and *NH.*

Reqd, bi, if words, in which *H* and *N* are together

$$= 2 \times \frac{6\,!}{3\,!} = 240.$$

Finally the number of words beginning with *H* and ending with *N* is same as the number of arrangements of ARYAA, which is equal to

$$\frac{5\,!}{3\,!} = 20$$

Example 15:

In how many ways can the letters of the word STRANGE be arranged so theat the vowels occupy only the odd places.

Solution:

Number of letters in STRANGE is 7 so there are 4 odd places namely First, Third, Fifth and Seventh.

Vowel are only 2. the number of ways in which 2 vowels can occupy 4 places is ${}^4P_2$. For each such way, the remaining 5 places are to ocuupied by five remaining consonants. This can be achieved in 5 ! different ways. These vowels can be arranged within themselves in two ways.

Hence the required number is ${}^4\sqrt{P}_2 \times 5\,! \times 2$

$$= 2 \times \frac{4\,!}{2\,!} \times 120 = 1440.$$

EXERCUSES

1. How many words can be formed from the letters of COURTESY ? How many of them will begin with C and end with Y?
2. How many numbers can be formed between 10 and 1000 with the help of digits 2, 3, 4, 0, 8, 9?
3. In how many ways, other arrangements can be formed out of the letters of DOGMATIC ?

4. How many different words can be formed out of the letters of word "TRIANGLE". How many begin with T ? How many begin with T and end with E.

5. How many numbers can be formed with the digits 1, 2, 3, 4, 3, 2, 1 so that odd digits always occupy odd places?

6. How many arrangements can be made out of the letters of the word DRAUGHT, the vowels never being separated?

7. There are 5 boys and 3 girls. In how many ways can they stand in a row so that no two girls are together.

8. How many different words can be made out of the letters of the letters of the word "ALLAHABAD". In how many of these will the vowels occupy the even places ?

9. How many numbers less thatn 1000 and divisible by 5 can be formed with the digits 0, 1, 2, 3, 4, 5, 6, 7, 8, 9, each digit not occurring more than once in each number?

10. A number of four different digits is formed by using the digits 1, 2, 3, 4, 5, 6, 7, in all possible ways, each digit occurring once only. Find (i) how many such numbers can be formed and (ii) how many of them are greater than 3400.

11. In how many ways can a consonant and a vowel be chosen out of the letters of each of the words (i) LOGRITHM (ii) EQUATION?

12. Find how many words can be formed of the letters of the word 'FAILURE' the four vowels always coming together.

13. Find the number of pernutations of the letters of the word SIGNAL such that the vowels may occupy only odd positions?

14. Find the number of all possible different words into which the word INTERFERE can be converted by changed of place of letters, it is being given that no two consonants are to be together.

15. How many numbers between 1000 and 10,000 can be formed with the digits 1, 2, 3, 4, 5, 6, 7, 8, 9? How many of them are odd?

ANSWERS

1. 40320, 720.
2. 125.
3. 40319.
4. 40320, 5040, 720.
5. 18.
6. 1440.
7. 14400
8. 7560,60

9. 155. 10 840, 520.

11.(*i*) 18 (*ii*) 15 12. 576

13.144. 14. 240.

15.3024, 1680.

COMBINATIONS

Suppose we are given a number of objects. Each collection of objects which can be made out of the above objects is called a *combination.*

Thus if $\{a, b, c\}$ is a given set of objects and two objects are to be chosen. The different combinations are given by $\{a, b\}$, $\{b, c\}$ and $\{c, a\}$. Note that pernutations are six namely *ab, ba, ca, ac, bc, cb.* Although *ab, ba* are two pernutations yet they form a single comnination. In combination, the order in which the elements are selected does not matter. similarly the number of combinations of three objects chosen from $\{a, b, c, d\}$ is 4 namely $\{a, b, c\}$, $\{b, c, d\}$, $\{c, d, a\}$, $\{a, b, d\}$.

Notation: If $0 \le r \le n$ then the number of combinations of r objects taken out of n objects is denoted by ${}^nC_r$.

TO PROVE THAT $${}^nC_r = \frac{{}^nP_r}{r!} = \frac{n!}{(n-r)!\,r!}$$

Proof. Let ${}^nC_r = x$. So, we get x groups each consisting of r objects chosen out of n given objects. But each of thse groups can be re–arranged within itself r ! times. Hence the total number of pernutations of r objects chosen from n objects is equal to $x \times r$!. In other words ${}^nP = x \times r$!

i.e. $$x = \frac{{}^nP_r}{r!}. \quad \text{Hence } {}^nC_r = \frac{{}^nP_r}{r!} = \frac{\underline{|n}}{\underline{|n-r}\,\underline{|r}}$$

Cor. ${}^nC_0 = {}^nC_n = 1.$

Proof. $${}^nC_0 = \frac{n!}{0!\,n!} = 1 \text{ and } {}^nC_n = \frac{n!}{0!\,n!} = 1.$$

Example 16:

In how many ways, a partly of 4 or more can be selected from 10 persons?

Solution:

We are to calculate

$${}^{10}C_4 + {}^{10}C_5 + {}^{10}C_6 + {}^{10}C_7 + {}^{10}C_8 + {}^{10}C_9 + {}^{10}C_{10}$$

Now $\quad {}^{10}C_4 = \frac{10!}{4!\,6!} = \frac{10.9.8.7}{4.3.2.1} = 210$

$${}^{10}C_5 = \frac{10!}{5!\,5!} = \frac{10.9.8.6}{5.4.3.2.1} = 252$$

$${}^{10}C_6 = \frac{10!}{4!\,5!} = \frac{10.9.8.7}{4.3.2.1} = 210$$

$${}^{10}C_7 = \frac{10!}{3!\,7!} = \frac{10.9.8}{3.2.1} = 120$$

$${}^{10}C_8 = \frac{10!}{2!\,8!} = \frac{10.9}{2.1} = 45$$

$${}^{10}C_9 = \frac{10!}{1!\,9!} = \frac{10}{1} = 10$$

$${}^{10}C_{10} = \frac{10!}{0!\,10!} = \frac{1}{1} = 1$$

Hence required number = 210 + 252 + 210 + 120 + 45 + 10 + 1

= 848.

Example 17:

If ${}^nC_{12} = {}^nC_8$, find ${}^nC_{17}$ and ${}^{22}C_n$.

Solution:

$${}^nC_{12} = {}^nC_8 \Rightarrow \frac{n!}{(n-12)!\,12!} = \frac{n!}{(n-8)!\,8!}$$

$\Rightarrow (n-8)!\,8! = (n-12)!\,12!$

$\Rightarrow (n-8)(n-9)(n-10)(n-11) = 12.11.10.9$

$\Rightarrow (n-8)(n-9)(n-10)(n-11) = 11880$

$\Rightarrow (n^2-19n+88)(n^2-19n+90) = 11880$

$\Rightarrow (x^2+88)(x+90) = 11880 \qquad$ where $x = n^2 - 19n$

$\Rightarrow x^2 + 178x + 7920 = 11880$

$\Rightarrow x^2 + 178x - 3960 = 0$

$\Rightarrow (x+198)(x-20) = 0$

$\Rightarrow x = -198,\ x = 20.$

In case $x = 20$, $n^2 - 19n - 20 = 0 \quad \Rightarrow \quad n$ is imaginary.

Since n is a positive integer, we get $n = 20$.

Hence $\quad {}^nC_{17} = {}^{20}C_{17} \dfrac{20.19.18}{3.2.1} = 1140$

$${}^{22}C_n = {}^{22}C_{20} \frac{22.21}{2.1} = 231.$$

COMPLEMENTARY COMBINATIONS

Prove that $\quad {}^nC_r = {}^nC_{n-r}$.

Proof. $\quad {}^nC_r = \dfrac{n!}{(n-r)!\,r!} = \dfrac{n!}{[n-(n-r)!\,(n-r)!}$

$$= \frac{n!}{(n-r)!\,[n-(n-r)]!}$$

$$= {}^nC_{n-1}$$

Two combinations C_x and ${}^nC_y$ are said to be complementary if $x + y = n$. Thus ${}^nC_r$ and ${}^nC_{n-r}$ are complementary combinations.

TO PROVE THAT ${}^nC_x = {}^nC_y \Rightarrow$ EITHER $x = y$ OR $x + y = n$

Proof. $\quad {}^nC_x = {}^nC_y \Rightarrow \dfrac{n!}{(n-x)!\,x!} = \dfrac{n!}{(n-y)!\,y!}$

$\Rightarrow \quad (n-x)!\,x! = (n-y)!\,y! \qquad \ldots(i)$

Suppose $x \neq y$ and let $x > y$.

Now (i) $\Rightarrow (n-x)!\,x(x-1)\ldots(y+1) = (n-y)$

$\Rightarrow x(x-1)\ldots(y+1) = (n-y)(n-y-1)\ldots(n-x+1) \quad \ldots$(ii)

Suppose further that $x + y \neq n$.

Then either $x + y > n \quad$ or $\quad x + y < n$.

If $x + y > n$, we have $x > n - y$, $x - 1 > n - y - 1$, ...,

$y + 1 = x - (x - y - 1) > n - y - (x - y - 1) = n - x + 1$

hence the product on LHS $\neq$ product on RHS of (ii).

This is absurd !

If $x + y < n$, we have $x < n - y$, $x - 1 < n - y - 1$, ...,

$y + 1 = x - (x - y - 1) < n - y - (x - y - 1) = n - x + 1$.

Thus LHS < RHS of (ii), which is agian absurd.

Hence if $x > y$, then we must have $x + y = n$.

It can be similarly shown that if $x < y$, then also $x + y = n$.

REMARK. The above result helps us in solving many problems in combination theory. Consider Example 17. We are given that ${}^nC_{12} = {}^nC_8$

Then $12 + 8 = n$ so that $n = 20$ and now rest of the problem can be done by usual calculations.

TO PROVE THAT ${}^nC_r + {}^nC_{r-1} = {}^{n+1}C_r$.

Proof. LHS $\dfrac{n!}{(n-r)!\,r!} + \dfrac{n!}{(n-r)!\,(n-r+1)!}$

$$= \frac{n!}{(n-r)!\,(n-r)!}\left\{\frac{1}{r} + \frac{1}{n-r+1}\right\}$$

$$= \frac{n!}{(n-1)!\,(n-r)!}\left\{\frac{n-r+1+r}{r(n-r+1)}\right\}$$

$$= \frac{(n+1)n!}{r(r-1)!\,(n-r+1)(n-r)!}$$

$$= \frac{(n+1)!}{r!\,(n+1-r)!}$$

$$= {}^{n+1}C_r.$$

TO PROVE THAT ${}^nC_0 + {}^nC_1 + {}^nC_2 + \dots \dots + {}^nC_n = 2^n$

Proof. For $n = 1$, LHS $= {}^1C_0 + {}^1C_1 = 1 + 1 = 2 = 2^1$.

So, our result holds for $n = 1$

Suppose the result is true for $n = k$

i. e. ${}^kC_0 + {}^kC_1 + \dots {}^kC_k = 2^k$

Now put $n = k = 1$

By § 10.13, ${}^{k+1}C_1 = {}^kC_1 + {}^kC_0$

${}^{k+1}C_2 = {}^nC_2 + {}^kC_1$

... ...

${}^{k+1}C_k = {}^nC_k + {}^kC_{k-1}$.

Adding we get

${}^{k+1}C_1 + {}^{k+1}C_2 + \dots \dots + {}^{k+1}C_k$

${}^kC_0 + 2\,({}^kC_1 + {}^nC_2 + \dots \dots + {}^kC_{k-1}) + {}^kC_k$

Thus ${}^{k+1}C_0 + {}^{k+1}C_1 + \dots \dots + {}^{k+1}C_{k+1}$

$^{k-1}C_0 + {}^kC_0 + 2\,({}^kC_1 + {}^nC_2 + \ldots\ldots + {}^kC_{k-1}) + {}^kC_k + {}^{k-1}C_{k+1}$

$= 2\,({}^kC_0 + {}^kC_1 + \ldots\ldots + {}^kC_k)$

as $^{k+1}C_0 = 1 = {}^kC_0$ and $^{k-1}C_{k+1} = 1 = {}^kC_k$.

Using (*i*) we get

$^{k-1}C_0 + {}^{k+1}C_1 + \ldots\ldots + {}^{k+1}C_{k+1} = 2.2^k = 2^{k+1}$.

Hence the result holds for $n = k + 1$. Thus, by Principle of Mathematical Induction, result is true for all Positive integers.

TO FIND THE NUMBER OF WAYS IN WHICH $m + n$ OBJECTS CAN BE DIRIDED INTO TWO GROUPS CONTAINING m AND n OBJECTS RESPECTIVELY

Let the required number of ways be equal to x. This is same as number of combinations of $m + n$ objects taken m at a time, for each combination of m objects gives rise to another combtnation of n objects.

Thud $\quad x = {}^{m+1}C_m = \dfrac{(m+n)!}{n!m!}$

REMARK. In case $m = n$, We shall get $x \dfrac{2m!}{(m!)^2}$. But in this counung, all ways are not different, Since it is possibie to interchange two groups without obtaining a new distribution, So, in such cases, the number of different combination is $\dfrac{2m}{2!(m!)^2}$.

Example 18:

Suppose we want to divide the collection {a, b, c, d} into two groups each having two elements.

Solution:

Possible groups are (*ab*), (*cd*); (*cd*), (*ab*); (*ac*), (*bd*); (*bd*), (*ac*); (*ad*), (*bc*); (*be*), (*ad*).

But if we look carefully, we note that first two groups, second two groups and third two groups are actually the same,placed in different order. So number of different groups is 3.

i.e. $\quad \dfrac{4!}{2!(2!)^2}$. and not $\dfrac{4!}{(2!)^2} = 6$

REMARK. The above result can be generalised as unber.

The number of ways, $a_1 + a_2 + \ldots\ldots + a_t$ objects can be divided into groups of $a_1 + a_2, \ldots\ldots, a_t$ objects respectively is equal to

$$\frac{a_1 + a_2 + \dots\dots + a_t)!}{a_1!a_2!a_3!\dots\dots a_t!}.$$

Further, if $a_1 = a_2 = \dots \dots = a_t$, the number of different groups

obtained $= \dfrac{(a_1 + a_2 + \dots\dots a_t)!}{t!(a_1!)^t}. = \dfrac{(ta_1)!}{t!(a_1!)^t}.$

Example 19:

If $^{18}C_r = {}^{18}C_{r+2,}$ *find* $^rC_5$.

Solution:

$^{18}C_r = {}^{18}C_{r+2} \quad \Rightarrow \quad r = r + 2 \quad$ or $\quad r = r + 2 = 18.$

Since $\quad r \neq r + 2, \quad$ we get $\quad 2r = 16 \quad$ i.e. $\quad r = 8.$

Hence $\quad {}^rC_5 = {}^8C_5 = \dfrac{8.7.6}{3.2.1} = 53.$

Example 20:

A boat is to be manned by 8 men, of whom 2 can only row on bow side and 1 can only row on stroke side, in how many ways can the crew be arranged?

Solution:

Let A, B be the men on bow side and C be on the stroke side. On bow side two mote men are to be selected out of remaining $8 - 3 = 5$ men. This can be done in $^bC_2$ ways. Once this selection is made, the remaining men are to occupy the stroke side. Further each combination of a side can be made in 4 ! ways.

So required number of arrangetnents $= 4\,! \times 4\,! \times {}^5C_2 = 5760.$

Example 21:

How many triangles can be made by joining the vertices of decagon? How many diagonals will it have?

Solution:

A triangle is obtained by joining any three vertices of a decagon. Thus number of triangles obtained from a deeagon

$$^{10}C_3 = \frac{10.9.8}{3.2.1} = 120.$$

Now a diagonal is obtained by joining any two vertices which are not adjacent.

Total number of ways in which two vertices o a decagon can be joined

$$= {}^{10}C_2 = \frac{10.9}{2} = 45.$$

Number of lines obtained by joining two adjacent vertices

=number of sides of decagon = 10.

Hence number of diagonals = 45 − 10 = 35.

Example 22:

A committee consisting of 5 member is to be formed out of 6 men and 4 women. How many committees can be formed so that at least one woman is always there in the committee?

Solution:

Total number of committees = ${}^{10}C_5 = 252$.

The number of committees in which no woman is there = ${}^{2}C_5 = {}^{6}C_1 = 6$.

Hence the number of committees in which at least one woman is included =252 − 6 = 246.

Example 23:

In an examination a candidate is required to answer 6 out of 10 questions which are divided inta two groups each containing 5 questions and not permitted to attempt more than 4 questions from each group. in how many ways, can he make up his choice?

Solution:

The candidate can have following three choices.

(*i*) 2 questions from group *A* and 4 from group *B*.

(*ii*) 3 questions from group *A* and 3 from group *B*.

(*iii*) 4 questions from group *A* and 2 from group *B*.

Now first choice can be made up in ${}^5C_2 \times {}^5C_4$, *i.e.* 50 ways. Secand choice can be made up in ${}^5C_3 \times {}^5C_3$ ways *i.e.* 100 ways and third choice can be made up ${}^5C_4 \times {}^5C_2$ ways *i.e.* 50 ways.

So, total number of ways, he can make up his choice

= 50 + 100 + 50 = 200.

Example 24:

Find out the number of ways in which a cricket team consisting of 11 players can be selected from 14 players. Also find out how many of these (a) will include captain (b) will not include captain?

Solution:

The number of ways in which 11 players can be selected from 14 players is $^{14}C_{11} = \frac{14.13.12}{3.2.1} = 14 \times 2 \times 13 = 28 \times 13 = 364.$

(*a*) Here we have to select 10 players from remaining 13 players as captain is to beincluded in each solution. Therefore, required number of ways $^{13}C_{10} = \frac{13.12.11}{3.2.1} = 13 \times 2 \times 11 = 26 \times 11 = 286.$

(*b*) Here we have to exclude captain in each selection. in other words, we have to select 11 players from 13 players. This can be done in $^{13}C_{11}$ ways $= \frac{13.12}{2.1} = 78$ ways.

TO FIND OUT THE NUMBER OF COMBINATIONS OF *r* THINGS TAKEN OUT OF *n* GIVEN THINGS WHEN THE GIVEN THINGS ARE NOT ALL DIFFERENT

A general formula for such a case may be quite complicated, but a particular case can be solved by methods discussed in following examples.

Example 25:

How many combinations can be made by taking 6 letters out of the word PROPORTION.

Solution:

There are 10 letters in PROPORTION. consisting of (*P P*), (*R R*), (*O O O*), T, I, N. Thus two *P*'s, two *R*'s, and 3 *O*'s are similar. In all the required combinations some may contain all sissimilar lettars, some may not contain all different letters. Following casearise:

(*i*) All the letters are different.

(*ii*) 4 letters are different and 2 are similar.

(*iii*) 3 letters are different and 3 are similar.

(*iv*) 2 letters are same 2 letters are same and remaining 2 letters are different.

(*v*) 3 letters are same, 2 letters are same and one letter different from the others.

(*vi*) 2 letters are same, 2 letters are same and remaining 2 letters are also same.

Case I. There are six different letters. The required number of combinations $= {}^6C_6 = 1.$

Case II. There are three pairs of similar things (*P P*), (*R R*) and two *O*'s from (*O O O*). One pair can be chosen in 3 ways. Remaining 4 diffreent letters are to be seleeted from remaining 5 different letters. This can be done in $^5C_4$= 5 ways. Hence the number of combinations of this type = 3 × 5 = 15.

Case III. The only letter which occurs 3 times in 'Proportion' is *O*. So three similar letters can be chosen in one way only. Remaining three different letters are to be chosen from the remaining 5 letters. This can be perfound in $^5C_3$ *i.e.* 10 ways. Hence the number of combinations of type III is 10 × 1 = 10.

Case IV . Two pairs of similar letters can be chosen in$^3C_2$ *i.e.* 3 ways and remaining 2 different letters can be chosen in $^4C_2$ *i.e.* 6 ways. Hence the number of cmobinations in this case is 3 × 6 = 18.

Case V. 3 similar letters can be chosen in one way, 2 sumilar letters can be chosen in 2 ways (either *P P* or *R R*) and the remaining one letter can be chosen in 4ways. So the required number of combinations is 1 × 2 × 4=8.

Case VI. Three pairs of similar letters can be chosen only in one way *i. e.* (*O O*), (*P P*), and (*R R*).

Hence the total number of required combinations

= 1 + 15 + 10 + 18 + 8 + 1 = 53.

Example 26:

From 6 boys and 4 girls, 5 are to be selected to admission for a particular house. In how many ways can this be done of there must be exactly 2 girls.

Solution:

Two girls can be selected out of 4 girls in $^4C_2 = \frac{4.3}{2.1} = 6$ ways.

The remaining three are to be boys. They are to be selected out of 6 boys. The number of ways in which 3 boys can be chosen out of 6 boys is

$$^6C_3 = \frac{6.5.4}{3.2.1} = 20.$$

Required number of ways = 6 × 20 = 120.

Example 27:

Out of 10 consonants and 4 vowels how many vords can be formed each containing 6 consonants and 3 vowels.

Solution:

6 consonants can be selected out of 10 in $^{10}C_6$ ways wh le 3 vowels can be selected out of 4 is *n* $^4C_3$ ways.

Total number of letters in each choice of 6 consonants and 3 vowels is 9. so each choice can be arranged in 9 ! different ways.

hence total number of words

$= {}^{10}C_6 \times 4\sqrt{}\ C_3 \times 9\ !$

$= \frac{10\times9\times8\times7}{4\times3\times.2.1}\times4\times9!$

$= 304819200.$

Example 28:

In how many ways can 12 Science students and 9 Commerce students sittin arow so that no two Commerce students may be together?

Solution:

Let the Science students be arranged as

$- S - S - S - \ldots. -$

(S standing for position occupied by a science student.) In order that no two Commerce students may be together, they must occupy odd positions which are 13 in number.

Hence we are to select a position out of these 13, which can be done in ${}^{13}C_9$ ways.

But $\quad {}^{13}C_9 = {}^{13}C_4 = \frac{13.12.11.10}{4.3.2.1} = 715.$

Consequently required number is 715.

Example 29:

Find the number of combinations if the letters of the word EXAMINATION taken our at a time.

Solution:

There are 11 letters in EXAMINATION consisting of (A A), (I I), (NN), E, X, T, O, M. Following cases arise:

(*i*) All the four letters are different.

(*ii*) two are different two are alike.

(*iii*) two pairs of alike letters.

Case I. There are 8 different letters, we are to choose 4.Number of combinaions formed in this manner = ${}^8C_4$ = 70.

Case II. One pair of alike ketters can be chosen in three ways and the remaining two different letters are to be chosen from 7.

Number of such combinations $= 3 \times {}^7C_2 = 3 \times 21 = 63$.

Case III. There are three pairs of alike letters:

so the number of combinations having two pairs of alike ketters

$= {}^3C_2 = 3$

Consequently total number of combinations

$= 70 + 63 + 3$

$= 136$.

EXERCISES

1. If ${}^{2n}C_r = {}^{2n}C_{r+2}$, prove that $r = n - 1$.
2. If $m = {}^nC_2$, prove that ${}^mC_2 = 3\ {}^{n+1}C_4$.
3. If ${}^{28}C_{2r} = {}^{24}C_{2r-4} = 125: 11$, find r.
4. Find the number of combinations of 50 things taking 46 at a time.
5. In how many ways can 12 things be divided equally among 4 persons?
6. From 4 officers and 8 privates, in how many ways can 6 be chosen (i) to include exactly on officer (ii) to include at least on officer ?
7. How many combinations of the letters of the word ALLITERATION can be made if fourletters are to be taken at a time ?
8. Find the number of combinations of the letters of the word COMMERCE taken four together.
9. How many words can be formed by taking three consonants and two vowels form a set of 10 consonants and 4 vowels?
10. How many words formed by taking 6 letters from CENTRIFUGAL will contain *C, T* and *F*?
11. How many words can be formed out of the letters of EXAMINATION taken 4 at a time.
12. How many words can be formed taking 2 vowels of the word 'DEVASTATION?' In how many of them will both *T*"s will be together?
13. How many combinations can be formed taking 4 letters at a time from the word SUCCEEDING.
14. How many words can be formed from the letters of CIRCUMFERENCE taking 3 consonants and 3 vowels?
15. A question paper contains 6 questions, each having an alterna tive. in how many ways can an examinee answer one or more questions?

4

Matrix

INTRODUCTION

Definition : *A system of any mn numbers arranged in a rectangular array of m rows and n columns is called a matrix of order m × n or an m × n matrix (which is read as m by n matrix).*

Or

A set of mn elements of a set S arranged in a rectangular array of m rows and n columns is called an m × n matrix over S.

A m × n matrix is usually written as

$$\begin{bmatrix} a_{11} & a_{12} & a_{13} & \cdots & a_{1n} \\ a_{21} & a_{22} & a_{22} & \cdots & a_{2n} \\ \cdots & \cdots & \cdots & \cdots & \cdots \\ a_{m1} & a_{m2} & a_{m3} & \cdots & a_{mn} \end{bmatrix} \text{is an } m \times n \text{ matrix.}$$

where the symbols a_{ij} represent any numbers (a_{ij} lies in the ith row and jth column).

Notes:

1. A compact form of the above matrix may be represented by the symbols $[a_{ij}]$, (a_{ij}) $\| a_{ij} \|$ or by a single capital letter A, say.

   The element a_{ij} belong to ith row and th column and sometime called the (;)th element of the matrix.

2. Each of the mn numbers constituting an m × n matrix is known as an *element of the matrix.*

   The elements of matrix may be scalar or vector quantities.

3. The plural of 'matrix' is 'matrices'.

Example:

Write down the orders of the matrices:

(a) $\begin{bmatrix} 2 & 3 & 5 \\ 1 & 0 & 3 \end{bmatrix}$; *(b)* $\begin{bmatrix} 2 \\ 3 \end{bmatrix}$.

(c) [3, 4, 5]; (d) [1].

Solution:

The order of the given matrix are:

(a) 2 × 3; (b) 2 × 1; (c) 1 × 3; (d) 1 × 1. **Ans.**

MULTIPLICATION OF MATRICES

Definition: *Let* $A = [a_{ij}]$ $m \times n$ *and* $B = B = [b_{ik}]$ $n \times p$ *be two matrices such that the number of column in A is equal to the number of row in B. Then the matrix* $[c_{ik}]$, $m \times p$ *such that* $c_{ik} = \sum_{j=1}^{n} a_{ij} b_{ik}$ *is called the product of the matrices A ande B in that order and we write = AB.*

As an example, consider the matrices

$$A = \begin{bmatrix} 1 & 2 & 3 \\ 4 & 5 & 6 \end{bmatrix} \text{ and } B \begin{bmatrix} 7 & 8 \\ 9 & 10 \\ 11 & 12 \end{bmatrix}$$

Here the number of columns in A = 3 = then number of rows in B and thus we can evaluate AB.

Let Ab = $[c_{ij}]$, where $[c_{ij}]$ is 2 × 2 matrix.

Now to write c_{11}, we take the elements of the first row of A *viz.*, 1, 2, 3 in this order and the elements of the first column of B *viz.*, 7, 9, 11 in this order and form the products 1.7, 2.9, 3.11 and finally add them.

i.e., $c_{11} = 1.7 + 2.9 + 3.11 = 58$

Similarly $c_{12} = 1.8 + 2.10 + 3.12 = 64$;

$c_{21} = 4.7 + 5.9 + 6.11 = 139$

and $c_{22} = 4.8 + 5.10 + 6.12 = 154$

Hence $AB = [c_{ij}] = \begin{bmatrix} c_{11} & c_{12} \\ c_{21} & c_{12} \end{bmatrix} = \begin{bmatrix} 58 & 64 \\ 139 & 154 \end{bmatrix}$.

Note: The product AB can be calculated only if the number of columns in A be equal to the number of rows in b. The two matrices A and B satisfying this condition are called conformable to multiplication.

POST-MULTIPLICATION AND PRE-MULTIPLICATION OF MATRICES

The matrix AB is the matrix A post-multiplied by B whereas the matrix BA is the matrix A pre-multiplied by B.

In the product AB, the matrix A is know as the pre-factor and the matrix B is know as the post-factor.

The product in both the above the above cases viz. AB and BA may or may not exist and may be equal or different.

i.e., we say AB ≠ BA in general.

The same is discussed on the next page:

Case 1: If the matrix A is $m \times n$ and the matrix B is $n \times k$, then the product AB exists whereas BA does not exist, since we know that AB can be calculated only if the numbers of columns in A is equal to the number of rows in B.

Case 2: If the matrix A is $m \times n$ and the matrix B is $n \times m$, then both AB and BA exist, but the matrix AB is $m \times m$ while the matrix BA is $n \times n$.

Hence AB ≠ BA thought AB and BA exist.

Case 3: If both A and B are square matrices of the same order, then AB as well as BA exist but are not necessarily equal *i.e.,* if

$$A = \begin{bmatrix} 1 & 2 \\ 3 & 4 \end{bmatrix} \text{ and } B = \begin{bmatrix} 3 & 1 \\ 4 & 7 \end{bmatrix}$$

then $$AB = \begin{bmatrix} 1 & 2 \\ 3 & 4 \end{bmatrix} \times \begin{bmatrix} 3 & 1 \\ 4 & 7 \end{bmatrix} = \begin{bmatrix} 1.3 + 2.4 & 1.1 + 2.7 \\ 3.3 + 4.4 & 3.1 + 4.7 \end{bmatrix}$$

$$= \begin{bmatrix} 11 & 25 \\ 25 & 31 \end{bmatrix}$$

and $$BA = \begin{bmatrix} 3 & 1 \\ 4 & 7 \end{bmatrix} \times \begin{bmatrix} 1 & 2 \\ 3 & 4 \end{bmatrix} = \begin{bmatrix} 3.1 + 1.3 & 3.2 + 1.4 \\ 4.1 + 7.3 & 4.2 + 7.4 \end{bmatrix}$$

$$= \begin{bmatrix} 6 & 10 \\ 25 & 36 \end{bmatrix}$$

$\therefore$ AB ≠ BA.

But if $A = \begin{bmatrix} 1 & 0 \\ 0 & -2 \end{bmatrix}$ and $B = \begin{bmatrix} 1 & 0 \\ 0 & 4 \end{bmatrix}$

then $AB = \begin{bmatrix} 1 & 0 \\ 0 & -2 \end{bmatrix} \times \begin{bmatrix} 1 & 0 \\ 0 & 4 \end{bmatrix} = \begin{bmatrix} 1.1 + 0.0 & 1.0 + 0.4 \\ 0.1 - 2.0 & 0.0 - 2.4 \end{bmatrix}$

$= \begin{bmatrix} 1 & 0 \\ 0 & -8 \end{bmatrix}$

and $BA = \begin{bmatrix} 1 & 0 \\ 0 & 4 \end{bmatrix} \times \begin{bmatrix} 1 & 0 \\ 0 & -2 \end{bmatrix} = \begin{bmatrix} 1.1 + 0.0 & 1.0 + 0\,(-2) \\ 0.1 + 4.0 & 0.0 + 4\,(-2) \end{bmatrix}$

$= \begin{bmatrix} 1 & 0 \\ 0 & -8 \end{bmatrix}$

$\therefore$ AB = BA.

Hence in general AB ≠ BA.

Note 1: If AB = BA, then matrices A and B are said to commit. If AB = – BA, the matrices A and B are said to anticommute.

Note 2: If $A = \begin{bmatrix} 1 & 1 \\ 1 & 1 \end{bmatrix}$ and $B = \begin{bmatrix} 1 & 0 \\ -1 & 0 \end{bmatrix}$,

then $AB = \begin{bmatrix} 1 & 1 \\ 1 & 1 \end{bmatrix} \times \begin{bmatrix} 1 & 0 \\ -1 & 0 \end{bmatrix}$

$= \begin{bmatrix} 1.1 + 1.(-1) & 1.0 + 1.0 \\ 1.1 + 1.(-1) & 1.0 + 1.0 \end{bmatrix} = \begin{bmatrix} 0 & 0 \\ 0 & 0 \end{bmatrix}$

i.e., AB is zero matrix (or null matrix) whereas neither A nor B is a zero matrix.

$\therefore$ AB = O does not imply that either A = 0 or B = O.

Here $BA = \begin{bmatrix} 1 & 0 \\ -1 & 0 \end{bmatrix} \times \begin{bmatrix} 1 & 1 \\ 1 & 1 \end{bmatrix}$

$= \begin{bmatrix} 1.1 + 0.1 & 1.1 + 0.1 \\ -1.1 + 0.1 & -1.1 + 0.1 \end{bmatrix} = \begin{bmatrix} 1 & 1 \\ -1 & -1 \end{bmatrix}$

i.e., BA ≠ O

If $A = \begin{bmatrix} 4 & 4 \\ 3 & 3 \end{bmatrix}$ and $B = \begin{bmatrix} -1 & 1 \\ 1 & -1 \end{bmatrix}$, then

$AB = \begin{bmatrix} 4 & 4 \\ 3 & 3 \end{bmatrix} \times \begin{bmatrix} -1 & 1 \\ 1 & -1 \end{bmatrix}$

$$= \begin{bmatrix} 4\,(-1) + 4\,(1) & 4\,(1) + 4\,(-1) \\ 3\,(-1) + 3\,(1) & 3\,(1) + 3\,(-1) \end{bmatrix} = \begin{bmatrix} 0 & 0 \\ 0 & 0 \end{bmatrix} = 0$$

i.e., the product of two non-zero square matrices can be a zero matrix.

and $\quad BA = \begin{bmatrix} -1 & 1 \\ 1 & -1 \end{bmatrix} \times \begin{bmatrix} 4 & 4 \\ 3 & 3 \end{bmatrix}$

$$= \begin{bmatrix} (-1).\,4 + 1.3 & (-1).\,4 + 1.3 \\ 1.4 + (-1).\,3 & 1.4 + (-1).\,3 \end{bmatrix} = \begin{bmatrix} -1 & -1 \\ 1 & 1 \end{bmatrix} \neq 0$$

NILPOTENT MATRIX

Definition: *A square matrix A si called Nilpotent matrix of order* ***m,*** *provided it satisfies the relation* $A^m = O$ *and* $A^{m-1} \neq O$, *where m is a positive integer and* ***O*** *is the null matrix of order m.*

For example, the matrix $A = \begin{bmatrix} 0 & 1 \\ 0 & 0 \end{bmatrix}$ is a nilpotent matrix, since

$$A = \begin{bmatrix} 0 & 1 \\ 0 & 0 \end{bmatrix} \neq \mathbf{O},$$

$$A^2 = \begin{bmatrix} 0 & 1 \\ 0 & 0 \end{bmatrix} + \begin{bmatrix} 0 & 1 \\ 0 & 0 \end{bmatrix} = \begin{bmatrix} 0.0 + 1.0 & 0.1 + 1.0 \\ 0.0 + 0.0 & 0.1 + 0.0 \end{bmatrix}$$

$$= \begin{bmatrix} 0 & 0 \\ 0 & 0 \end{bmatrix} = \mathbf{O},$$

$$A^3 = \lambda^2 \cdot A = O \cdot A = O.$$

i.e., A is a matrix which is not itself a zero matrix though its powers are zero matrices and so it is a nilpotent matrix *(Another definition of nilpotent matrix).*

Example:

Show that $A = \begin{bmatrix} 1 & 2 & 3 \\ 1 & 2 & 3 \\ -1 & -2 & -3 \end{bmatrix}$ *is a nilpotent matrix of order 2.*

Solution:

Given $A = \begin{bmatrix} 1 & 2 & 3 \\ 1 & 2 & 3 \end{bmatrix} \neq O$

$$\therefore A^2 = \begin{bmatrix} 1 & 2 & 3 \\ 1 & 2 & 3 \\ -1 & -2 & -3 \end{bmatrix} \times \begin{bmatrix} 1 & 2 & 3 \\ 1 & 2 & 3 \\ -1 & -2 & -3 \end{bmatrix}$$

$$= \begin{bmatrix} 1.1 + 2.2 + 3(-1) & 1.2 + 2.2 + 3(-2) & 1.3 + 2.3 + 3(-3) \\ 1.1 + 2.1 + 3(-1) & 1.2 + 2.2 + 3(-2) & 1.3 + 2.3 + 3(-3) \\ -1.1 - 2.1 - 3(-1) & -1.2 - 2.2 - 3(-2) & -1.3 - 2.3 - 3(-3) \end{bmatrix}$$

$$= \begin{bmatrix} 0 & 0 & 0 \\ 0 & 0 & 0 \\ 0 & 0 & 0 \end{bmatrix} = \mathbf{O}$$, where **O** is the null matrix of order 3.

i.e., $\mathbf{A^2 = O}$ $\mathbf{A \neq O}$. Hence **A** is a nilpotent matrix of order 2.

INVOLUTORY MATRIX

Definition: *A square matrix A is called Involutory provided it satisfies the relation $A^2 = \mathbf{I}$, where **I** is the identity matrix.*

For example, the matrix $A = \begin{bmatrix} 1 & 0 \\ 1 & -1 \end{bmatrix}$ is involutory matrix,

since $A^2 = \begin{bmatrix} 1 & 0 \\ 1 & -1 \end{bmatrix} \times \begin{bmatrix} 1 & 0 \\ 1 & -1 \end{bmatrix}$

$$= \begin{bmatrix} 1.1 + 0.0 & 1.0 + 0.(-1) \\ 0.1 + (-1).0 & 0.0 + (-1).(-1) \end{bmatrix} = \begin{bmatrix} 1 & 0 \\ 0 & 1 \end{bmatrix} = I$$

Example 1:

If A is any square martix of order n and I_n is the identity martix of order n, such that $(I_n - A) = O$, then show that A is an involutory martix.

Solution:

Given that $(I_n - A)(I_n + A) = O$

$\Rightarrow$ $I_n^2 + I_n \cdot A - A \cdot I_n - A^2 = O$

$\Rightarrow$ $I_n + A - A - A^2 = O$, $\quad \because I_n^2 = I_n, I_n \cdot A = A = A \cdot I_n$.

$\Rightarrow$ $I_n - A^2 = O$

$\Rightarrow$ $A^2 = I_n$

i.e., A is involutory by definition.

ADDITION OF MATRICES

If there be two m × n matrices given by $A = [a_{ij}]$ and $B = [b_{ij}]$, then the matrix A + B is defined as the matrix each element of which is the sum of the corresponding elements of A and B *i.e.,*

$$A + B = [a_{ij} + b_{ij}],$$

where i = 1, 2, 3,..., m and j = 1, 2, 3,..., n.

For example: If $A = \begin{bmatrix} a_1 & b_1 & c_1 \\ a_2 & b_2 & c_2 \end{bmatrix}$ and $B = \begin{bmatrix} a_3 & b_3 & c_3 \\ a_4 & b_4 & c_4 \end{bmatrix}$

then $A = B = \begin{bmatrix} a_1 + a_3 & b_2 + b_3 & c_1 + c_3 \\ a_2 + a_4 & b_2 + b_4 & c_2 + c_4 \end{bmatrix}$

NILPOTENT MATRIX

Definition: *A square matrix A si called Nilpotent matrix of order* ***m****, provided it satisfies the relation* $A^m = O$ *and* $A^{m-1} \neq O$*, where m is a positive integer and* ***O*** *is the null matrix of order m.*

For example, the matrix $A = \begin{bmatrix} 0 & 1 \\ 0 & 0 \end{bmatrix}$ is a nilpotent matrix, since

$$A = \begin{bmatrix} 0 & 1 \\ 0 & 0 \end{bmatrix} \neq \mathbf{O},$$

$$A^2 = \begin{bmatrix} 0 & 1 \\ 0 & 0 \end{bmatrix} + \begin{bmatrix} 0 & 1 \\ 0 & 0 \end{bmatrix} = \begin{bmatrix} 0.0 + 1.0 & 0.1 + 1.0 \\ 0.0 + 0.0 & 0.1 + 0.0 \end{bmatrix}$$

$$= \begin{bmatrix} 0 & 0 \\ 0 & 0 \end{bmatrix} = \mathbf{O},$$

$$A^3 = \lambda^2 \cdot A = O \cdot A = O.$$

i.e., A is a matrix which is not itself a zero matrix though its powers are zero matrices and so it is a nilpotent matrix *(Another definition of nilpotent matrix).*

Example:

Show that $A = \begin{bmatrix} 1 & 2 & 3 \\ 1 & 2 & 3 \\ -1 & -2 & -3 \end{bmatrix}$ *is a nilpotent matrix of order 2.*

Solution:

Given $A = \begin{bmatrix} 1 & 2 & 3 \\ 1 & 2 & 3 \\ -1 & -2 & -3 \end{bmatrix} \neq O$

$$\therefore A^2 = \begin{bmatrix} 1 & 2 & 3 \\ 1 & 2 & 3 \\ -1 & -2 & -3 \end{bmatrix} \times \begin{bmatrix} 1 & 2 & 3 \\ 1 & 2 & 3 \\ -1 & -2 & -3 \end{bmatrix}$$

$$= \begin{bmatrix} 1.1 + 2.2 + 3(-1) & 1.2 + 2.2 + 3(-2) & 1.3 + 2.3 + 3(-3) \\ 1.1 + 2.1 + 3(-1) & 1.2 + 2.2 + 3(-2) & 1.3 + 2.3 + 3(-3) \\ -1.1 - 2.1 - 3(-1) & -1.2 - 2.2 - 3(-2) & -1.3 - 2.3 - 3(-3) \end{bmatrix}$$

$= \begin{bmatrix} 0 & 0 & 0 \\ 0 & 0 & 0 \\ 0 & 0 & 0 \end{bmatrix} = \mathbf{O}$, where $\mathbf{O}$ is the null matrix of order 3.

i.e., $\mathbf{A^2 = O \; A \neq O}$. Hence **A** is a nilpotent matrix of order 2.

TRIANGULAR MATRICES

It every element above or below the leading diagonal is zero, then the matrix is called a *Triangular Matrix.*

(a) Upper Triangular Matrix: A square matrix A whose elements $a_{ij} = 0$ for $i < j$ is called an upper triangular matrix.

For example $\begin{bmatrix} a_{11} & a_{12} & a_{13} \ldots\ldots a_{1n} \\ 0 & a_{22} & a_{23} \ldots\ldots a_{2n} \\ 0 & 0 & a_{33} \ldots\ldots a_{3n} \\ \ldots & \ldots & \ldots\ldots\ldots \\ 0 & 0 & 0 \ldots\ldots a_{nn} \end{bmatrix}$

(b) Lower Triangular Matrix: A square matrix A whose element $a_{ij} = 0$ for $i < j$ is called a lower triangular matrix.

For example $\begin{bmatrix} a_{11} & 0 & 0 \ldots\ldots 0 \\ a_{21} & a_{22} & 0 \ldots\ldots 0 \\ a_{31} & a_{32} & a_{33} \ldots\ldots 0 \\ \ldots & \ldots & \ldots \; \ldots\ldots 0 \\ a_{n1} & a_{n2} & a_{n3} \ldots\ldots a_{nn} \end{bmatrix}$

TYPES OF MATRICES

(a) **Horizontal Matrix:** If in a matrix the number of columns is more than the number of rows then it is called a horizontal matrix.

For example $\begin{bmatrix} 1 & 3 & 2 & 3 \\ 2 & 5 & 7 & 9 \end{bmatrix}$ is a horizontal matrix.

(b) **Vertical Matrix:** If in a matrix the number of rows is more than the number of columns it is called a vertical matrix.

For example $\begin{bmatrix} 2 & 3 \\ 3 & 5 \\ 4 & 6 \\ 5 & 7 \end{bmatrix}$ is a vertical matrix.

(c) **Column Matrix:** If there if only one column in a matrix, it is called a column matrix.

For example $\begin{bmatrix} 2 \\ 3 \\ 4 \end{bmatrix}$ This is also called column vector.

(d) **Square Matrix:** If m = n *i.e.,* the number of rows and columns of a matrix are equal, then the matrix is of order n × n and is called a square matrix of order n.

For example $\begin{bmatrix} 2 & 3 & 1 \\ 1 & 5 & 2 \\ 7 & 6 & 9 \end{bmatrix}$ is a square matrix and $\begin{bmatrix} 1 & 3 & 2 & 3 \\ 2 & 5 & 7 & 9 \end{bmatrix}$

is a rectangular matrix.

(e) **Rectangular Matrices:** When the number of rows and columns of the array are not equal, then the matrix is known as a rectangular matrix.

(f) **Row Matrix:** If in a matrix, there is only one row it is called a row matrix. For example [1, 2, 3]. This is also called a row vector.

(g) **Null (or zero) Matrix:** If all the elements of an m × n matrix are zero, then it is called a null or zero matrix and denoted by $O_{m \times n}$ or simply O.

For example $\begin{bmatrix} 0 & 0 & 0 \\ 0 & 0 & 0 \end{bmatrix}$ is the 2 × 3 mull matrix.

(h) **Unit Matrix:** A square matrix having unity for its elements in the leading diagonal and all the other elements as zero is called an unit matrix.

For example $\begin{bmatrix} 1 & 0 & 0 & 0 \\ 0 & 0 & 0 & 0 \\ 0 & 0 & 1 & 0 \\ 0 & 0 & 0 & 1 \end{bmatrix}$ is unit matrix of order 4 × 4 denote it byI_4.

$\therefore$ an a-rowed square matrix $[a_{ij}]$ is called a unit matrix provided

$$a_{ij} = 1, \text{ whenever } i = j$$
$$= 0, \text{ whenever } i \neq j.$$

(i) **Equal Matrix:** Two matrices A $[(a_{ij})]$ are said to be equal if (a) they are of the same type *i.e.,* if they have same number of rows and columns and (b) the elements in the corresponding positions of the two matrices are equal.

From the definition given above it is evident that

1. If A = B, then B = A (Symmetry)
2. A = A, where A is any matrix. (Reflexivity
3. If A = B and B = C, then A = C (Transitivity)

i.e., the relation of equality in the set of all matrices is an equivalence relation.

(j) **Diagonal Matrix:** A square matrix in which all elements except those in the main (or leading) diagonal are zero is know as a diagonal matrix.

For example $\begin{bmatrix} 2 & 0 & 0 \\ 0 & 3 & 0 \\ 0 & 0 & 7 \end{bmatrix}$ is a 3-rowed diagonal matrix.

The um of the diagonal elements of a square matrix A (say) is called the trace of the matrix A.

(k) **Sub-Matrix:** A matrix which is obtained from a given matrix by deleting any number of rows and number of columns is called a sub-matrix of the given matrix.

For example $\begin{bmatrix} 1 & 2 \\ 3 & 4 \end{bmatrix}$ is a sub-matrix of $\begin{bmatrix} 5 & 3 & 2 \\ 1 & 1 & 2 \\ 7 & 3 & 4 \end{bmatrix}$

(l) **Diagonal Element and Orinciple Diagonal:** Those elements a_{ij} of any matrix $[a_{ij}]$ are called diagonal elements for which i = j.

The line along which the above elements lie is called the Principal diagonal or the Diagonal of the matrix.

DIAGONAL MATRIX

Defintion: A square matrix in which all element except those element there in the leading diagonal are zero is called a diagonal matrix. Thus, an n-rowed square matrix $[a_{ij}]$ is a diagonal matrix iff $a_{ij} = 0$ wherever $i \neq j$. If A $[a_{ij}]$ is a diagonal matrix of order n. It must be in the following form.

For eample
$$\begin{bmatrix} a_{11} & 0 & 0 \ldots\ldots & 0 \\ 0 & a_{22} & 0 \ldots\ldots & 0 \\ 0 & 0 & a_{33} \ldots\ldots & \ldots \\ \ldots & \ldots & \ldots\ldots\ldots & \ldots \\ 0 & 0 & 0 \ldots\ldots & a_{nn} \end{bmatrix}$$

Theorem 1:

Any two diangonal matrices of the same order commute under mltiplication.

Proof:

Let any two diagonal matrices be

$$A = \begin{bmatrix} a_1 & 0 & 0 & \ldots & 0 \\ 0 & a_2 & 0 & \ldots & 0 \\ \ldots & \ldots & \ldots & \ldots & \ldots \\ 0 & 0 & 0 & \ldots & a_n \end{bmatrix} \text{ and } B = \begin{bmatrix} b_1 & 0 & 0 & \ldots & 0 \\ 0 & b_2 & 0 & \ldots & 0 \\ \ldots & \ldots & \ldots & \ldots & \ldots \\ 0 & 0 & 0 & \ldots & b_n \end{bmatrix}$$

Then we have

$$AB = \begin{bmatrix} a_1 & 0 & 0 & \ldots & 0 \\ 0 & a_2 & 0 & \ldots & 0 \\ \ldots & \ldots & \ldots & \ldots & \ldots \\ 0 & 0 & 0 & \ldots & a_n \end{bmatrix} \times \begin{bmatrix} b_1 & 0 & 0 & \ldots & 0 \\ 0 & b_2 & 0 & \ldots & 0 \\ \ldots & \ldots & \ldots & \ldots & \ldots \\ 0 & 0 & 0 & \ldots & b_n \end{bmatrix}$$

$$\Rightarrow \quad AB = \begin{bmatrix} a_1b_1 & 0 & 0 & \ldots & 0 \\ 0 & a_2b_2 & 0 & \ldots & 0 \\ \ldots & 0 & \ldots & \ldots & \ldots \\ 0 & 0 & 0 & \ldots & a_nb_n \end{bmatrix} \qquad \ldots(1)$$

$$\text{and}\quad BA = \begin{bmatrix} b_1 & 0 & 0 & \dots & 0 \\ 0 & b_2 & 0 & \dots & 0 \\ \dots & \dots & \dots & \dots & 0 \\ 0 & 0 & 0 & \dots & b_n \end{bmatrix} \times \begin{bmatrix} a_1 & 0 & 0 & \dots & 0 \\ 0 & a_2 & 0 & \dots & 0 \\ \dots & \dots & \dots & \dots & \dots \\ 0 & 0 & 0 & \dots & a_n \end{bmatrix}$$

$$= \begin{bmatrix} b_1a_1 & 0 & 0 & \dots & 0 \\ 0 & b_2a_1 & 0 & \dots & 0 \\ \dots & \dots & \dots & \dots & 0 \\ 0 & 0 & 0 & \dots & b_na_n \end{bmatrix} \qquad \dots\text{(ii)}$$

$\therefore$ From (1) and (2), we find that AB = BA and each one of them is a diagonal matrix. **Hence proved.**

Theorme 2:

Product of any two diagonal matrices of order n is a diagonal matrix of order n.

Proof:

The proof is same as of theorem 1 above. (Do yourself).

Theorem 3:

Sum of any two diagonal matrices of order n is a diagonal matrix of order n and commute under addition.

Proof:

Let any two diagonal matrices be

$$A = \begin{bmatrix} a_1 & 0 & 0 & \dots & 0 \\ 0 & a_2 & 0 & \dots & 0 \\ \dots & \dots & \dots & \dots & \dots \\ 0 & 0 & 0 & \dots & a_n \end{bmatrix} \text{ and } B = \begin{bmatrix} b_1 & 0 & 0 & \dots & 0 \\ 0 & b_2 & 0 & \dots & 0 \\ \dots & \dots & \dots & \dots & \dots \\ 0 & 0 & 0 & \dots & b_n \end{bmatrix}$$

$$\therefore A + B = \begin{bmatrix} a_1 + b_1 & 0 & 0 & \dots & 0 \\ 0 & a_2 + b_2 & 0 & \dots & \dots \\ \dots & \dots & \dots & \dots & 0 \\ 0 & 0 & 0 & \dots & a_n + b_n \end{bmatrix} \qquad \dots(1)$$

$$\text{And } B + A = \begin{bmatrix} b_1 + a_1 & 0 & 0 & \ldots & 0 \\ 0 & b_2 + a_2 & 0 & \ldots & 0 \\ \ldots & \ldots & \ldots & \ldots & 0 \\ 0 & 0 & 0 & \ldots & b_n + a_n \end{bmatrix} \quad \ldots(2)$$

∴ From (1) and (2), we get A + B = B + A and each one of them is a diagonal matrix of odrer n.

SEALAR MATRIX

Definition: If in a square matrix A all the diagonal elements are equal to a (where a ≠ 0) and all the remaining elements are equal to zero then it is called a scalar matrix. Thus, n × n square matrix $[a_{ij}]$ is called a sealar matrix iff for some member a.

For example $\begin{bmatrix} a & 0 & 0 & 0 \\ 0 & a & 0 & 0 \\ 0 & 0 & a & 0 \\ 0 & 0 & 0 & a \end{bmatrix}$ is a scalar matrix of order 4 × 4.

Commutative Matrices

Definition: If A and B are two square matrices such that AB = BA, then A and B are called *commutative* or are said to *commute.*

If **AB = – BA,** the matrices **A** and **B** are said to *anti-commute.*

Example 1:

Show that the matrices A and B anti-commute, where

$$A = \begin{bmatrix} 1 & -1 \\ 2 & -1 \end{bmatrix} \text{ and } B = \begin{bmatrix} 1 & 1 \\ 4 & -1 \end{bmatrix}$$

Solution:

$$\text{Here } AB = \begin{bmatrix} 1 & -1 \\ 2 & -1 \end{bmatrix} \times \begin{bmatrix} 1 & 1 \\ 4 & -1 \end{bmatrix}$$

$$= \begin{bmatrix} 1.1 - 1.4 & 1.1 + (-1).(-1) \\ 2.1 - 1.4 & 2.1 + (-1).(-1) \end{bmatrix} = \begin{bmatrix} -3 & 2 \\ -2 & 3 \end{bmatrix}$$

$$\text{And } BA = \begin{bmatrix} 1 & 1 \\ 4 & -1 \end{bmatrix} \times \begin{bmatrix} 1 & -1 \\ 2 & -1 \end{bmatrix} \quad \ldots(1)$$

$$= \begin{bmatrix} 1.1 + 1.2 & 1.(-1) + 1.(-1) \\ 4.1 + (-1).2 & 4.(-1) + (-1).(-1) \end{bmatrix} = \begin{bmatrix} 3 & -2 \\ 2 & -3 \end{bmatrix}$$

$$= - \begin{bmatrix} -3 & 2 \\ -2 & 3 \end{bmatrix} \qquad ...(2)$$

∴ From (1) and (2) we find that **AB** = – **BA**.

Hence **A** and **B** anti-commute.

Example 2:

If $A = \begin{bmatrix} a & 0 & 0 \\ 0 & a & 0 \\ 0 & 0 & a \end{bmatrix}$ *and* $B = \begin{bmatrix} a_{11} & a_{12} & a_{13} \\ a_{21} & a_{22} & a_{23} \\ a_{31} & a_{32} & a_{33} \end{bmatrix}$

Then prove that AB = BA = aB.

Solution:

$$AB = \begin{bmatrix} a & 0 & 0 \\ 0 & a & 0 \\ 0 & 0 & a \end{bmatrix} \times \begin{bmatrix} a_{11} & a_{12} & a_{13} \\ a_{21} & a_{22} & a_{23} \\ a_{31} & a_{32} & a_{33} \end{bmatrix}$$

$$= \begin{bmatrix} aa_{11} & aa_{12} & aa_{13} \\ aa_{21} & aa_{22} & aa_{23} \\ aa_{31} & aa_{32} & aa_{33} \end{bmatrix} = a \begin{bmatrix} a_{11} & a_{12} & a_{13} \\ a_{21} & a_{22} & a_{23} \\ a_{31} & a_{32} & a_{33} \end{bmatrix}$$

= aB.

$$\text{Similarly } BA = \begin{bmatrix} a_{11} & a_{12} & a_{13} \\ a_{21} & a_{22} & a_{23} \\ a_{31} & a_{32} & a_{33} \end{bmatrix} \times \begin{bmatrix} a & 0 & 0 \\ 0 & a & 0 \\ 0 & 0 & a \end{bmatrix}$$

$$= \begin{bmatrix} aa_{11} & aa_{12} & aa_{13} \\ aa_{21} & aa_{22} & aa_{23} \\ aa_{31} & aa_{32} & aa_{33} \end{bmatrix} = a \begin{bmatrix} a_{11} & a_{12} & aa_{13} \\ a_{21} & a_{22} & a_{23} \\ a_{31} & a_{32} & a_{33} \end{bmatrix}$$

= aB.

Hence AB = BA = aB.

UNIT MATRIX OR IDENTITY MATRIX

Definition: *If in a scalar martix the diagonal element a = 1, then the matrix is called the unit matrix or identity matrix and is denoted by* I_n *in* *the case of n × n matrix. Thus a square matrix $A = [a_{ij}]$ n × n is called an identity or unit matrix iff* $a_{ij}\begin{cases}1 & \text{when } i=j\\ 0 & \text{when } i\neq j\end{cases}$

For example $I_4 = \begin{bmatrix}1 & 0 & 0 & 0\\ 0 & 1 & 0 & 0\\ 0 & 0 & 1 & 0\\ 0 & 0 & 0 & 1\end{bmatrix}$

Example 1:

Prove that $I^m = I^{m-1} = ... = I^2 = I$, where m is any positive integer and I_n is the unit matrix of order n × n.

Solution:

Let A be any n × n matrix and I be the nuit martrix of order n × n i.e., $I = I_n$.

Now we know that $AI_n = I_nA = A$

But $I_n = I$. ...(1)

$\therefore AI = IA = A$

Taking A = I, we have $I \cdot I = I \Rightarrow I^2 = I$ ...(2)

Again form (1), taking $A = I^2$, where $I^2 = I$ (proved), we get

$I^2 \cdot I = I^3 \quad \Rightarrow \quad I^3 = I^2 = I$, from (2).

Proceeding in this way, we can prove that

$I^m = I^{m-1} = ... = I^3 = I^3 = I$, where m is any positive integer.

Example 2:

If A be any n × n matrix and I_n is the identity martix of order n × n, then prove that $AI_n = I_nA = A$.

Solution:

Let us suppose that

$A = \begin{bmatrix}a_{11} & a_{12} & ... & a_{1n}\\ a_{21} & a_{22} & ... & a_{2n}\\ ... & ... & ... & ...\\ a_{n1} & a_{n2} & ... & a_{nn}\end{bmatrix}$ and $I_n = \begin{bmatrix}1 & 0 & ... & 0\\ 0 & 1 & ... & 0\\ ... & ... & ... & 0\\ 0 & 0 & ... & 1\end{bmatrix}$

$$\therefore A \cdot I_n = \begin{bmatrix} a_{11} & a_{12} & \cdots & a_{1n} \\ a_{21} & a_{22} & \cdots & a_{2n} \\ \cdots & \cdots & \cdots & \cdots \\ a_{n1} & a_{n2} & \cdots & a_{nn} \end{bmatrix} \times \begin{bmatrix} 1 & 0 & \cdots & 0 \\ 0 & 1 & \cdots & 0 \\ \cdots & \cdots & \cdots & 0 \\ 0 & 0 & \cdots & 1 \end{bmatrix}$$

$$= \begin{bmatrix} a_{11}.1 + a_{12}.0 + \ldots + a_{1n}.0 & a_{11}.0 + a_{12}.1 + \ldots + a_{1n}.0 & \cdots & \cdots & a_{11}.0 + a_{12}.0 + \ldots + a_{1n}.1 \\ a_{21}.1 + a_{22}.0 + \ldots + a_{2n}.0 & a_{21}.0 + a_{22}.1 + \ldots + a_{2n}.0 & \cdots & \cdots & a_{21}.0 + a_{22}.0 + \ldots \div a_{1n}.1 \\ \cdots & \cdots & \cdots & \cdots & \cdots \\ a_{n1}.1 + a_{n2}.0 + \ldots + a_{nn}.0 & a_{n1}.0 + a_{n2}.1 + \ldots + a_{nn}.0 & \cdots & \cdots & a_{n1}.0 + a_{n2}.0 + \ldots + a_{nn}.1 \end{bmatrix}$$

$$= \begin{bmatrix} a_{11} & a_{12} & \cdots & a_{1n} \\ a_{21} & a_{22} & \cdots & a_{2n} \\ \cdots & \cdots & \cdots & \cdots \\ a_{n1} & a_{n2} & \cdots & a_{nn} \end{bmatrix} = A$$

Similarly we can show that $I_n \cdot A = A$.

Hence we have $A \cdot I_n = I_n \cdot A = A$.

PERIODIC MATRIX

Definition: *A square matrix A is called periodic, if* $A^{k+1} = A$, *where k is a positive integer.*

If k is the least positive integer for which $A^{k+1} = A$, *then A is said to be of* **period k.**

Idempotent Matrix

Definition: A square matrix A is called idempotent provided it satisfies the relation $A^2 = A$.

SYMMETRIC IDEMPOTENT MATRIX

Definition: *A square matrix A is called symmetric idempotent if* $A = A'$ *and* $A^2 = A$, *where A' is the transposed matrix of A.*

Example:

Show that if A and B are matrices of order $n \times n$ *and such that* $AB = A$ *and* $BA = B$, *then A and B are idempotent martices.*

Solution:

We have ABA = (AB) A = (A) A, $\because$ AB = A (given)

$\Rightarrow$ ABA = A^2 ...(1)

Also ABA = A (BA) = A (B) $\because$ BA = B (given)

= AB = A, $\because$ AB = A (given)

$\Rightarrow$ ABA = A ...(2)

From (1) and (2), we have $A^2 = A$ i.e., A is idempotent.

In a similar manner, we can prove that

BAB = B (AB) = B (A), $\because$ AB = A (given)

= BA = B, $\because$ BA = B (given)

$\Rightarrow$ BAB = B ...(3)

Also BAB = (BA) B = (B) B, $\because$ BA = B (given)

$\Rightarrow$ BAB = B^2 ...(4)

$\therefore$ From (3) and (4), we have $B^2 = B$ *i.e.,* B is idempotent.

SOLVED EXAMPLES

Example 1:

If $A = \begin{bmatrix} \cos\theta & -\sin\theta \\ \sin\theta & \cos\theta \end{bmatrix}$, $B = \begin{bmatrix} \cos\phi & -\sin\phi \\ \sin\phi & \cos\phi \end{bmatrix}$ *show that AB = BA.*

Solution:

$$AB = \begin{bmatrix} \cos\theta & -\sin\theta \\ \sin\theta & \cos\theta \end{bmatrix} \times \begin{bmatrix} \cos\phi & -\sin\phi \\ \sin\phi & \cos\phi \end{bmatrix}$$

$$= \begin{bmatrix} \cos\theta\cos\phi - \sin\theta\sin\phi & -\cos\theta\sin\phi - \sin\theta\cos\phi \\ \sin\theta\cos\phi + \cos\theta\sin\phi & -\sin\theta\sin\phi + \cos\theta\cos\phi \end{bmatrix}$$

$$= \begin{bmatrix} \cos(\theta+\phi) & -\sin(\theta+\phi) \\ \sin(\theta+\phi) & \cos(\theta+\phi) \end{bmatrix} \quad ...(1)$$

$$\text{And } BA = \begin{bmatrix} \cos\phi & -\sin\phi \\ \sin\phi & \cos\phi \end{bmatrix} \times \begin{bmatrix} \cos\theta & -\sin\theta \\ \sin\theta & \cos\theta \end{bmatrix}$$

$$= \begin{bmatrix} \cos\phi\cos\theta - \sin\phi\sin\theta & -\cos\phi\sin\theta - \sin\phi\cos\theta \\ \sin\phi\cos\theta + \cos\phi\sin\theta & -\sin\phi\sin\theta + \cos\phi\cos\theta \end{bmatrix}$$

$$= \begin{bmatrix} \cos(\theta+\phi) & -\sin(\theta+\phi) \\ \sin(\theta+\phi) & \cos(\theta+\phi) \end{bmatrix}$$

$\therefore$ From (1) and (2) we get AB = BA. **Hence proved.**

Example 2:

If $A = \begin{bmatrix} 1 & 5 & 6 \\ -6 & 7 & 0 \end{bmatrix}$ and $B = \begin{bmatrix} 1 & -5 & 7 \\ 8 & -7 & 7 \end{bmatrix}$ *then find A + B and A – B.*

Solution:

$$A + B = \begin{bmatrix} 1 & 5 & 6 \\ -6 & 7 & 0 \end{bmatrix} + \begin{bmatrix} 1 & -5 & 7 \\ 8 & -7 & 7 \end{bmatrix}$$

$$= \begin{bmatrix} 1+1 & 5-5 & 6+7 \\ -6+8 & 7-7 & 0+7 \end{bmatrix} = \begin{bmatrix} 2 & 0 & 13 \\ 2 & 0 & 7 \end{bmatrix}$$ **Ans.**

$$\text{and } A - B = \begin{bmatrix} 1 & 5 & 6 \\ -6 & 7 & 0 \end{bmatrix} - \begin{bmatrix} 1 & -5 & 7 \\ 8 & -7 & 7 \end{bmatrix}$$

$$= \begin{bmatrix} 1-1 & 5-(-5) & 6-7 \\ -6-7 & 8-7-(-7) & 0-7 \end{bmatrix} = \begin{bmatrix} 0 & 10 & -1 \\ -14 & 14 & -7 \end{bmatrix}$$ **Ans.**

Example 3:

Find the product of the following two matrices

$$\begin{bmatrix} 0 & c & -b \\ -c & 0 & a \\ b & -a & 0 \end{bmatrix} \times \begin{bmatrix} a^2 & ab & ac \\ ab & b^2 & bc \\ ac & bc & c^2 \end{bmatrix}$$

Solution:

The required product

$$= \begin{bmatrix} 0 & c & -b \\ -c & 0 & a \\ b & -a & 0 \end{bmatrix} \times \begin{bmatrix} a^2 & ab & ac \\ ab & b^2 & bc \\ ac & bc & c^2 \end{bmatrix}$$

$$= \begin{bmatrix} 0.a^2 + c.ab - b.ac & 0.ab + c.b^2 - b.bc & 0ac + c.bc - b.c^2 \\ -ca^2 + 0.ab\ a.ac & -c.ab + 0b^2 + a.bc & -c.ac + 0.bc + a.c^2 \\ b.a^2 - a.ab + 0.ac & b.ab - a.b^2 + 0.bc & b.ac - a.bc + 0.c^2 \end{bmatrix}$$

$$= \begin{bmatrix} 0 & 0 & 0 \\ 0 & 0 & 0 \\ 0 & 0 & 0 \end{bmatrix}$$

Example 4:

Evaluate A^3 if $A = \begin{bmatrix} \cosh\theta & \sinh\theta \\ \sinh\theta & \cosh\theta \end{bmatrix}$

Solution:

$$A^2 = \begin{bmatrix} \cosh\theta & \sinh\theta \\ \sinh\theta & \cosh\theta \end{bmatrix} \times \begin{bmatrix} \cosh\theta & \sinh\theta \\ \sinh\theta & \cosh\theta \end{bmatrix}$$

$$= \begin{bmatrix} \cosh^2\theta + \sinh^2\theta & \cosh\theta\sinh\theta + \sinh\theta\cosh\theta \\ \sinh\theta\cosh\theta + \cosh\theta\sinh\theta & \sinh^2\theta + \cosh\theta \end{bmatrix}$$

$$= \begin{bmatrix} \cosh 2\theta & \sinh 2\theta \\ \sinh 2\theta & \cosh 2\theta \end{bmatrix} \because \begin{matrix} \cosh^2\theta + \sinh^2\theta = \cosh 2\theta \\ 2\cosh\theta\cosh\theta = \sinh 2\theta \end{matrix}$$

$$\therefore \quad A^3 = A^2A = \begin{bmatrix} \cosh 2\theta & \sinh 2\theta \\ \sinh 2\theta & \cosh 2\theta \end{bmatrix} \begin{bmatrix} \cosh 2\theta & \sinh 2\theta \\ \sinh 2\theta & \cosh 2\theta \end{bmatrix}$$

$$= \begin{bmatrix} \cosh 2\theta\cosh\theta + \sinh 2\theta\sinh\theta & \cosh 2\theta\sinh\theta + \sinh 2\theta\cosh\theta \\ \sinh 2\theta\cosh\theta + \cosh 2\theta\sinh\theta & \sinh 2\theta\sinh\theta + \cosh 2\theta\cosh\theta \end{bmatrix}$$

$$= \begin{bmatrix} \cosh(2\theta + \theta) & \sinh(2\theta + \theta) \\ \sinh(2\theta + \theta) & \cosh(2\theta + \theta) \end{bmatrix},$$

$\because$ sinh (A + B) = sinh A cosh B + cosh A sin B

sinh (A + B) = cosh A cosh B + sinh A sin B

$$= \begin{bmatrix} \cosh 3\theta & \sinh 2\theta \\ \sinh 3\theta & \cosh 2\theta \end{bmatrix}$$

Example 5(a):

If A, B, C are three matrices such that

$A = [x, y, z]$, $B = \begin{bmatrix} a & h & g \\ h & b & f \\ g & f & c \end{bmatrix}$, $C = \begin{bmatrix} x \\ y \\ z \end{bmatrix}$, *evaluate ABC.*

Solution:

$$B = [x, y, z] \times \begin{bmatrix} a & h & g \\ h & b & f \\ g & f & c \end{bmatrix}$$

$$= [x.a + y.h + z.g \quad x.h + y.b + z.f \quad x.g + y.f + z.c]$$

$$\Rightarrow \quad ABC = [ax + hy + gz \quad hx + by + fz \quad gx + fy + cz \times \begin{bmatrix} x \\ y \\ z \end{bmatrix}$$

$$= [x (ax + hy + gz) + y (hx + by + fz) + z (gx + fy + cz)] \text{ (Note)}$$

$$= [ax^2 + by^2 + cz^2 + 2hxy + 2gzx + 2fyz]. \qquad \textbf{Ans.}$$

Example 5(b):

Find the values of x, y, z in the following equation

$$\begin{bmatrix} 1 & 2 & 3 \\ 3 & 1 & 2 \\ 2 & 3 & 1 \end{bmatrix} \times \begin{bmatrix} x \\ y \\ z \end{bmatrix} = \begin{bmatrix} 4 & -2 \\ 0 & -6 \\ -1 & 2 \end{bmatrix} \times \begin{bmatrix} 2 \\ 1 \end{bmatrix}$$

Solution:

$$\begin{bmatrix} 1 & 2 & 3 \\ 3 & 1 & 2 \\ 2 & 3 & 1 \end{bmatrix} \times \begin{bmatrix} x \\ y \\ z \end{bmatrix} = \begin{bmatrix} 1.x + 2y + 2z \\ 3x + 1.y + 2z \\ 2x + 3y + 1.z \end{bmatrix} \qquad ...(1)$$

$$\text{And} \begin{bmatrix} 4 & -2 \\ 0 & -6 \\ -1 & 2 \end{bmatrix} \times \begin{bmatrix} 2 \\ 1 \end{bmatrix} = \begin{bmatrix} 4.2 + (-2).1 \\ 0.2 + (-6).1 \\ -1.2 + \quad 2.1 \end{bmatrix} = \begin{bmatrix} 6 \\ -6 \\ 0 \end{bmatrix} \qquad .(2)$$

With the help of (1) and (2), the given equation reduces to

$$\begin{bmatrix} x + 2y + 3z \\ 3x + \ y + 2z \\ 2x + 3y + \ z \end{bmatrix} = \begin{bmatrix} 6 \\ -6 \\ 0 \end{bmatrix}$$

From this on comparing the corresponding elements on both sides we get $x + 2y + 3z = 6$; $3x + y + 2z = -6$ and $2x + 3y + z = 0$.

Solving these we get $x = -4$, $y = 2$, $z = 2$. **Ans.**

Example 5(c):

If $A = \begin{bmatrix} 1 & 2 & 3 \\ 0 & 1 & 2 \\ 0 & 0 & 1 \end{bmatrix}$; $B = \begin{bmatrix} x \\ y \\ z \end{bmatrix}$ and $AB = \begin{bmatrix} 6 \\ 3 \\ 1 \end{bmatrix}$ *find the values of* x, y, z.

Solution:

$$AB = \begin{bmatrix} 1 & 2 & 3 \\ 0 & 1 & 2 \\ 0 & 0 & 1 \end{bmatrix} \times \begin{bmatrix} x \\ y \\ z \end{bmatrix}$$

$$\Rightarrow \quad \begin{bmatrix} 6 \\ 3 \\ 1 \end{bmatrix} = \begin{bmatrix} 1.x + 2.y + 3z \\ 0.x + 1.y + 2z \\ 0.x + 0.y + 1.z \end{bmatrix}$$

$\Rightarrow \quad 6 = x + 2y + 3z,\ 3 = y + 2z,\ 1 = z,$ **(Note)**

comparing the corresponding elements of the matrices on both sides.

Solving these we get x = 1, y = 1, z = 1. **Ans.**

Example 6:

If $A = \begin{bmatrix} i & 0 \\ 0 & -i \end{bmatrix}$, $B = \begin{bmatrix} 0 & -1 \\ 1 & 0 \end{bmatrix}$, $C = \begin{bmatrix} 0 & i \\ i & 0 \end{bmatrix}$, *prove that*

$A^2 = B^2 = C^2 = -I$ *and* $AB = -C = -BA$, *where* $I = \begin{bmatrix} 1 & 1 \\ 0 & 1 \end{bmatrix}$

Solution:

$$A^2 = \begin{bmatrix} i & 0 \\ 0 & -i \end{bmatrix} \times \begin{bmatrix} i & 0 \\ 0 & -i \end{bmatrix}$$

$$= \begin{bmatrix} i.i + 0.0 & i.0 + (-1) \\ 0.i - i.0 & 0.0 + (-i)(-i) \end{bmatrix} = \begin{bmatrix} -1 & 0 \\ 0 & -1 \end{bmatrix}$$

$$= \begin{bmatrix} 1 & 0 \\ 0 & 1 \end{bmatrix}$$

$$= -I.$$

and $B^2 = \begin{bmatrix} 1 & -1 \\ 0 & 0 \end{bmatrix} \times \begin{bmatrix} 1 & -1 \\ 0 & 0 \end{bmatrix} = \begin{bmatrix} 0.0 - 1.1 & 0.(-1) + (-1).0 \\ 1.0 + 0.1 & 1(-1) + 0.0 \end{bmatrix}$

$$= \begin{bmatrix} -1 & 0 \\ 0 & -1 \end{bmatrix} = -\begin{bmatrix} 1 & 0 \\ 0 & 1 \end{bmatrix} = -1$$

Similarly we can prove that $C^2 = -I$. Hence $A^2 = B^2 = C^2 = -I$.

Again $AB = \begin{bmatrix} i & 0 \\ 0 & -i \end{bmatrix} \times \begin{bmatrix} 0 & -1 \\ 1 & 0 \end{bmatrix}$

$$= \begin{bmatrix} i.0 + 0.1 & i(-1) + 0.0 \\ 0.0 - i(1) & 0(-1) - i.0 \end{bmatrix}$$

$$= \begin{bmatrix} 0 & -i \\ -i & 0 \end{bmatrix} = -\begin{bmatrix} 0 & i \\ i & 0 \end{bmatrix} = -C$$

and $BA = \begin{bmatrix} 0 & -1 \\ 1 & 0 \end{bmatrix} \times \begin{bmatrix} i & 0 \\ 0 & -i \end{bmatrix}$

$$= \begin{bmatrix} 0.i - i.o & 0.0 - 1(-i) \\ 1.i + 0.0 & 1.0 + 0(-1) \end{bmatrix}$$

$$= \begin{bmatrix} 0 & i \\ i & 0 \end{bmatrix} = C$$

Hence $AB = -C = -BA$.

Example 7:

If $\begin{bmatrix} 4 \\ 1 \\ 3 \end{bmatrix} A = \begin{bmatrix} -4 & 8 & 4 \\ -1 & 2 & 1 \\ -3 & 6 & 3 \end{bmatrix}$, *find A.*

Solution:

We know that if X is an $m \times n$ matrix, Y is an $n \times k$ matrix, then the product XY is an $m \times k$ matrix.

Hence $\begin{bmatrix} 4 \\ 1 \\ 3 \end{bmatrix}$ is 3×1 matrix and $\begin{bmatrix} -4 & 8 & 4 \\ -1 & 2 & 1 \\ -3 & 6 & 3 \end{bmatrix}$

is 3×3 matrix, os A must be a 1×3 matrix *i.e.*, a row matrix. **(Note)**

$\therefore$ Let A = [a, b, c]

Then $\begin{bmatrix} 4 \\ 1 \\ 3 \end{bmatrix} \times [a\ b\ c] = \begin{bmatrix} -4 & 8 & 4 \\ -1 & 2 & 1 \\ -3 & 6 & 3 \end{bmatrix}$

Which gives $\begin{bmatrix} 4a & 4b & 4c \\ a & b & c \\ 3a & 3b & 3c \end{bmatrix} = \begin{bmatrix} -4 & 8 & 4 \\ -1 & 2 & 1 \\ -3 & 6 & 3 \end{bmatrix}$

Comparing corresponding elements we have

$$4a = -4,\ a = -1,\ 3a = -3,\ 4b = 8,\ b = 2,\ 3b = 6$$

and $4c = 4,\ c = 1,\ 3c = 3.$

All these are satisfied by $a = -1,\ b = 2,\ c = 1.$

Hence from (1) we have

$A = [a\ b\ c] = [-1, 2, 1].$ **Ans.**

Example 8:

Given $A_i = \begin{bmatrix} 0 & 0 & 0 & 1 \\ 0 & 0 & 1 & 0 \\ 0 & 1 & 0 & 0 \\ 1 & 0 & 0 & 0 \end{bmatrix}$; $A_2 = \begin{bmatrix} 0 & 0 & 0 & i \\ 0 & 0 & -i & 0 \\ 0 & i & 0 & 0 \\ -i & 0 & 0 & 0 \end{bmatrix}$

$A_2 = \begin{bmatrix} 0 & 0 & 1 & 0 \\ 0 & 0 & 0 & -1 \\ 1 & 0 & 0 & 0 \\ 0 & -1 & 0 & 0 \end{bmatrix}$ and $A_4 = \begin{bmatrix} 1 & 0 & 0 & 0 \\ 0 & 1 & 0 & 0 \\ 0 & 0 & -1 & 0 \\ 0 & 0 & 0 & -1 \end{bmatrix}$

Show that $A_i A_k + A_k A_i = 2I$ or O according as $i = k$ or $i \neq k$ and I is the matrix of order 4 and i and k take the values 1, 2, 3 and 4.

Solution:

Let $i = k = 1$ (say). Then

$A_iA_k = A_1A_2 = A_kA_i$

$\therefore A_iA_k = A_1A_1 = \begin{bmatrix} 0 & 0 & 0 & 1 \\ 0 & 0 & 1 & 0 \\ 0 & 1 & 0 & 0 \\ 1 & 0 & 0 & 0 \end{bmatrix} \times \begin{bmatrix} 0 & 0 & 0 & 1 \\ 0 & 0 & 1 & 0 \\ 0 & 1 & 0 & 0 \\ 1 & 0 & 0 & 0 \end{bmatrix}$

$$= \begin{bmatrix} 0+0+0+1 & 0+0+0+0 & 0+0+0+0 & 0+0+0+0 \\ 0+0+0+0 & 0+0+1+0 & 0+0+0+0 & 0+0+0+0 \\ 0+0+0+0 & 0+0+0+0 & 0+1+0+0 & 0+0+0+0 \\ 0+0+0+0 & 0+0+0+0 & 0+0+0+0 & 1+0+0+0 \end{bmatrix}$$

$$= \begin{bmatrix} 1 & 0 & 0 & 0 \\ 0 & 1 & 0 & 0 \\ 0 & 0 & 1 & 0 \\ 0 & 0 & 0 & 1 \end{bmatrix} = I$$

$\therefore\quad A_iA_k + A_kA_i = I + I = 2I$ **Hence proved.**

If i ≠ k, let i = 3 and k = 2

Then $A_iA_k = A_3A_2 = \begin{bmatrix} 0 & 0 & 1 & 0 \\ 0 & 1 & 0 & -1 \\ 1 & 0 & 0 & 0 \\ 0 & -1 & 0 & 0 \end{bmatrix} \times \begin{bmatrix} 0 & 0 & 0 & i \\ 0 & 0 & -i & 0 \\ 0 & i & 0 & 0 \\ -i & 0 & 0 & 0 \end{bmatrix}$

$$= \begin{bmatrix} 0+0+0+0 & 0+0+i+0 & 0+0+0+0 & 0+0+0+0 \\ 0+0+0+i & 0+0+0+0 & 0+0+0+0 & 0+0+0+0 \\ 0+0+0+0 & 0+0+0+0 & 0+0+0+0 & i+0+0+0 \\ 0+0+0+0 & 0+0+0+0 & 0+i+0+0 & 0+0+0+0 \end{bmatrix}$$

$$= \begin{bmatrix} 0 & i & 0 & 0 \\ i & 0 & 0 & 0 \\ 0 & 0 & 0 & i \\ 0 & 0 & i & 0 \end{bmatrix} = i \begin{bmatrix} 0 & 1 & 0 & 0 \\ 1 & 0 & 0 & 0 \\ 0 & 0 & 0 & 1 \\ 0 & 0 & 1 & 0 \end{bmatrix}$$

And $A_kA_i = A_2A_3 = \begin{bmatrix} 0 & 0 & 0 & i \\ 0 & 0 & -i & 0 \\ 0 & i & 0 & 0 \\ -i & 0 & 0 & 0 \end{bmatrix} \times \begin{bmatrix} 0 & 0 & 1 & 0 \\ 0 & 0 & 0 & -1 \\ 1 & 0 & 0 & 0 \\ 0 & -1 & 0 & 0 \end{bmatrix}$

$= \begin{bmatrix} 0 & -i & 0 & 0 \\ -i & 0 & 0 & 0 \\ 0 & 0 & 0 & -i \\ 0 & 0 & -i & 0 \end{bmatrix}$ Multiplying in the usual way

$$= -i \begin{bmatrix} 0 & 1 & 0 & 0 \\ 1 & 0 & 0 & 0 \\ 0 & 0 & 0 & 1 \\ 0 & 0 & 1 & 0 \end{bmatrix}$$

$$\therefore A_iA_k + A_kA_i = i \begin{bmatrix} 0 & 1 & 0 & 0 \\ 1 & 0 & 0 & 0 \\ 0 & 0 & 0 & 1 \\ 0 & 0 & 1 & 0 \end{bmatrix} - i \begin{bmatrix} 0 & 1 & 0 & 0 \\ 1 & 0 & 0 & 0 \\ 0 & 0 & 0 & 1 \\ 0 & 0 & 1 & 0 \end{bmatrix}$$

$$= O$$

Hence proved.

We can in a similar way prove the above result by giving i and k other values also.

Example 9(a):

If a is an idempotent matrix, then the matrix $B = I - A$ is idempotent and $AB = O = BA$.

Solution:

We know $IA = AI = A$. ...(1)

Also A being an idempotent matrix, we have $A^2 = A$. ...(2)

Since I and A are square matrices, so I – A is also a square matrix and therefore we have

$$(I - A)^2 = (I - A)(I - A)$$
$$= (I - A) I - (I - A) A, \text{ by distributive law}$$
$$= I^2 - AI - IA + A^2$$
$$= I - A - A + A, \text{ from (1), (2) and } I^2 = I$$

$\Rightarrow$ $(I - A)^2 = I - A$, i.e., I – A or B is an idempotent matrix by definition.

Again $AB = A(I - A) = AI - A^2$, by distributive law

$= A - A$, from (1) and (2)

i.e., $AB = O$.

And $BA = (I - A) A = IA - A^2$, by distributive law

$= A - A$, from (1) and (2)

$\Rightarrow$ $BA = O$. **Hence proved.**

Example 9(b):

If A and B are idempotent martices, then show that AB is idempotent if A and B commute.

Solution:

If A is idempotent, then $A^2 = A$ and if B is idempotent, then $B^2 = B$. ...(1)

And if A and B commute, then $AB = BA$ ...(2)

Now $(AB)^2 = (AB).(AB)$

$= A(BA)B$, by associative law

$= A\,(AB)\,B$, from (2)

$= (AA)\,(BB)$, by associative law

$= A^2B^2$

$= AB$, by (1)

Hence ÀB is idempotent.

Example 10:

Show that the matrix $A = \begin{bmatrix} 2 & -2 & -4 \\ -1 & 3 & 4 \\ 1 & -2 & -3 \end{bmatrix}$ *is idempotent.*

Solution:

$$A^2 = A \cdot A = \begin{bmatrix} 2 & -2 & -4 \\ -1 & 3 & 4 \\ 1 & -2 & -3 \end{bmatrix} \times \begin{bmatrix} 2 & -2 & -4 \\ -1 & 3 & 4 \\ 1 & -2 & -3 \end{bmatrix}$$

$$= \begin{bmatrix} 2.2 - 2(-1) - 4.1 & 2(-2) - 2.3 - 4(-2) & 2(-4) - 2.4 - 4(-3) \\ -1.2 + 3(-1) + 4.1 & -1(-2) + 3.3 + 4(-2) & -1(-4) + 3.4 \div 4(-3) \\ 1.2 - 2(-1) - 3.1 & 1(-2) - 2.3 - 3(-2) & 1(-4) - 2.4 - 3(-3) \end{bmatrix}$$

$$= \begin{bmatrix} 2 & -2 & -4 \\ -1 & 3 & 4 \\ 1 & -2 & -3 \end{bmatrix} = A$$

Hence the matrix A is idempotent.

Example 11:

Show that the matrix $A = \begin{bmatrix} -5 & -8 & 0 \\ 3 & 5 & 0 \\ 1 & 2 & -1 \end{bmatrix}$ *is involutory,*

Solution:

$$A^2 = \begin{bmatrix} -5 & -8 & 0 \\ 3 & 5 & 0 \\ 1 & 2 & -1 \end{bmatrix} \times \begin{bmatrix} -5 & -8 & 0 \\ 3 & 5 & 0 \\ 1 & 2 & -1 \end{bmatrix}$$

$$= \begin{bmatrix} (-5).(-5)+(-8).3+0.1 & (-5).(-8)+(-8)5+0.2 & (-5).0+(-8).0+0\,(-1) \\ 3.(-5)+5.3+0.1 & 3.(-8)+5.5+0.2 & 3.0+5.0+0.(-1) \\ 1.(-5)+2.3+(-1).1 & 1.(-8)+2.5+(-1).2 & 1.0+2.0+(-1)\,(-1) \end{bmatrix}$$

$$= \begin{bmatrix} 25-24+0 & -40-40+0 & 0+0+0 \\ -15+15+0 & -24+25+0 & 0+0+0 \\ -5+6-1 & -8+10-2 & 0+0+1 \end{bmatrix} = \begin{bmatrix} 1 & 0 & 0 \\ 0 & 1 & 0 \\ 0 & 0 & 1 \end{bmatrix} = I$$

Hence the given matrix a is involutory.

Example 12:

If $A = \begin{bmatrix} 2 & 3 & 1 \\ 0 & -1 & 5 \end{bmatrix}$ and $B = \begin{bmatrix} 1 & 2 & -6 \\ 0 & -1 & 3 \end{bmatrix}$ *evaluate (a)* $A^2 - B^2$ *and (b) AB and BA.*

Solution:

(a) $A^2 = \begin{bmatrix} 2 & 3 & 1 \\ 0 & -1 & 5 \end{bmatrix} \times \begin{bmatrix} 2 & 3 & 1 \\ 0 & -1 & 5 \end{bmatrix}$, which does not exist as number of columns in the first matrix is not equal to number of rows in the second matrix.

Similarly B^2 does not exist.

(b) AB and BA both do not exist, the reason being the same as in part (a) above.

Example 13:

Solve the following equations for A and B:

$2A - B = \begin{bmatrix} 3 & -3 & 0 \\ 3 & 3 & 2 \end{bmatrix}$, $2B + A = \begin{bmatrix} 4 & 1 & 5 \\ -1 & 4 & -4 \end{bmatrix}$

Solution:

Given $2A - B = \begin{bmatrix} 3 & -3 & 0 \\ 3 & 3 & 2 \end{bmatrix}$

Multiplying both sides by 2, we get

$$4A - 2B = 2\begin{bmatrix} 3 & -3 & 0 \\ 3 & 3 & 2 \end{bmatrix} = \begin{bmatrix} 6 & -6 & 0 \\ 6 & 6 & 4 \end{bmatrix} \quad ...(1)$$

Also given that $2B + A = \begin{bmatrix} 4 & 1 & 5 \\ -1 & 4 & -4 \end{bmatrix}$ ...(2)

Adding (1) and (2) we get

$$5A = \begin{bmatrix} 6 & -6 & 0 \\ 6 & 6 & 4 \end{bmatrix} + \begin{bmatrix} 4 & 1 & 5 \\ -1 & 4 & -4 \end{bmatrix}$$

$$= \begin{bmatrix} 6+4 & -6+1 & 0+5 \\ 6-1 & 6+4 & 4+4 \end{bmatrix} = \begin{bmatrix} 10 & -5 & 5 \\ 5 & 10 & 0 \end{bmatrix}$$

$$\Rightarrow \quad A = \frac{1}{2}\begin{bmatrix} 10 & -5 & 5 \\ 5 & 10 & 0 \end{bmatrix} = \begin{bmatrix} 2 & -1 & 1 \\ 1 & 2 & 0 \end{bmatrix}$$

Again from (2) we get

$$2B = \begin{bmatrix} 4 & 1 & 5 \\ -1 & 4 & -4 \end{bmatrix} - A$$

$$\Rightarrow \quad 2B = \begin{bmatrix} 4 & 1 & 5 \\ -1 & 4 & -4 \end{bmatrix} - \begin{bmatrix} 2 & -1 & 1 \\ 1 & 2 & 0 \end{bmatrix}$$

$$= \begin{bmatrix} 4-2 & 1+1 & 5-1 \\ -1-1 & 4-2 & -4-0 \end{bmatrix} = \begin{bmatrix} 2 & 2 & 4 \\ -2 & 2 & -4 \end{bmatrix}$$

$$\Rightarrow \quad B = \frac{1}{2}\begin{bmatrix} 2 & 2 & 4 \\ -2 & 2 & -4 \end{bmatrix} = \begin{bmatrix} 1 & 1 & 2 \\ -1 & 1 & -2 \end{bmatrix}.$$ **Ans.**

Example 14(a):

Given $A = \begin{bmatrix} 1 & 2 & -3 \\ 5 & 0 & 2 \\ 1 & -1 & 1 \end{bmatrix}$ and $B = \begin{bmatrix} 3 & -1 & 2 \\ 4 & 2 & 5 \\ 2 & 0 & 3 \end{bmatrix}$,

find the matrix C such that A + 2C = B.

Solution:

Given that A + 2C = B or 2= B – A

$$\Rightarrow \quad 2C = \begin{bmatrix} 3 & -1 & 2 \\ 4 & 2 & 5 \\ 2 & 0 & 3 \end{bmatrix} - \begin{bmatrix} 1 & 2 & -3 \\ 5 & 0 & 2 \\ 1 & -1 & 1 \end{bmatrix}$$

$$= \begin{bmatrix} 3-1 & -1-2 & 2-(-3) \\ 4-5 & 2-0 & 5-2 \\ 2-1 & 0-(-1) & 3-1 \end{bmatrix} = \begin{bmatrix} 2 & -3 & 5 \\ -1 & 2 & 3 \\ 1 & 1 & 2 \end{bmatrix}$$

$$\Rightarrow \quad C = \frac{1}{2}\begin{bmatrix} 2 & -3 & 0 \\ -1 & 2 & 3 \\ 1 & 1 & 2 \end{bmatrix} = \begin{bmatrix} 1 & -\frac{3}{2} & \frac{5}{2} \\ -\frac{1}{2} & 1 & \frac{3}{2} \\ \frac{1}{2} & \frac{1}{2} & 1 \end{bmatrix}$$

Example 14(b):

If $A = \begin{bmatrix} 2 & 3 & 1 \\ 0 & -1 & 5 \end{bmatrix}$ and $B = \begin{bmatrix} 1 & 2 & -6 \\ 0 & -1 & 3 \end{bmatrix}$ *evaluate 3A – 4B.*

Solution:

$$2A - 4B = 3\begin{bmatrix} 2 & 3 & 1 \\ 0 & -1 & 5 \end{bmatrix} - 4\begin{bmatrix} 1 & 2 & -6 \\ 0 & -1 & 3 \end{bmatrix}$$

$$= \begin{bmatrix} 6 & 9 & 3 \\ 0 & -3 & 15 \end{bmatrix} - \begin{bmatrix} 4 & 8 & -24 \\ 0 & -4 & 12 \end{bmatrix}$$

$$= \begin{bmatrix} 6-4 & 9-8 & 3-(-24) \\ 0-0 & -3-(-4) & 15-12 \end{bmatrix}$$

$$= \begin{bmatrix} 2 & 1 & 27 \\ 0 & 1 & 3 \end{bmatrix}.$$

Example 14(c):

Calculate AB and BA if

$$A = \begin{bmatrix} 1 & -1 & 1 \\ -3 & 2 & -1 \\ -2 & 1 & 0 \end{bmatrix}, B = \begin{bmatrix} 1 & 2 & 3 \\ 2 & 4 & 6 \\ 1 & 2 & 3 \end{bmatrix}$$

Solution:

$$AB = \begin{bmatrix} 1 & -1 & 1 \\ -3 & 2 & -1 \\ -2 & 1 & 0 \end{bmatrix} \times \begin{bmatrix} 1 & 2 & 3 \\ 2 & 4 & 6 \\ 1 & 2 & 3 \end{bmatrix}$$

$$= \begin{bmatrix} 1.1 + (-1).2 + 1.1 & 1.2 \ (-1).4 + 1.2 & 1.3 + (-1).6 + 1.3 \\ (-3).1 + 2.2 + (-1).1 & (-3).2 + 2.4 + (-1).2 & (-3).3 + 2.6 + (-1).3 \\ (-2).1 + 1.2 + 0.1 & (-2).2 + 1.4 + 0.2 & (-2).3 + 1.6 + 0.3 \end{bmatrix}$$

$$= \begin{bmatrix} 0 & 0 & 0 \\ 0 & 0 & 0 \\ 0 & 0 & 0 \end{bmatrix} = O, \text{ where O is } 3 \times 3 \text{ null matrix.}$$ **Ans.**

$$\text{And BA} = \begin{bmatrix} 1 & 2 & 3 \\ 2 & 4 & 6 \\ 1 & 2 & 3 \end{bmatrix} \times \begin{bmatrix} 1 & -1 & 1 \\ -3 & 2 & -1 \\ -2 & 1 & 0 \end{bmatrix}$$

$$= \begin{bmatrix} 1.1 + 2.(-3) + 3(-2) & 1(-1) + 2.2 + 3.1 & 1.1 + 2.(-1) + 3.0 \\ 2.1 + 4.(-3) + 6(-2) & 2(-1) + 4.2 + 6.1 & 2.1 + 4.(-1) + 6.0 \\ 1.1 + 2.(-3) + 3(-2) & 1(-1) + 2.2 + 3.1 & 1.1 + 2.(-1) + 30 \end{bmatrix}$$

$$= \begin{bmatrix} -11 & 6 & -1 \\ -22 & 12 & -2 \\ -11 & 6 & -1 \end{bmatrix}$$ **Ans.**

[**Note:** AB $\neq$ BA].

Example 14(d):

If $A = \begin{bmatrix} 2 & 3 & 4 \\ 1 & 2 & 3 \\ -1 & 1 & 2 \end{bmatrix}$ and $B = \begin{bmatrix} 1 & 3 & 0 \\ -1 & 2 & 1 \\ 0 & 0 & 2 \end{bmatrix}$ *evaluate AB, BA or which-ever exists.*

Solution:

$$AB = \begin{bmatrix} 2 & 3 & 4 \\ 1 & 2 & 3 \\ -1 & 1 & 2 \end{bmatrix} \times \begin{bmatrix} 1 & 3 & 0 \\ -1 & 2 & 1 \\ 0 & 0 & 2 \end{bmatrix}$$

$$= \begin{bmatrix} 2.1 + 3(-1) + 4.0 & 2.3 + 3.2 + 4.0 & 2.0 + 3.1 + 4.2 \\ 1.1 + 2(-1) + 3.0 & 1.3 + 2.2 + 3.0 & 1.0 + 2.1 + 3.2 \\ -1.1 + 1(-1) + 2.0 & 1.3 + 1.2 + 2.0 & -1.0 + 1.1 + 2.2 \end{bmatrix}$$

$$= \begin{bmatrix} -1 & 12 & 11 \\ -1 & 7 & 8 \\ -2 & -1 & 5 \end{bmatrix}$$

$$\text{And BA} = \begin{bmatrix} 1 & 3 & 0 \\ -1 & 2 & 1 \\ 0 & 0 & 2 \end{bmatrix} \times \begin{bmatrix} 2 & 3 & 4 \\ 1 & 2 & 3 \\ -1 & 1 & 2 \end{bmatrix}$$

$$= \begin{bmatrix} 1.2 + 3.1 + 0\,(-1) & 1.3 + 3.2 + 0.1 & 1.4 + 3.3 + 0.2 \\ -1.2 + 2.1 + 1\,(-1) & -1.3 + 2.2 + 1.1 & -1.4 + 2.3 + 1.2 \\ 0.2 + 0.1 + 2\,(-1) & 0.3 + 0.2 + 2.1 & 0.4 + 0.3 + 2.2 \end{bmatrix}$$

$$= \begin{bmatrix} 5 & 9 & 13 \\ -1 & 2 & 4 \\ -2 & 2 & 4 \end{bmatrix}$$

Example 15:

Find the product matrix of the matrices

$$\begin{bmatrix} 2 & 1 & 2 & 1 \\ 1 & 1 & 1 & 1 \end{bmatrix} \text{ and } \begin{bmatrix} 2 & -1 & 0 \\ 0 & 4 & 1 \\ -2 & 1 & 0 \\ 1 & -3 & 2 \end{bmatrix}$$

Solution:

The required matrix

$$= \begin{bmatrix} 2 & 1 & 2 & 1 \\ 1 & 1 & 1 & 1 \end{bmatrix} \times \begin{bmatrix} 2 & -1 & 0 \\ 0 & 4 & 1 \\ -2 & 1 & 0 \\ 1 & -3 & 2 \end{bmatrix}$$

$$= \begin{bmatrix} 2.2 + 1.0 + 2\,(-2) + 1.1 & 2\,(-1) + 1.4 + 2.1 + 1\,(-3) & 2.0 + 1.1 + 2.0 + 1.2 \\ 1.2 + 1.0 + 1\,(-2) + 1.1 & 1\,(-1) + 1.4 + 1.1 + 1\,(-3) & 1.0 + 1.1 + 1.0 + 1.2 \end{bmatrix}$$

$$= \begin{bmatrix} 1 & 1 & 3 \\ 1 & 1 & 3 \end{bmatrix}.$$

Ans.

Example 16:

$$A = \begin{bmatrix} 1 & 1 & -1 \\ 2 & -3 & 4 \\ 3 & -2 & 3 \end{bmatrix};\ B = \begin{bmatrix} -1 & -2 & -1 \\ 6 & 12 & 6 \\ 5 & 10 & 5 \end{bmatrix},\ C = \begin{bmatrix} -1 & -1 & 1 \\ 2 & 2 & -2 \\ -3 & -3 & 3 \end{bmatrix}$$

show that AB and CA are null matrices but $AB \neq O$, $AC \neq O$.

Solution:

$$AB = \begin{bmatrix} 1 & 1 & -1 \\ 2 & -3 & 4 \\ 3 & -2 & 3 \end{bmatrix} \times \begin{bmatrix} -1 & -2 & -1 \\ 6 & 12 & 6 \\ 5 & 10 & 5 \end{bmatrix}$$

$$= \begin{bmatrix} 1(-1) + 1.6 + (-1).5 & 1(-2) + 1.12 + (-1).10 & 1(-1) + 1.6 + (-1).5 \\ 2(-1) - 3.6 + 4.5 & 2(-2) - 3.12 + 4.10 & 2(-1) - 3.6 + 4.5 \\ 3(-1) - 2.6 + 3.5 & 3(-2) - 2.12 + 3.10 & 3(-1) - 2.6 + 3.5 \end{bmatrix}$$

$$= \begin{bmatrix} 0 & 0 & 0 \\ 0 & 0 & 0 \\ 0 & 0 & 0 \end{bmatrix}$$, which is null matrix.

This is known as 'unusual property' of matrix multiplication

$$CA = \begin{bmatrix} -1 & -1 & 1 \\ 2 & 2 & -2 \\ -3 & -3 & 3 \end{bmatrix} \times \begin{bmatrix} 1 & 1 & -1 \\ 2 & -3 & 4 \\ 3 & -2 & 3 \end{bmatrix}$$

$$= \begin{bmatrix} 1.1 - 1.2 + 1.3 & -1.1 - 1(-3) + 1(-2) & -1(-1) - 1.4 + 1.3 \\ 2.1 + 2.2 - 2.3 & 2.1 + 2(-3) - 2(-2) & 2(-1) + 2.4 - 2.3 \\ -3.1 - 3.2 + 3.3 & -3.1 - 3(-3) + 3(-2) & -3(-1) - 3.4 + 3.3 \end{bmatrix}$$

$$= \begin{bmatrix} 0 & 0 & 0 \\ 0 & 0 & 0 \\ 0 & 0 & 0 \end{bmatrix}$$, which is a null matrix. **Hence proved.**

We can prove in a similar way that BA ≠ O and AC ≠ O.

Example 17:

If $A = \begin{bmatrix} 1 & -2 & 3 \\ -4 & 2 & 5 \end{bmatrix}$ and $B = \begin{bmatrix} 2 & 3 \\ 4 & 5 \\ 2 & 1 \end{bmatrix}$ *find AB and show that AB ≠ BA.*

Solution:

$$AB = \begin{bmatrix} 1 & -2 & 3 \\ -4 & 2 & 5 \end{bmatrix} \times \begin{bmatrix} 2 & 3 \\ 4 & 5 \\ 2 & 1 \end{bmatrix}$$

$$= \begin{bmatrix} 1.2 + (-2).4 + 3.2 & 1.3 + (-2).5 + 3.1 \\ -4.2 + 2.4 + 5.2 & -4.3 + 2.5 + 5.1 \end{bmatrix}$$

$$= \begin{bmatrix} 0 & -4 \\ 10 & 3 \end{bmatrix}$$

and $\quad BA = \begin{bmatrix} 2 & 3 \\ 4 & 5 \\ 2 & 1 \end{bmatrix} \times \begin{bmatrix} 1 & -2 & 3 \\ -4 & 2 & 5 \end{bmatrix}$

$$= \begin{bmatrix} 2.1 + 3(-4) & 2(-2) + 3(2) & 2(3) + 3(5) \\ 4.1 + 5(-4) & 4(-2) + 5(2) & 4(3) + 5(5) \\ 2.1 + 1(-4) & 2(-2) + 1(2) & 2(3) + 1(5) \end{bmatrix}$$

$$= \begin{bmatrix} -10 & 2 & 21 \\ -16 & 2 & 37 \\ -2 & -2 & 11 \end{bmatrix}$$

Hence $AB \neq BA$.

Example 18(a):

If $A = \begin{bmatrix} 1 & 2 \\ 3 & 0 \\ 4 & 1 \end{bmatrix}$ and $B = \begin{bmatrix} 0 & 1 & 0 \\ 0 & 2 & 1 \\ 2 & 3 & 0 \end{bmatrix}$, *find BA.*

Solution:

$$BA = \begin{bmatrix} 0 & 1 & 0 \\ 0 & 2 & 1 \\ 2 & 3 & 0 \end{bmatrix} \times \begin{bmatrix} 1 & 2 \\ 3 & 0 \\ 4 & 1 \end{bmatrix}$$

$$= \begin{bmatrix} 0.1 + 1.3 + 0.4 & 0.2 + 1.0 + 0.1 \\ 0.1 + 2.3 + 1.4 & 0.2 + 2.0 + 1.1 \\ 2.1 + 3.3 + 0.4 & 2.2 + 3.0 + 0.1 \end{bmatrix}$$

$$= \begin{bmatrix} 3 & 0 \\ 10 & 1 \\ 11 & 4 \end{bmatrix}$$

Example 18(b):

Multiply [3 – 1 4] and $\begin{bmatrix} -2 \\ 6 \\ 3 \end{bmatrix}$

Solution:

$$[3 \ -1 \ 4] \times \begin{bmatrix} -2 \\ 6 \\ 3 \end{bmatrix}$$

$$= [3\,(-2) + (-1)\,.\,6 + 4.3] = [0]$$ **Ans.**

Example 19:

If $A = \begin{bmatrix} 0 & 1 \\ 1 & 2 \end{bmatrix}$ and $B = \begin{bmatrix} 1 \\ 2 \end{bmatrix}$, *find AB and BA, if they exist.*

Solution:

$$AB = \begin{bmatrix} 0.1 + 1.2 \\ 1.1 + 2.2 \end{bmatrix} = \begin{bmatrix} 2 \\ 5 \end{bmatrix}$$ **Ans.**

Also BA does not exist as the number of columns in B and number of rows in A are not equal.

Example 20:

If $A = \begin{bmatrix} 1 & 4 & 0 \\ 2 & 5 & 0 \\ 3 & 6 & 0 \end{bmatrix}$, $B = \begin{bmatrix} 3 & 2 & 1 \\ 1 & 2 & 3 \\ 4 & 5 & 6 \end{bmatrix}$ and $C = \begin{bmatrix} 3 & 2 & 1 \\ 1 & 2 & 3 \\ 7 & 8 & 9 \end{bmatrix}$

then evaluate AB – AC.

Solution:

$$AB = \begin{bmatrix} 1 & 4 & 0 \\ 2 & 5 & 0 \\ 3 & 6 & 0 \end{bmatrix} \times \begin{bmatrix} 3 & 2 & 1 \\ 1 & 2 & 3 \\ 4 & 5 & 6 \end{bmatrix}$$

$$= \begin{bmatrix} 1.3 + 4.1 + 0.4 & 1.2 + 4.2 + 0.5 & 1.1 + 4.3 + 0.6 \\ 2.3 + 5.1 + 0.4 & 2.2 + 5.2 + 0.5 & 2.1 + 5.3 + 0.6 \\ 3.3 + 6.1 + 0.4 & 3.2 + 6.2 + 0.5 & 3.1 + 6.3 + 0.6 \end{bmatrix}$$

$$= \begin{bmatrix} 7 & 10 & 13 \\ 11 & 14 & 17 \\ 15 & 18 & 21 \end{bmatrix} \quad ...(1)$$

And $AC = \begin{bmatrix} 1 & 4 & 0 \\ 2 & 5 & 0 \\ 3 & 6 & 0 \end{bmatrix} \times \begin{bmatrix} 3 & 2 & 1 \\ 1 & 2 & 3 \\ 7 & 8 & 9 \end{bmatrix}$

$$= \begin{bmatrix} 1.3 + 4.1 + 0.7 & 1.2 + 4.2 + 0.8 & 1.1 + 4.3 + 0.9 \\ 2.3 + 5.1 + 0.7 & 2.2 + 5.2 + 0.8 & 2.1 + 5.3 + 0.2 \\ 3.3 + 6.1 + 0.7 & 3.2 + 6.2 + 0.8 & 3.1 + 6.3 + 0.9 \end{bmatrix}$$

$$= \begin{bmatrix} 7 & 10 & 13 \\ 11 & 14 & 17 \\ 15 & 18 & 21 \end{bmatrix} \quad ...(2)$$

$\therefore$ From (1) and (2) we get AB – AC

$$= \begin{bmatrix} 7 & 10 & 13 \\ 11 & 14 & 17 \\ 15 & 18 & 21 \end{bmatrix} - \begin{bmatrix} 7 & 10 & 13 \\ 11 & 14 & 17 \\ 15 & 18 & 21 \end{bmatrix} = \begin{bmatrix} 0 & 0 & 0 \\ 0 & 0 & 0 \\ 0 & 0 & 0 \end{bmatrix} \quad \textbf{Ans.}$$

Example 21:

Find the square of the matrix

$$\begin{bmatrix} -1 & 1 & 1 & 1 \\ 1 & -1 & 1 & 1 \\ 1 & 1 & -1 & 1 \\ 1 & 1 & 1 & -1 \end{bmatrix}$$

Solution:

$$\begin{bmatrix} -1 & 1 & 1 & 1 \\ 1 & -1 & 1 & 1 \\ 1 & 1 & -1 & 1 \\ 1 & 1 & 1 & -1 \end{bmatrix}^2$$

$$= \begin{bmatrix} -1 & 1 & 1 & 1 \\ 1 & -1 & 1 & 1 \\ 1 & 1 & -1 & 1 \\ 1 & 1 & 1 & -1 \end{bmatrix} \times \begin{bmatrix} -1 & 1 & 1 & 1 \\ 1 & -1 & 1 & 1 \\ 1 & 1 & -1 & 1 \\ 1 & 1 & 1 & -1 \end{bmatrix}$$

$$= \begin{bmatrix} (-1)(-1) + 1.1 + 1.1 + 1.1 & (-1).1 + 1(-1) + 1,1 + 1.1 & (-1).1 + 1.1 + 1(-1) + 1.1 & (-1).1 + 1.1 + 1.1 + 1(-1) \\ 1.(-1) + (-1).1 + 1.1 + 1.1 & 1.1 + (-1)(-1) + 1.1 + 1.1 & 1.1 + (-1).1 + 1(-1) + 1.1 & 1.1 + (-1).1 + 1.1 + 1(-1) \\ 1.(-1) + 1.1 + (-1).1 + 1.1 & 1.1 + 1(-1) + (-1).1 + 1.1 & 1.1 + 1.1 + (-1).(-1) + 1.1 & 1.1 + 1.1 + (-1).1 + 1(-1) \\ 1.(-1) + 1.1 + 1.1 + (-1).1 & 1.1 + 1.(-1) + 1.1 + (-1).1 & 1.1 + 1.1 + 1,(-1) + (-1).1 & 1.1 + 1.1 + 1.1 + (-1)(-1) \end{bmatrix}$$

$$= \begin{bmatrix} 4 & 0 & 0 & 0 \\ 0 & 4 & 0 & 0 \\ 0 & 0 & 4 & 0 \\ 0 & 0 & 0 & 4 \end{bmatrix} = 4 \begin{bmatrix} 1 & 0 & 0 & 0 \\ 0 & 1 & 0 & 0 \\ 0 & 0 & 1 & 0 \\ 0 & 0 & 0 & 1 \end{bmatrix}$$ **Ans.**

Example 12:

I*if* $A = \begin{bmatrix} 1 & 1 & 3 \\ 2 & 2 & 6 \\ -1 & -1 & -3 \end{bmatrix}$, *show that* $A^2 = O$.

Solution:

$$A^2 = \begin{bmatrix} 1 & 1 & 3 \\ 2 & 2 & 6 \\ -1 & -1 & -3 \end{bmatrix} \times \begin{bmatrix} 1 & 1 & 3 \\ 2 & 2 & 6 \\ -1 & -1 & -3 \end{bmatrix}$$

$$= \begin{bmatrix} 1.1 + 1.2 + 3.(-1) & 1.1 + 1.2 + 3.(-1) & 1.3 + 1.6 + 3.(-3) \\ 2.1 + 2.2 + 6.(-1) & 2.1 + 2.2 + 6.(-1) & 2.3 + 2.6 + 6.(-3) \\ -1.1 - 1.2 - 3.(-1) & -1.1 + 1.2 - 3(-1) & -1.3 + 1.6 - 3.(-3) \end{bmatrix}$$

$$= \begin{bmatrix} 0 & 0 & 0 & 0 \\ 0 & 0 & 0 & 0 \\ 0 & 0 & 0 & 0 \end{bmatrix}$$ = O, where O is 3 × 3 null matrix. **Hence proved**

ADJOINT OF A MATRIX

Definition: *If* C_{ij} *be the cofactor of the element* a_{ij} *in* $|a_{ij}|$ *of the n × n matrix* $A = [a_{ij}]$, *then*

$$\text{adjoint of } A = \begin{bmatrix} C_{11} & C_{21} \ldots C_{n1} \\ C_{12} & C_{22} \ldots C_{n2} \\ \ldots\ldots & \ldots\ldots \\ \ldots\ldots & \ldots\ldots \\ C_{1n} & C_{2n} \ldots C_{nn} \end{bmatrix}$$

This is also written as Adj A.

or adjoint of A = transposed of C, where $C = \begin{bmatrix} C_{11} & C_{21} \ldots C_{1n} \\ C_{12} & C_{22} \ldots C_{2n} \\ \ldots\ldots & \ldots\ldots \\ C_{n1} & C_{n2} \ldots C_{nn} \end{bmatrix}$

While solving problems we generally use this definition.

Here students should note carefully that the cofactors of the elements of the first row of $[a_{ij}]$ are the elements of the first column of Adj. A.

Similarly, the cofactors of the elements of the first column of $|a_{ij}|$ are the element of first row of Adj. A.

Example 1:

Find the adjoint of matrix $A = \begin{bmatrix} 1 & 2 & 3 \\ 0 & 5 & 0 \\ 2 & 4 & 3 \end{bmatrix}$

Solution:

For the given matrix A, we have

$$C_{11} = \begin{vmatrix} 5 & 0 \\ 4 & 3 \end{vmatrix} = 16; \; C_{12} = -\begin{vmatrix} 0 & 0 \\ 2 & 3 \end{vmatrix}$$

$$= 0; \; C_{13} = \begin{vmatrix} 0 & 5 \\ 2 & 4 \end{vmatrix} = -10;$$

$$C_{21} = -\begin{vmatrix} 2 & 3 \\ 4 & 3 \end{vmatrix} = -15; \; C_{22}$$

$$= \begin{vmatrix} 1 & 3 \\ 2 & 3 \end{vmatrix} = -3; \; C_{23} = -\begin{vmatrix} 1 & 2 \\ 2 & 4 \end{vmatrix} = 0;$$

$$C_{31} = \begin{vmatrix} 2 & 3 \\ 5 & 0 \end{vmatrix} = -15; \; C_{32} = -\begin{vmatrix} 1 & 3 \\ 0 & 0 \end{vmatrix}$$

$$= 0; \; C_{33} = \begin{vmatrix} 1 & 2 \\ 2 & 5 \end{vmatrix} = 5;$$

$$\therefore \; C = \begin{bmatrix} 15 & 0 & -10 \\ 6 & -3 & 0 \\ -15 & 0 & 5 \end{bmatrix}$$

$$\therefore \; \text{Adj. A} = C'(\text{i.e. transpose of C}) = \begin{bmatrix} 15 & 6 & -15 \\ 0 & -3 & 0 \\ -10 & 0 & 5 \end{bmatrix}$$

Ans.

Example 2:

Find the adjoint of $A = \begin{bmatrix} -1 & -2 & 3 \\ -2 & 1 & 1 \\ -4 & -5 & 2 \end{bmatrix}$

Solution:

For the given matrix A, we have

$C_{11} = \begin{vmatrix} 1 & 1 \\ -5 & 2 \end{vmatrix} = 7;\ C_{12} = -\begin{vmatrix} -2 & 1 \\ -4 & 2 \end{vmatrix} = 0;\ C_{13} = \begin{vmatrix} -2 & 1 \\ -4 & -5 \end{vmatrix} = 14;$

$C_{21} = -\begin{vmatrix} -2 & 3 \\ -5 & 2 \end{vmatrix} = -11;\ C_{22} = \begin{vmatrix} -1 & 3 \\ -4 & 2 \end{vmatrix} = 10;\ C_{23} = -\begin{vmatrix} -1 & -2 \\ -4 & -5 \end{vmatrix} = 3;$

$C_{31} = \begin{vmatrix} -2 & 3 \\ 1 & 1 \end{vmatrix} = -5;\ C_{32} = \begin{vmatrix} -1 & 3 \\ -2 & 1 \end{vmatrix} = -5;\ C_{33} = \begin{vmatrix} -1 & -2 \\ -2 & 1 \end{vmatrix} = -5;$

$\therefore\ C = \begin{bmatrix} 7 & 0 & 14 \\ -11 & 10 & 3 \\ -5 & -5 & -5 \end{bmatrix}$

$\therefore\ \text{Adj. } A = C' = \begin{bmatrix} 7 & -11 & -5 \\ 0 & 10 & -5 \\ 14 & 3 & -5 \end{bmatrix}$ **Ans.**

Example 3:

Find the adjoint of the matrix $A = \begin{bmatrix} 1 & 1 & 1 \\ 1 & 2 & -3 \\ 2 & -1 & 3 \end{bmatrix}$

Solution:

For the given matrix A, we have

$C_{11} = \begin{vmatrix} 2 & -3 \\ -1 & 3 \end{vmatrix} = 3;\ C_{12} = -\begin{vmatrix} 1 & -3 \\ 2 & 3 \end{vmatrix} = 9;\ C_{13} = \begin{vmatrix} 1 & 2 \\ 2 & -1 \end{vmatrix} = -5;$

$C_{21} = -\begin{vmatrix} 1 & 1 \\ -1 & 3 \end{vmatrix} = -4;\ C_{22} = \begin{vmatrix} 1 & 1 \\ 2 & 3 \end{vmatrix} = 1;\ C_{23} = -\begin{vmatrix} 1 & 1 \\ 2 & -1 \end{vmatrix} = 3;$

$C_{31} = \begin{vmatrix} 1 & 1 \\ 2 & -3 \end{vmatrix} = -5; \ C_{32} = -\begin{vmatrix} 1 & 1 \\ 1 & -3 \end{vmatrix} = 4; \ C_{33} = \begin{vmatrix} 1 & 1 \\ 1 & 2 \end{vmatrix} = 5;$

$\therefore \quad C = \begin{bmatrix} 3 & -9 & -5 \\ -4 & 1 & 3 \\ -5 & 4 & 1 \end{bmatrix}$

$\therefore \quad \text{Adj. } A = C' = \begin{bmatrix} 3 & -4 & -5 \\ -9 & 1 & 4 \\ -5 & 3 & 1 \end{bmatrix}$ **Ans.**

Example 4:

Find the adjoint of the matrix $A = \begin{bmatrix} 1 & 0 & -1 \\ 3 & 4 & 5 \\ 0 & -6 & -7 \end{bmatrix}$

Solution:

For the given matrix A, we have

$C_{11} = \begin{vmatrix} 4 & 5 \\ -6 & -7 \end{vmatrix} = 2; \ C_{12} = -\begin{vmatrix} 3 & 5 \\ 0 & -7 \end{vmatrix} = 21; \ C_{13} = \begin{vmatrix} 3 & 4 \\ 0 & -6 \end{vmatrix} = 18;$

$C_{21} = -\begin{vmatrix} 0 & -1 \\ -6 & -7 \end{vmatrix} = -6; \ C_{22} = \begin{vmatrix} 1 & -1 \\ 0 & -7 \end{vmatrix} = -7; \ C_{23} = -\begin{vmatrix} 1 & -0 \\ 0 & -0 \end{vmatrix} = 6;$

$C_{31} = \begin{vmatrix} 0 & -1 \\ 4 & 5 \end{vmatrix} = 5; \ C_{32} = -\begin{vmatrix} 1 & -1 \\ 3 & 5 \end{vmatrix} = -8; \ C_{33} = \begin{vmatrix} 1 & 0 \\ 3 & 4 \end{vmatrix} = 4;$

$\therefore \quad C = \begin{bmatrix} 2 & 21 & -18 \\ -6 & -7 & 6 \\ 4 & -8 & 4 \end{bmatrix}$

$\therefore \quad \text{Adj. } A = C' = \begin{bmatrix} 2 & -6 & 4 \\ 21 & -7 & 6 \\ 4 & -8 & 4 \end{bmatrix}$ **Ans.**

THEOREMS ON ADJOINT OF A MATRIX

Theorem 1:

If $A = [a_{ij}]$ be an $n \times n$ matrix, then

$$|\,Adj\ A\,| = |\,A\,|^{n-1}, \text{ if } |\,A\,| \neq 0$$

Proof:

We know that $|A| . |B| = |AB|$

$\therefore \quad |A| . |\text{Adj } A| = |A . \text{Adj } A|$

$$= \begin{bmatrix} |A| & 0 & 0 & .. & 0 \\ 0 & |A| & 0 & .. & 0 \\ 0 & 0 & |A| & .. & 0 \\ .. & .. & .. & .. & .. \\ 0 & 0 & 0 & .. & |A| \end{bmatrix},$$

$\Rightarrow \quad |A| . |\text{Adj } A| = \{ |A| \}^n$ **(Note)**

Dividing both sides by $|A|$, since $|A| \neq 0$, we get

$|\text{Adj } A| = |A|^{n-1}$ **Hence proved.**

Theorem 2:

If A and B are two $n \times n$ matrices, then

Adj (AB) = (Adj B) . (Adj A).

Proof:

We know A . (Adj A) = | A | . 1

So we have (AB) . (Adj AB) = | AB | . 1 ...(i)

Now (AB) . (Adj B). (Adj A)

= A . B , Adj B . Adj A

= A . (B . Adj B) . (Adj A) (Note)

= A . | B | . I . Adj A, $\because$ B . Adj B = | B | . I

= A . | B | . Adj A, $\because$ I . Adj A = Adj A as I . A = A always

= | B | . A . Adj A (Note)

= | B | . | A | . I, $\because$ A . Adj A = | A | . I

= | A | . | B | . I

= | AB | . I $\because$ | A | . | B | = | AB | ...(ii)

∴ From (i) and (ii) we get

$$(AB) . (\text{Adj } AB) = (AB) . (\text{Adj } B) . (\text{Adj } A)$$

$\Rightarrow$ $\text{Adj } (AB) = (\text{Adj } B) . (\text{Adj } A).$ **Hence proved.**

Theorem 3:

If $A = [a_{ij}]$ be an $n \times n$ matrix, then

$A . (\text{Adj } A) = (\text{Adj } A) . A = |A| . I$, where I is on $n \times n$ identity matrix.

Proof:

We know Adj $A = [C'_{ik}]$,

where C_{kj} is the cofactor of a_{kj} in $|A|$ and $C'_{jk} = C_{kj}$,

Therefore $A . (\text{Adj } A) = [a_{ij}] [C'_{ik}]$

$$= [B_{ik}], \text{ say,} \qquad \text{...(ii)}$$

where
$$B_{ik} = \sum_{j=1}^{n} a_{ij} C'_{jk}$$

$$= \sum_{j=1}^{n} a_{ij} C'_{kj} \qquad \because C'_{jk} = C_{kj}$$

$$\left.\begin{aligned} &= |A|, \text{ if } i = k \\ &= 0, \quad \text{if } i \neq k \end{aligned}\right\}$$

∴ From (i), (i, k)th element of A . (Adj. A) = | A |, or 0 according as i = k or i ≠ k.

i.e.,, all diagonal terms of A . (Adj . A) are | A | and non-diagonal terms are zero.

i.e.,
$$A . (\text{Adj} . A) = \begin{bmatrix} |A| & 0 & 0 & .. & 0 \\ 0 & |A| & 0 & .. & 0 \\ 0 & 0 & |A| & .. & 0 \\ .. & .. & .. & .. & .. \\ 0 & 0 & 0 & .. & |A| \end{bmatrix}$$

$$= |A| \begin{bmatrix} 1 & 0 & 0 & .. & 0 \\ 0 & 1 & 0 & .. & 0 \\ 0 & 0 & 1 & .. & 0 \\ .. & .. & .. & .. & .. \\ 0 & 0 & 0 & .. & 1 \end{bmatrix}$$

$$= |A| . I. \quad ...(i)$$

Similarly we can prove that (Adj. A) . A = | A | . I ...(ii)

Hence from (i) and (ii), we get

$$A . (Adj\ A) = (Adj\ A) . A = |A| . I$$

$$\Rightarrow \quad A . \frac{(Adj\ A)}{|A|} = \frac{(Adj\ A)}{|A|} . A = I$$

$$\Rightarrow \quad A^{-1} = \frac{(Adj\ A)}{|A|}, \quad \because AA^{-1} = I = A^{-1}A$$

i.e.,, the inverse of $A = \frac{Adj\ A}{|A|}$. ...(iii)

Note: The result (iii) gives us another method of finding the inverse of a given matrix.

Example 1:

How will you use the notion of determinant to compute the inverse of a non-singular square matrix? Compute the inverse of the matrix

$$A = \begin{bmatrix} 1 & 2 & 3 \\ 4 & 5 & 6 \\ 7 & 8 & 10 \end{bmatrix}$$

Solution:

We have for the matrix A,

$$C_{11} = \begin{vmatrix} 5 & 6 \\ 8 & 10 \end{vmatrix} = 2; \; C_{12} = -\begin{vmatrix} 4 & 6 \\ 7 & 10 \end{vmatrix} = 2; \; C_{13} = \begin{vmatrix} 4 & 5 \\ 7 & 8 \end{vmatrix} = -3;$$

$$C_{21} = -\begin{vmatrix} 2 & 3 \\ 8 & 10 \end{vmatrix} = 4; \; C_{22} = \begin{vmatrix} 1 & 3 \\ 7 & 10 \end{vmatrix} = -11; \; C_{23} = -\begin{vmatrix} 1 & 2 \\ 7 & 8 \end{vmatrix} = 6;$$

$$C_{31} = \begin{vmatrix} 2 & 3 \\ 5 & 6 \end{vmatrix} = -3; \; C_{32} = -\begin{vmatrix} 1 & 3 \\ 4 & 6 \end{vmatrix} = 6; \; C_{33} = \begin{vmatrix} 1 & 2 \\ 4 & 5 \end{vmatrix} = -3;$$

$$\therefore \quad C = \begin{bmatrix} 2 & 2 & -3 \\ 4 & -11 & 6 \\ -3 & 6 & -3 \end{bmatrix}$$

$$\therefore \quad \text{Adj. A} = C' = \begin{bmatrix} 2 & 4 & -3 \\ 2 & -11 & 6 \\ -3 & 6 & -3 \end{bmatrix}$$

Also $|A| = \begin{vmatrix} 1 & 2 & 3 \\ 4 & 5 & 6 \\ 7 & 8 & 10 \end{vmatrix} = \begin{vmatrix} 1 & 0 & 0 \\ 4 & -3 & -6 \\ 7 & -6 & -11 \end{vmatrix}$, replacing C_2, C_3 by $C_2 - 2C_1, C_3 - 3C_1$ respectively

$$\therefore \quad A^{-1} = \frac{\text{Adj. A}}{|A|} = \frac{1}{3}\begin{bmatrix} 2 & 4 & -3 \\ 2 & -11 & 6 \\ -3 & 6 & -3 \end{bmatrix}$$

$$= \begin{bmatrix} -\frac{2}{3} & -\frac{4}{3} & 1 \\ -\frac{2}{3} & \frac{11}{3} & -2 \\ 1 & -2 & 1 \end{bmatrix}$$

Ans.

Example 2(a):

Find the inverse of the matrix A, where $A = \begin{bmatrix} 1 & 0 & -4 \\ -2 & 2 & 5 \\ 3 & -1 & 2 \end{bmatrix}$

Solution:

Here $|A| = \begin{vmatrix} 1 & 0 & -4 \\ -2 & 2 & 5 \\ 3 & -1 & 2 \end{vmatrix} = \begin{vmatrix} 1 & 0 & 0 \\ -2 & 2 & -3 \\ 3 & -1 & 14 \end{vmatrix}$ applying $C_3 + 4C_1$

$$= \begin{vmatrix} 2 & -3 \\ -1 & 14 \end{vmatrix} = 28 - 3 = 25 \neq 0$$

Also we have

$C_{11} = \begin{vmatrix} 2 & 5 \\ -1 & 2 \end{vmatrix} = 9$; $C_{12} = -\begin{vmatrix} -2 & 5 \\ 3 & 2 \end{vmatrix} = 19$; $C_{13} = \begin{vmatrix} -2 & 2 \\ 3 & -1 \end{vmatrix} = -4$;

$C_{21} = -\begin{vmatrix} 0 & -4 \\ -1 & 2 \end{vmatrix} = 4$; $C_{22} = \begin{vmatrix} 1 & -4 \\ 3 & 2 \end{vmatrix} = 14$; $C_{23} = -\begin{vmatrix} 1 & 0 \\ 3 & -1 \end{vmatrix} = 1$;

$$C_{31} = \begin{vmatrix} 0 & -4 \\ 2 & 5 \end{vmatrix} = 8;\ C_{32} = -\begin{vmatrix} 1 & -4 \\ -2 & 5 \end{vmatrix} = 3;\ C_{33} = \begin{vmatrix} 1 & 0 \\ -2 & 2 \end{vmatrix} = 2;$$

$$\therefore\ C = \begin{bmatrix} 9 & 19 & -4 \\ 4 & 14 & 1 \\ 8 & 3 & 2 \end{bmatrix}$$

$$\therefore\ \text{Adj. } A = C' = \begin{bmatrix} 9 & 4 & 8 \\ 19 & 14 & 3 \\ -4 & 1 & 2 \end{bmatrix}$$

$$\therefore\ A^{-1} = \frac{\text{Adj. } A}{|A|} = \frac{1}{25}\begin{bmatrix} 9 & 4 & 8 \\ 19 & 14 & 3 \\ -4 & 1 & 2 \end{bmatrix}$$

Ans.

Example 2(b):

If $A = \begin{bmatrix} 1 & 1 & 1 \\ 2 & 2 & 3 \\ 1 & 4 & 9 \end{bmatrix}$, *find* A^{-1}.

Solution:

$$\text{Here } |A| = \begin{vmatrix} 1 & 1 & 1 \\ 2 & 2 & 3 \\ 1 & 4 & 9 \end{vmatrix} = \begin{vmatrix} 1 & 0 & 0 \\ 2 & 0 & 1 \\ 1 & 3 & 8 \end{vmatrix}$$

replacing C_2, C_3 by $C_2 - C_1$ and $C_3 - C_1$

$$= \begin{vmatrix} 0 & 1 \\ 3 & 8 \end{vmatrix} = 0 \qquad \text{...(i)}$$

Also for the matrix A, we have

$$C_{11} = \begin{vmatrix} 2 & 3 \\ 4 & 9 \end{vmatrix} = 6;\ C_{12} = -\begin{vmatrix} 2 & 3 \\ 1 & 9 \end{vmatrix} = -15;\ C_{13} = \begin{vmatrix} 2 & 2 \\ 1 & 4 \end{vmatrix} = 6;$$

$$C_{21} = -\begin{vmatrix} 1 & 1 \\ 4 & 9 \end{vmatrix} = -5;\ C_{22} = \begin{vmatrix} 1 & 1 \\ 1 & 9 \end{vmatrix} = 8;\ C_{23} = -\begin{vmatrix} 1 & 1 \\ 1 & 4 \end{vmatrix} = -3;$$

$$C_{31} = \begin{vmatrix} 1 & 1 \\ 2 & 3 \end{vmatrix} = 1;\ C_{32} = -\begin{vmatrix} 1 & 1 \\ 2 & 3 \end{vmatrix} = -1;\ C_{33} = \begin{vmatrix} 1 & 1 \\ 2 & 2 \end{vmatrix} = 0;$$

$$\therefore \quad C = \begin{bmatrix} 6 & -15 & 6 \\ -5 & 8 & -3 \\ 1 & -1 & 0 \end{bmatrix}$$

$$\therefore \quad \text{Adj. } A = C' = \begin{bmatrix} 6 & -5 & 1 \\ -15 & 8 & -1 \\ 6 & -3 & 0 \end{bmatrix}$$

$$\therefore \quad A^{-1} = \frac{\text{Adj. } A}{|A|} = -\frac{1}{3}\begin{bmatrix} 6 & -5 & 1 \\ -15 & 8 & -1 \\ 6 & -3 & 0 \end{bmatrix}$$

$$= \begin{bmatrix} 2 & 5/3 & -1/3 \\ 5 & -8/3 & 1/2 \\ -2 & 1 & 0 \end{bmatrix}$$

Ans.

Example 3:

Find the adjoin and inverse of the matrix $A = \begin{bmatrix} \cos\alpha & -\sin\alpha \\ \sin\alpha & \cos\alpha \end{bmatrix}$

Solution:

Here $|A| = \begin{bmatrix} \cos\alpha & -\sin\alpha \\ \sin\alpha & \cos\alpha \end{bmatrix}$

$$= \cos^2\alpha + \sin^2\alpha = 1 \neq 0 \qquad \text{...(i)}$$

Also we have

$C_{11} = \cos\alpha$, $C_{12} = -\sin\alpha$, $C_{21} = -(-\sin\alpha) = \sin\alpha$ and $C_{22} = \cos\alpha$

(Note)

$$\therefore \quad C = \begin{bmatrix} \cos\alpha & -\sin\alpha \\ \sin\alpha & \cos\alpha \end{bmatrix}$$

$$\therefore \quad \text{Adj } A = C' = \begin{bmatrix} \cos\alpha & \sin\alpha \\ -\sin\alpha & \cos\alpha \end{bmatrix}$$

$$\therefore \quad A^{-1} = \frac{\text{Adj A}}{|A|} = \begin{bmatrix} \cos\alpha & \sin\alpha \\ -\sin\alpha & \cos\alpha \end{bmatrix}$$

substituting values from (i) and (ii). **Ans.**

Example 4(a):

Find the adjoint and inverse of the matrix $A = \begin{bmatrix} \cos\theta & -\sin\theta & 0 \\ \sin\theta & \cos\theta & 0 \\ 0 & 0 & 1 \end{bmatrix}$

Solution:

For the given matrix A, we have

$$C_{11} = \begin{vmatrix} \cos\theta & 0 \\ 0 & 1 \end{vmatrix} = \cos\theta; \; C_{12} = -\begin{vmatrix} \sin\theta & 0 \\ 0 & 1 \end{vmatrix} = -\sin\theta;$$

$$C_{13} = \begin{vmatrix} \sin\theta & \cos\theta \\ 0 & 0 \end{vmatrix} = \theta;$$

$$C_{21} = -\begin{vmatrix} -\sin\theta & 0 \\ 0 & 1 \end{vmatrix} = \sin\theta; \; C_{22} = \begin{vmatrix} \cos\theta & 0 \\ 0 & 1 \end{vmatrix} = \cos\theta;$$

$$C_{23} = -\begin{vmatrix} \cos\theta & -\sin\theta \\ 0 & 0 \end{vmatrix} = \theta;$$

$$C_{31} = \begin{vmatrix} -\sin\theta & 0 \\ \cos\theta & 0 \end{vmatrix} = 0; \; C_{32} = -\begin{vmatrix} \cos\theta & 0 \\ \sin\theta & 0 \end{vmatrix} = 0;$$

$$C_{33} = \begin{vmatrix} \cos\theta & -\sin\theta \\ \sin\theta & \cos\theta \end{vmatrix} = 1;$$

$$\therefore \quad C = \begin{bmatrix} \cos\theta & -\sin\theta & 0 \\ \sin\theta & \cos\theta & 0 \\ 0 & 0 & 1 \end{bmatrix}$$

$$\therefore \quad \text{Adj. A} = C' = \begin{bmatrix} \cos\theta & \sin\theta & 0 \\ -\sin\theta & \cos\theta & 0 \\ 0 & 0 & 1 \end{bmatrix}$$

$$\text{Also } |A| = \begin{vmatrix} \cos\theta & -\sin\theta & 0 \\ \sin\theta & \cos\theta & 0 \\ 0 & 0 & 1 \end{vmatrix} = \begin{vmatrix} \cos\theta & -\sin\theta \\ \sin\theta & \cos\theta \end{vmatrix}$$

$$= \cos^2\theta + \sin^2\theta = 1 \neq 0$$

$\therefore$ The inverse of A $= \dfrac{\text{Adj. A}}{|A|} = \begin{vmatrix} \cos\theta & \sin\theta & 0 \\ -\sin\theta & \cos\theta & 0 \\ 0 & 0 & 1 \end{vmatrix}$ **Ans.**

Example 4(b):

Find the inverse of A $= \begin{bmatrix} 0 & 1 & 2 \\ 1 & 2 & 3 \\ 3 & 1 & 1 \end{bmatrix}$

Solution:

For the given matrix A, we have

$C_{11} = \begin{vmatrix} 2 & 3 \\ 1 & 1 \end{vmatrix} = -1;\ C_{12} = -\begin{vmatrix} 1 & 3 \\ 3 & 1 \end{vmatrix} = 8;\ C_{13} = \begin{vmatrix} 1 & 2 \\ 3 & 1 \end{vmatrix} = -5;$

$C_{21} = -\begin{vmatrix} 1 & 2 \\ 1 & 1 \end{vmatrix} = 1;\ C_{22} = \begin{vmatrix} 0 & 2 \\ 3 & 1 \end{vmatrix} = -6;\ C_{23} = -\begin{vmatrix} 0 & 1 \\ 3 & 1 \end{vmatrix} = 3;$

$C_{31} = \begin{vmatrix} 1 & 2 \\ 2 & 3 \end{vmatrix} = -1;\ C_{32} = -\begin{vmatrix} 0 & 2 \\ 1 & 3 \end{vmatrix} = 2;\ C_{33} = \begin{vmatrix} 0 & 1 \\ 1 & 2 \end{vmatrix} = -1;$

$\therefore \quad C = \begin{bmatrix} -1 & 8 & -5 \\ 1 & -6 & 3 \\ -1 & 2 & -1 \end{bmatrix}$

$\therefore$ Adj. A $= C' = \begin{bmatrix} -1 & 1 & -1 \\ 8 & -6 & 2 \\ -5 & 3 & -1 \end{bmatrix}$

and $\quad |A| = \begin{vmatrix} 0 & 1 & 2 \\ 1 & 2 & 3 \\ 3 & 1 & 1 \end{vmatrix} = \begin{vmatrix} 0 & 1 & 0 \\ 1 & 2 & -1 \\ 3 & 1 & 1 \end{vmatrix}$, replacing C_3 by $C_3 - 2C_2$

$\Rightarrow \quad |A| = -\begin{vmatrix} 1 & -1 \\ 3 & -1 \end{vmatrix} = -2 \neq 0$

$\therefore$ Inverse of A $= \dfrac{\text{Adj. A}}{|A|}$

$$= -\frac{1}{2}\begin{bmatrix} -1 & 1 & -1 \\ 8 & -6 & 2 \\ -5 & 3 & -1 \end{bmatrix} = \frac{1}{2}\begin{bmatrix} 1 & -1 & 1 \\ -8 & 6 & -2 \\ 5 & -3 & 1 \end{bmatrix}$$ **Ans.**

Example 4(c):

Given $A = \begin{bmatrix} 1 & -2 & -1 \\ 2 & 3 & 1 \\ 0 & 5 & -2 \end{bmatrix}$, *compute det. A, Adj. A and* A^{-1}.

Solution:

Here $|A| = \begin{vmatrix} 1 & -2 & -1 \\ 2 & 3 & 1 \\ 0 & 5 & -2 \end{vmatrix}$

$= \begin{vmatrix} 1 & -2 & -1 \\ 0 & 7 & 3 \\ 0 & 5 & -2 \end{vmatrix}$, replacing R_2 by $R_2 - 2R_1$

$= \begin{vmatrix} 7 & 3 \\ 5 & -2 \end{vmatrix}$, expanding with respect to C_1

$= 7(-2) - 5(3) = -(29) \neq 0.$ **Ans.**

For we have

$C_{11} = \begin{vmatrix} 3 & 1 \\ 5 & -2 \end{vmatrix} = -11$; $C_{12} = -\begin{vmatrix} 2 & 1 \\ 0 & -2 \end{vmatrix} = 4$; $C_{13} = \begin{vmatrix} 2 & 3 \\ 0 & 5 \end{vmatrix} = 10$;

$C_{21} = -\begin{vmatrix} -2 & -1 \\ 3 & -2 \end{vmatrix} = -9$; $C_{22} = \begin{vmatrix} 1 & -1 \\ 0 & -2 \end{vmatrix} = -2$; $C_{23} = -\begin{vmatrix} 1 & -2 \\ 0 & 5 \end{vmatrix} = -5$;

$C_{31} = \begin{vmatrix} -2 & -1 \\ 3 & 1 \end{vmatrix} = 1$; $C_{32} = -\begin{vmatrix} 1 & -1 \\ 2 & 1 \end{vmatrix} = -3$; $C_{33} = \begin{vmatrix} 1 & -2 \\ 2 & 3 \end{vmatrix} = 7$;

$\therefore \quad C = \begin{bmatrix} -11 & 4 & 10 \\ -9 & -2 & -5 \\ 1 & -3 & 7 \end{bmatrix}$

$$\therefore \quad \text{Adj. } A = C' = \begin{bmatrix} -11 & -9 & 1 \\ 4 & -2 & -3 \\ 10 & -5 & 7 \end{bmatrix}$$ **Ans.**

$$\therefore \quad A^{-1} = \frac{\text{Adj. } A}{|A|}$$

$$= -\frac{1}{29}\begin{bmatrix} -11 & -9 & 1 \\ 4 & -2 & -3 \\ 10 & -5 & 7 \end{bmatrix} = \frac{1}{29}\begin{bmatrix} 11 & -9 & -1 \\ -4 & 2 & 3 \\ -10 & 5 & -7 \end{bmatrix}$$ **Ans.**

Example 5:

Find the inverse of $A = \begin{bmatrix} 1 & 2 & 3 \\ 2 & 4 & 5 \\ 3 & 5 & 6 \end{bmatrix}$

Solution:

For the given matrix A, we have

$C_{11} = \begin{vmatrix} 4 & 5 \\ 5 & 6 \end{vmatrix} = -1$; $C_{12} = -\begin{vmatrix} 2 & 5 \\ 3 & 6 \end{vmatrix} = 3$; $C_{13} = \begin{vmatrix} 2 & 4 \\ 3 & 5 \end{vmatrix} = -2$;

$C_{21} = -\begin{vmatrix} 2 & 3 \\ 5 & 6 \end{vmatrix} = 3$; $C_{22} = \begin{vmatrix} 1 & 3 \\ 3 & 6 \end{vmatrix} = -3$; $C_{23} = -\begin{vmatrix} 1 & 2 \\ 3 & 5 \end{vmatrix} = 1$;

$C_{31} = \begin{vmatrix} 2 & 3 \\ 4 & 5 \end{vmatrix} = -2$; $C_{32} = -\begin{vmatrix} 1 & 3 \\ 2 & 5 \end{vmatrix} = 1$; $C_{33} = \begin{vmatrix} 1 & 2 \\ 2 & 4 \end{vmatrix} = 0$;

$$\therefore \quad C = \begin{bmatrix} -1 & 3 & -2 \\ 3 & -3 & 1 \\ -2 & 1 & 0 \end{bmatrix}$$

$$\therefore \quad \text{Adj. } A = C' = \begin{bmatrix} -1 & 3 & -2 \\ 3 & -3 & 1 \\ -2 & 1 & 0 \end{bmatrix}$$

Also $|A| = \begin{bmatrix} 1 & 2 & 3 \\ 2 & 4 & 5 \\ 3 & 5 & 6 \end{bmatrix} = \begin{bmatrix} 1 & 0 & 0 \\ 2 & 0 & -1 \\ 3 & -1 & -3 \end{bmatrix}$, replacing C_2, C_3 by $C_2 - 2C_1$, $C_3 - 3C_1$

$$= \begin{vmatrix} 0 & -1 \\ -1 & -3 \end{vmatrix} = -1$$

$$\therefore \quad A^{-1} = \frac{\text{Adj. A}}{|A|} = -\begin{bmatrix} -1 & 3 & -2 \\ 3 & -3 & 1 \\ -2 & 1 & 0 \end{bmatrix}$$ (Note)

$$= \begin{bmatrix} 1 & -3 & 2 \\ -3 & 3 & -1 \\ 2 & -1 & 0 \end{bmatrix}$$ **Ans.**

Example 6:

Find the adjoint of the matrix A and evaluate A^{-1}, where

$$A = \begin{bmatrix} 2 & 2 & 2 \\ 2 & 5 & 5 \\ 2 & 5 & 11 \end{bmatrix}$$

Solution:

Hence for the matrix A, we have

$$C_{11} = \begin{vmatrix} 5 & 5 \\ 5 & 11 \end{vmatrix} = 30; \; C_{12} = -\begin{vmatrix} 2 & 5 \\ 2 & 11 \end{vmatrix} = -12; \; C_{13} = \begin{vmatrix} 2 & 5 \\ 2 & 5 \end{vmatrix} = 0;$$

$$C_{21} = -\begin{vmatrix} 2 & 2 \\ 5 & 11 \end{vmatrix} = -12; \; C_{22} = \begin{vmatrix} 2 & 2 \\ 2 & 11 \end{vmatrix} = 18; \; C_{23} = -\begin{vmatrix} 2 & 2 \\ 2 & 5 \end{vmatrix} = -6;$$

$$C_{31} = \begin{vmatrix} 2 & 2 \\ 5 & 5 \end{vmatrix} = 0; \; C_{32} = -\begin{vmatrix} 2 & 2 \\ 2 & 5 \end{vmatrix} = -6; \; C_{33} = \begin{vmatrix} 2 & 2 \\ 2 & 5 \end{vmatrix} = 6;$$

$$\therefore \quad C = \begin{bmatrix} 30 & -12 & 0 \\ -12 & 18 & -6 \\ 0 & -6 & 6 \end{bmatrix}$$

$$\therefore \quad \text{Adj. A} = C' = \begin{bmatrix} 30 & -12 & 0 \\ -12 & 18 & -6 \\ 0 & -6 & 6 \end{bmatrix}$$ **Ans.**

Also $|A| = \begin{bmatrix} 2 & 2 & 2 \\ 2 & 5 & 5 \\ 2 & 5 & 11 \end{bmatrix} = \begin{bmatrix} 2 & 0 & 0 \\ 2 & 3 & 3 \\ 2 & 3 & 9 \end{bmatrix}$, applying $C_2 - C_1$, $C_3 - C_1$

$$= 2\begin{vmatrix} 3 & 3 \\ 3 & 9 \end{vmatrix} 2\,[27 - 9] = 36$$

$$\therefore\ A^{-1} = \frac{\text{Adj. A}}{|A|} = \frac{1}{36}\begin{bmatrix} 30 & -12 & 0 \\ -12 & 18 & -6 \\ 0 & -6 & 6 \end{bmatrix}$$

$$= \frac{1}{36} \times 6\begin{bmatrix} 5 & -2 & 0 \\ -2 & 3 & -1 \\ 0 & -1 & 1 \end{bmatrix} = \begin{bmatrix} 5/6 & -1/3 & 0 \\ -1/3 & 1/2 & -1/6 \\ 0 & -1/6 & 1/6 \end{bmatrix}$$ **Ans.**

Example 7(a):

Find the inverse of $A = \begin{bmatrix} 1 & 2 & 3 \\ 0 & 5 & 0 \\ 2 & 0 & 3 \end{bmatrix}$

Solution:

Here $|A| = \begin{vmatrix} 1 & 2 & 3 \\ 0 & 5 & 0 \\ 2 & 0 & 3 \end{vmatrix} = 5\begin{vmatrix} 1 & 3 \\ 2 & 3 \end{vmatrix} = -15$ ...(i)

Also for the matrix A, we have

$C_{11} = \begin{vmatrix} 5 & 0 \\ 0 & 3 \end{vmatrix} = 15;\ C_{12} = -\begin{vmatrix} 0 & 0 \\ 2 & 3 \end{vmatrix} = 0;\ C_{13} = \begin{vmatrix} 0 & 5 \\ 2 & 0 \end{vmatrix} = -10;$

$C_{21} = -\begin{vmatrix} 2 & 3 \\ 0 & 3 \end{vmatrix} = -6;\ C_{22} = \begin{vmatrix} 1 & 3 \\ 2 & 3 \end{vmatrix} = -3;\ C_{23} = -\begin{vmatrix} 1 & 2 \\ 2 & 0 \end{vmatrix} = 4;$

$C_{31} = \begin{vmatrix} 2 & 3 \\ 5 & 0 \end{vmatrix} = 15;\ C_{32} = -\begin{vmatrix} 1 & 3 \\ 0 & 0 \end{vmatrix} = 0;\ C_{33} = \begin{vmatrix} 1 & 2 \\ 0 & 5 \end{vmatrix} = 5;$

$$\therefore\ C = \begin{bmatrix} 15 & 0 & -10 \\ -6 & -3 & 4 \\ -15 & 0 & 5 \end{bmatrix}.$$

$$\therefore\ A^{-1} = \frac{\text{Adj. A}}{|A|} = -\frac{1}{15}\begin{bmatrix} 15 & -6 & -15 \\ 0 & -3 & 0 \\ -10 & 4 & 5 \end{bmatrix}$$ **Ans.**

Example 7(b):

Find the inverse of $A = \begin{bmatrix} 1 & 2 & 4 \\ 5 & 7 & 8 \\ 9 & 10 & 12 \end{bmatrix}$

Solution:

Here $|A| = \begin{vmatrix} 1 & 2 & 4 \\ 5 & 7 & 8 \\ 9 & 10 & 12 \end{vmatrix} = \begin{vmatrix} 1 & 0 & 0 \\ 5 & -3 & -12 \\ 9 & -8 & -24 \end{vmatrix}$

replacing C_2, C_3 by $C_2 - 2C_1$, $C_3 - 4C_1$

$$\Rightarrow \quad |A| = \begin{vmatrix} -3 & -12 \\ -8 & -24 \end{vmatrix} = 72 - 96 = -24 \quad \ldots(i)$$

Also for the matrix A, we have

$C_{11} = \begin{vmatrix} 7 & 8 \\ 10 & 12 \end{vmatrix} = 4$; $C_{12} = -\begin{vmatrix} 5 & 8 \\ 9 & 12 \end{vmatrix} = 12$; $C_{13} = \begin{vmatrix} 5 & 7 \\ 9 & 10 \end{vmatrix} = -13$;

$C_{21} = -\begin{vmatrix} 2 & 4 \\ 10 & 12 \end{vmatrix} = 16$; $C_{22} = \begin{vmatrix} 1 & 4 \\ 9 & 12 \end{vmatrix} = -24$; $C_{23} = -\begin{vmatrix} 1 & 2 \\ 9 & 10 \end{vmatrix} = 8$;

$C_{31} = \begin{vmatrix} 2 & 4 \\ 7 & 8 \end{vmatrix} = -12$; $C_{32} = -\begin{vmatrix} 1 & 4 \\ 5 & 8 \end{vmatrix} = 12$; $C_{33} = \begin{vmatrix} 1 & 2 \\ 5 & 7 \end{vmatrix} = -3$;

$$\therefore \quad C = \begin{bmatrix} 4 & 12 & -13 \\ 16 & -24 & 8 \\ -12 & 12 & -3 \end{bmatrix}$$

$$\therefore \quad \text{Adj. } A = C' = \begin{bmatrix} 4 & 16 & -12 \\ 12 & -24 & 12 \\ -13 & 8 & -3 \end{bmatrix}$$

$$\therefore \quad A^{-1} = \frac{\text{Adj. } A}{|A|} = \frac{1}{24}\begin{bmatrix} 4 & 16 & -12 \\ 12 & -24 & 12 \\ -13 & 8 & -3 \end{bmatrix}, \text{ from (i), (ii)}$$

$$= \begin{bmatrix} -\frac{4}{24} & -\frac{16}{24} & \frac{12}{24} \\ -\frac{12}{24} & 1 & -\frac{12}{24} \\ \frac{13}{24} & -\frac{8}{24} & \frac{8}{24} \end{bmatrix} = \begin{bmatrix} -\frac{1}{6} & -\frac{2}{3} & \frac{1}{2} \\ -\frac{1}{2} & 1 & -\frac{1}{2} \\ \frac{13}{24} & -\frac{1}{3} & \frac{1}{3} \end{bmatrix}$$ **Ans.**

Example 7(c):

If $A = \begin{bmatrix} 1 & 4 & 0 \\ -1 & 2 & 2 \\ 0 & 0 & 2 \end{bmatrix}$, *find* A^{-1}.

Solution:

Here $|A| = \begin{vmatrix} 1 & 4 & 0 \\ -1 & 2 & 2 \\ 0 & 0 & 2 \end{vmatrix} = \begin{vmatrix} 1 & 4 & 0 \\ 0 & 6 & 2 \\ 0 & 0 & 2 \end{vmatrix}$, replacing R_2 by $R_2 + R_2$

$$= \begin{vmatrix} 6 & 2 \\ 0 & 2 \end{vmatrix} = 12$$

Also for the matrix A, we have

$C_{11} = \begin{vmatrix} 2 & 2 \\ 0 & 2 \end{vmatrix} = 4$; $C_{12} = -\begin{vmatrix} -1 & 2 \\ 0 & 2 \end{vmatrix} = 2$; $C_{13} = \begin{vmatrix} -1 & 2 \\ 0 & 0 \end{vmatrix} = 0$;

$C_{21} = -\begin{vmatrix} 4 & 0 \\ 0 & 2 \end{vmatrix} = -8$; $C_{22} = \begin{vmatrix} 1 & 0 \\ 0 & 2 \end{vmatrix} = 2$; $C_{23} = -\begin{vmatrix} 1 & 4 \\ 0 & 0 \end{vmatrix} = 0$;

$C_{31} = \begin{vmatrix} 4 & 0 \\ 2 & 2 \end{vmatrix} = 8$; $C_{32} = -\begin{vmatrix} 1 & 0 \\ -1 & 2 \end{vmatrix} = -2$; $C_{33} = \begin{vmatrix} 1 & 4 \\ -1 & 2 \end{vmatrix} = 6$;

$$\therefore \quad C = \begin{bmatrix} 4 & 2 & 0 \\ -8 & 2 & 0 \\ 8 & -2 & 6 \end{bmatrix}$$

$$\therefore \quad \text{Adj. } A = C' = \begin{bmatrix} 4 & -8 & 8 \\ 2 & 2 & -2 \\ 0 & 0 & 6 \end{bmatrix} = 2\begin{bmatrix} 2 & -4 & 4 \\ 1 & 1 & -1 \\ 0 & 0 & 3 \end{bmatrix}$$

$$\therefore\ A^{-1} = \frac{\text{Adj. A}}{|A|} = -\frac{1}{2} \times 2\begin{bmatrix} 2 & -4 & 4 \\ 1 & 1 & -1 \\ 0 & 0 & 3 \end{bmatrix} = \frac{1}{6}\begin{bmatrix} 2 & -4 & 4 \\ 1 & 1 & -1 \\ 0 & 0 & 3 \end{bmatrix}$$

Example 8:

If $A = \begin{bmatrix} 1 & 0 & -1 \\ 3 & 4 & 5 \\ 0 & -6 & -7 \end{bmatrix}$, *find adj. A and* A^{-1}.

Solution:

Here $|A| = \begin{vmatrix} 1 & 0 & -1 \\ 3 & 4 & 5 \\ 0 & -6 & -7 \end{vmatrix} = \begin{vmatrix} 1 & 0 & -1 \\ 0 & 4 & 8 \\ 0 & -6 & -7 \end{vmatrix}$, replacing R_2 by $R_2 - 3R_1$

$\Rightarrow \quad |A| = \begin{vmatrix} 4 & 8 \\ -6 & -7 \end{vmatrix} = -28 + 48 = 20$

Also for the matrix A, we have

$C_{11} = \begin{vmatrix} 4 & 5 \\ -6 & -7 \end{vmatrix} = 2;\ C_{12} = -\begin{vmatrix} 3 & 5 \\ 0 & -7 \end{vmatrix} = 21;\ C_{13} = \begin{vmatrix} 3 & 4 \\ 0 & -6 \end{vmatrix} = -18;$

$C_{21} = -\begin{vmatrix} 0 & -1 \\ -6 & -7 \end{vmatrix} = 6;\ C_{22} = \begin{vmatrix} 1 & -1 \\ 0 & -7 \end{vmatrix} = -7;\ C_{23} = -\begin{vmatrix} 1 & 0 \\ 0 & -6 \end{vmatrix} = 6;$

$C_{31} = \begin{vmatrix} 0 & -1 \\ 4 & 5 \end{vmatrix} = 4;\ C_{32} = -\begin{vmatrix} 1 & -1 \\ 3 & 5 \end{vmatrix} = -8;\ C_{33} = \begin{vmatrix} 1 & 0 \\ 3 & 4 \end{vmatrix} = 4;$

$\therefore\ C = \begin{bmatrix} 2 & 21 & -18 \\ 6 & -7 & 6 \\ 4 & -8 & 4 \end{bmatrix}$

$\therefore\ \text{Adj. A} = C' = \begin{bmatrix} 2 & 6 & 4 \\ 21 & -7 & -8 \\ -18 & 6 & 4 \end{bmatrix}$

$$\therefore \quad A^{-1} = \frac{\text{Adj. A}}{|A|} = \frac{1}{20}\begin{bmatrix} 2 & 6 & 4 \\ 21 & -7 & -8 \\ -18 & 6 & 4 \end{bmatrix}$$

Ans.

Example 9:

Find the reciprocal of the matrix $A = \begin{bmatrix} 2 & 1 & 2 \\ 2 & 2 & 1 \\ 1 & 2 & 2 \end{bmatrix}$

Solution:

Here $|A| = \begin{vmatrix} 2 & 1 & 2 \\ 2 & 2 & 1 \\ 1 & 2 & 2 \end{vmatrix} = \begin{vmatrix} 0 & -3 & -2 \\ 0 & -2 & -3 \\ 1 & 2 & 2 \end{vmatrix}$, applying $R_1 - 2R_3$, $R_2 - 2R_3$

$$= \begin{vmatrix} -3 & -2 \\ -2 & -3 \end{vmatrix} = 9 - 4 = 5$$

Also we have

$C_{11} = \begin{vmatrix} 2 & 1 \\ 2 & 2 \end{vmatrix} = 2$; $C_{12} = -\begin{vmatrix} 2 & 1 \\ 1 & 2 \end{vmatrix} = -3$; $C_{13} = \begin{vmatrix} 2 & 2 \\ 1 & 2 \end{vmatrix} = 2$;

$C_{21} = -\begin{vmatrix} 1 & 2 \\ 2 & 2 \end{vmatrix} = 2$; $C_{22} = \begin{vmatrix} 2 & 2 \\ 1 & 2 \end{vmatrix} = 2$; $C_{23} = -\begin{vmatrix} 2 & 1 \\ 1 & 2 \end{vmatrix} = -3$;

$C_{31} = \begin{vmatrix} 1 & 2 \\ 2 & 1 \end{vmatrix} = -3$; $C_{32} = -\begin{vmatrix} 2 & 2 \\ 2 & 1 \end{vmatrix} = 2$; $C_{33} = \begin{vmatrix} 2 & 1 \\ 2 & 2 \end{vmatrix} = 2$;

$$\therefore \quad C = \begin{bmatrix} 2 & -3 & 2 \\ 2 & 2 & -3 \\ -3 & 2 & 2 \end{bmatrix}$$

$$\therefore \quad \text{Adj. A} = C' = \begin{bmatrix} 2 & 2 & -3 \\ -3 & 2 & 2 \\ 2 & -3 & 2 \end{bmatrix}$$

$\therefore$ Reciprocal of A = A^{-1}

$$= \frac{\text{Adj. A}}{|A|} = \frac{1}{5}\begin{bmatrix} 2 & 2 & -3 \\ -3 & 2 & 2 \\ 2 & -3 & 2 \end{bmatrix}$$ **Ans.**

Example 10:

Find the inverse of $A = \begin{bmatrix} 1 & 2 & 1 \\ 3 & 2 & 3 \\ 1 & 1 & 2 \end{bmatrix}$

Solution:

For the given matrix A, we have

$C_{11} = \begin{vmatrix} 2 & 3 \\ 1 & 2 \end{vmatrix} = 1$; $C_{12} = -\begin{vmatrix} 3 & 3 \\ 1 & 2 \end{vmatrix} = -3$; $C_{13} = \begin{vmatrix} 3 & 2 \\ 1 & 1 \end{vmatrix} = 1$;

$C_{21} = -\begin{vmatrix} 2 & 1 \\ 1 & 2 \end{vmatrix} = -3$; $C_{22} = \begin{vmatrix} 1 & 1 \\ 1 & 2 \end{vmatrix} = 1$; $C_{23} = -\begin{vmatrix} 1 & 2 \\ 1 & 1 \end{vmatrix} = 1$;

$C_{31} = \begin{vmatrix} 2 & 1 \\ 1 & 3 \end{vmatrix} = 4$; $C_{32} = -\begin{vmatrix} 1 & 1 \\ 3 & 3 \end{vmatrix} = 0$; $C_{33} = \begin{vmatrix} 1 & 2 \\ 3 & 2 \end{vmatrix} = -4$;

$\therefore \quad C = \begin{bmatrix} 1 & -3 & 1 \\ -3 & 1 & 1 \\ 4 & 0 & -4 \end{bmatrix}$

$\therefore \quad \text{Adj. A} = C' = \begin{bmatrix} 1 & -3 & 4 \\ -3 & 1 & 0 \\ 1 & 1 & -4 \end{bmatrix}$

and $|A| = \begin{vmatrix} 1 & 2 & 1 \\ 3 & 2 & 3 \\ 1 & 1 & 2 \end{vmatrix} = \begin{vmatrix} 1 & 2 & 0 \\ 3 & 2 & 0 \\ 1 & 1 & 1 \end{vmatrix}$, replcaing C_2 by $C_3 - C_1$

$= \begin{vmatrix} 1 & 2 \\ 3 & 2 \end{vmatrix} = -4$ **Ans.**

Example 11(a):

Find the adjoint and rank of $A = \begin{bmatrix} 1 & 2 & 3 \\ 2 & 3 & 2 \\ 3 & 3 & 4 \end{bmatrix}$

Solution:

Here $|A| = \begin{vmatrix} 1 & 2 & 3 \\ 2 & 3 & 2 \\ 3 & 3 & 4 \end{vmatrix} = \begin{vmatrix} 1 & 0 & 0 \\ 2 & -1 & -4 \\ 3 & -3 & -5 \end{vmatrix}$

replacing C_2, C_3 by $C_2 - 2C_1$, $C_3 - 3C_1$

$$\Rightarrow \quad |A| = \begin{vmatrix} -1 & -4 \\ -3 & -5 \end{vmatrix} = (-1)(-5) - (-4)(-3) = -7 \quad \text{...(i)}$$

Also for the matrix A, we have

$C_{11} = \begin{vmatrix} 3 & 2 \\ 3 & 4 \end{vmatrix} = 6$; $C_{12} = -\begin{vmatrix} 2 & 2 \\ 3 & 4 \end{vmatrix} = -2$; $C_{13} = \begin{vmatrix} 2 & 3 \\ 3 & 3 \end{vmatrix} = -3$;

$C_{21} = -\begin{vmatrix} 2 & 3 \\ 3 & 4 \end{vmatrix} = 1$; $C_{22} = \begin{vmatrix} 1 & 3 \\ 3 & 4 \end{vmatrix} = -5$; $C_{23} = -\begin{vmatrix} 1 & 2 \\ 3 & 3 \end{vmatrix} = 3$;

$C_{31} = \begin{vmatrix} 1 & 3 \\ 3 & 2 \end{vmatrix} = -5$; $C_{32} = -\begin{vmatrix} 1 & 3 \\ 2 & 2 \end{vmatrix} = 4$; $C_{33} = \begin{vmatrix} 1 & 2 \\ 2 & 3 \end{vmatrix} = -1$;

$$\therefore \quad C = \begin{bmatrix} 6 & -2 & -3 \\ 1 & -5 & 3 \\ -5 & 4 & -1 \end{bmatrix}$$

$$\therefore \quad \text{Adj. } A = C' = \begin{bmatrix} 6 & 1 & -5 \\ -2 & -5 & 4 \\ -3 & 3 & -1 \end{bmatrix} \quad \text{...(ii) } \textbf{Ans.}$$

$$\therefore \quad A^{-1} = \frac{\text{Adj. } A}{|A|} = \frac{\begin{bmatrix} 6 & 1 & -5 \\ -2 & -5 & 4 \\ -3 & 3 & -1 \end{bmatrix}}{-7}, \text{ from (i), (ii)}$$

$$= \frac{1}{7}\begin{bmatrix} -6 & -1 & 5 \\ 2 & 5 & -4 \\ 3 & -3 & 1 \end{bmatrix} = \begin{bmatrix} -\frac{6}{7} & -\frac{1}{7} & \frac{5}{7} \\ \frac{2}{7} & \frac{5}{7} & -\frac{4}{7} \\ \frac{3}{7} & -\frac{3}{7} & \frac{1}{7} \end{bmatrix} \quad \textbf{Ans.}$$

Example 11(b):

Find A^{-1}, *if* $A = \begin{bmatrix} 2 & 3 & 1 \\ 1 & 2 & 3 \\ 3 & 1 & 2 \end{bmatrix}$

Solution:

Here $|A| = \begin{vmatrix} 2 & 3 & 1 \\ 1 & 2 & 3 \\ 3 & 1 & 2 \end{vmatrix} = \begin{vmatrix} 0 & 0 & 1 \\ -5 & -7 & 3 \\ -1 & -5 & 2 \end{vmatrix}$,

replacing C_1 and C_2 by $C_1 - 2C_3$ and $C_2 - 3C_3$ respectively.

$= \begin{vmatrix} -5 & -7 \\ -1 & -5 \end{vmatrix}$, expanding with respect to R_1

$= (-5)(-5) - (-7)(-1) = 25 - 7 = 18.$

Also we have

$C_{11} = \begin{vmatrix} 2 & 3 \\ 1 & 2 \end{vmatrix} = 1;\ C_{12} = -\begin{vmatrix} 1 & 3 \\ 3 & 2 \end{vmatrix} = 7;\ C_{13} = \begin{vmatrix} 1 & 2 \\ 3 & 1 \end{vmatrix} = -5;$

$C_{21} = -\begin{vmatrix} 3 & 1 \\ 1 & 2 \end{vmatrix} = 5;\ C_{22} = \begin{vmatrix} 2 & 1 \\ 3 & 2 \end{vmatrix} = 1;\ C_{23} = -\begin{vmatrix} 2 & 3 \\ 3 & 2 \end{vmatrix} = 7;$

$C_{31} = \begin{vmatrix} 3 & 1 \\ 2 & 3 \end{vmatrix} = 7;\ C_{32} = -\begin{vmatrix} 2 & 1 \\ 1 & 3 \end{vmatrix} = -5;\ C_{33} = \begin{vmatrix} 2 & 3 \\ 1 & 1 \end{vmatrix} = 1;$

$\therefore\ C = \begin{bmatrix} 1 & 7 & -5 \\ -5 & 1 & 7 \\ 7 & -5 & 1 \end{bmatrix}$

$\therefore\ \text{Adj. } A = C' = \begin{bmatrix} 1 & -5 & 7 \\ 7 & 1 & -5 \\ -5 & 7 & 1 \end{bmatrix}$

$\therefore\ A^{-1} = \dfrac{\text{Adj. } A}{|A|} = \dfrac{1}{18}\begin{bmatrix} 1 & -5 & 7 \\ 7 & 1 & -5 \\ -5 & 7 & 1 \end{bmatrix}$ **Ans.**

Example 12:

If A' denotes the transpose of a matrix A and

$$A = \begin{bmatrix} 1 & -2 & 3 \\ 0 & -1 & 4 \\ -2 & 2 & 1 \end{bmatrix} \text{ find } (A')^{-1}$$

Solution:

$$A' = \begin{bmatrix} 1 & 0 & -2 \\ -2 & -1 & 2 \\ 3 & 4 & 1 \end{bmatrix}, \text{ by definition of transpose of a matrix}$$

$$= B \text{ (say)}$$

$$\text{Now } |B| = \begin{vmatrix} 1 & 0 & -2 \\ -2 & -1 & 2 \\ 3 & 4 & 1 \end{vmatrix} = \begin{vmatrix} 1 & 0 & 0 \\ -2 & -1 & -2 \\ 3 & 4 & 7 \end{vmatrix}, \text{ replacing } C_3 \text{ by } C_3 + 2C_1$$

$$= \begin{vmatrix} -1 & -2 \\ 4 & 7 \end{vmatrix}, \text{expanding with respect to } R_1$$

$$= (-1)(7) - (-2)(4) = -7 + 8 = 1 \neq 0.$$

Also we have

$$C_{11} = \begin{vmatrix} -1 & 2 \\ 4 & 1 \end{vmatrix} = -9; \; C_{12} = -\begin{vmatrix} -2 & 2 \\ 3 & 1 \end{vmatrix} = 8; \; C_{13} = \begin{vmatrix} -2 & 1 \\ 3 & 4 \end{vmatrix} = -5;$$

$$C_{21} = -\begin{vmatrix} 0 & -2 \\ 4 & 1 \end{vmatrix} = -8; \; C_{22} = \begin{vmatrix} 1 & -2 \\ 3 & 1 \end{vmatrix} = 7; \; C_{23} = -\begin{vmatrix} 1 & 0 \\ 3 & 4 \end{vmatrix} = -4;$$

$$C_{31} = \begin{vmatrix} 0 & -2 \\ -1 & 2 \end{vmatrix} = -2; \; C_{31} = -\begin{vmatrix} 1 & -2 \\ -2 & 2 \end{vmatrix} = 2; \; C_{33} = \begin{vmatrix} 1 & 0 \\ -2 & -1 \end{vmatrix} = -1;$$

$$\therefore \quad C = \begin{bmatrix} -9 & 8 & -5 \\ -8 & 7 & -4 \\ -2 & 2 & -1 \end{bmatrix}$$

$$\therefore \quad \text{Adj. } B = C' = \begin{bmatrix} -9 & -8 & -2 \\ 8 & 7 & 2 \\ -5 & -4 & -1 \end{bmatrix}$$

$$\therefore \; B^{-1} = \frac{\text{Adj. B}}{|B|} = \begin{bmatrix} -9 & -8 & -2 \\ 8 & 7 & 2 \\ -5 & -4 & -1 \end{bmatrix}$$

$$\Rightarrow \quad (A')^{-1} = B^{-1} = \begin{bmatrix} -9 & -8 & -2 \\ 8 & 7 & 2 \\ -5 & -4 & -1 \end{bmatrix}$$ **Ans.**

Example 13:

Verify the theorem A . (Adj . A) = (Adj . A) . A = | A | I_2, when

$$A = \begin{bmatrix} 1 & 2 \\ 3 & -5 \end{bmatrix}.$$

Solution:

Here $|A| = \begin{vmatrix} 1 & 2 \\ 3 & -5 \end{vmatrix} = -5 - 6 = -11; \; I_2 = \begin{bmatrix} 1 & 0 \\ 0 & 1 \end{bmatrix}$ ...(i)

Also for the given matrix A, we have

$C_{11} = -5, C_{12} = -3; C_{21} = -2, C_{22} = 1$

$$\therefore \quad C = \begin{bmatrix} -5 & -3 \\ -2 & 1 \end{bmatrix} \text{ and as Adj. A} = C' = \begin{bmatrix} -5 & -2 \\ -3 & 1 \end{bmatrix}$$

$$\therefore \quad A\,.\,(\text{Adj. A}) = \begin{bmatrix} 1 & 2 \\ 3 & -5 \end{bmatrix} . \begin{bmatrix} -5 & -2 \\ -3 & 1 \end{bmatrix}$$

$$= \begin{bmatrix} 1.(-5) + 2(-3) & 1(-2) + 2.1 \\ 3(-5) - 5(-3) & 3(-2) - 5.1 \end{bmatrix}$$

$$= \begin{bmatrix} -11 & 0 \\ 0 & -11 \end{bmatrix}$$

$$\text{And (Adj. A) . A} = \begin{bmatrix} -5 & -2 \\ -3 & 1 \end{bmatrix} . \begin{bmatrix} 1 & 2 \\ 3 & -5 \end{bmatrix}$$

$$= \begin{bmatrix} -5.1 - 2.3 & -5.2 - 2(-5) \\ -3.1 + 1.3 & -3.2 + 1(-5) \end{bmatrix} = \begin{bmatrix} -11 & 0 \\ 0 & -11 \end{bmatrix}$$

$\therefore \quad$ A . (Adj. A) = (Adj. A) . A

$$= \begin{bmatrix} -11 & 0 \\ 0 & -11 \end{bmatrix} = -11 \begin{bmatrix} 1 & 0 \\ 0 & 1 \end{bmatrix}$$

$= |A| I_2$, from (i) **Hence verified.**

Example 14:

Find the inverse of $A = \begin{bmatrix} 1 & 2 & 3 \\ 1 & 3 & 4 \\ 1 & 4 & 3 \end{bmatrix}$

Solution:

Here $|A| = \begin{bmatrix} 1 & 2 & 3 \\ 1 & 3 & 4 \\ 1 & 4 & 3 \end{bmatrix} = \begin{bmatrix} 1 & 2 & 3 \\ 0 & 1 & 1 \\ 0 & 2 & 0 \end{bmatrix}$

replacing R_2, R_3 by $R_3 - R_1$, $R_3 - R_1$

$\Rightarrow$ $|A| = \begin{vmatrix} 1 & 1 \\ 2 & 0 \end{vmatrix} = -2$ ...(i)

Also for the matrix A, we have

$C_{11} = \begin{vmatrix} 3 & 4 \\ 4 & 3 \end{vmatrix} = -7$; $C_{12} = -\begin{vmatrix} 1 & 4 \\ 1 & 3 \end{vmatrix} = 1$; $C_{13} = \begin{vmatrix} 1 & 3 \\ 1 & 4 \end{vmatrix} = 1$;

$C_{21} = -\begin{vmatrix} 2 & 3 \\ 4 & 3 \end{vmatrix} = 6$; $C_{22} = \begin{vmatrix} 1 & 3 \\ 1 & 3 \end{vmatrix} = 0$; $C_{23} = -\begin{vmatrix} 1 & 2 \\ 1 & 4 \end{vmatrix} = -2$;

$C_{31} = \begin{vmatrix} 2 & 3 \\ 3 & 4 \end{vmatrix} = -1$; $C_{32} = -\begin{vmatrix} 1 & 3 \\ 1 & 4 \end{vmatrix} = -1$; $C_{33} = \begin{vmatrix} 1 & 2 \\ 1 & 3 \end{vmatrix} = 5$;

$\therefore$ $C = \begin{bmatrix} -7 & 1 & 1 \\ 6 & 0 & -2 \\ -1 & -1 & 1 \end{bmatrix}$

$\therefore$ Adj. $A = C' = \begin{bmatrix} -7 & 6 & -1 \\ 1 & 0 & -1 \\ 1 & -2 & 1 \end{bmatrix}$

$$\therefore\ A^{-1} = \frac{\text{Adj. A}}{|A|} = -\frac{1}{2}\begin{bmatrix} -7 & 6 & -1 \\ 1 & 0 & -1 \\ 1 & -2 & 1 \end{bmatrix}$$

$$= \begin{bmatrix} \frac{7}{2} & -3 & \frac{1}{2} \\ -\frac{1}{2} & 0 & \frac{1}{2} \\ -\frac{1}{2} & 1 & -\frac{1}{2} \end{bmatrix}$$

Ans.

EXISTENCE OF INVERSE

Theorem:

The necessary and sufficient condition that a square matrix may possess an inverse is that it be non-singular.

Proof:

The condition is necessary: If A is an $n \times n$ matrix and B is its inverse then by definition of the inverse we have

$$AB = I_n$$

Taking the determinants of both sides we get

$$|AB| = |I_n| \qquad \ldots(i)$$

But $|AB| = |A|.|B|$

and $|I_n| = 1$, where I_n is the $n \times n$ identity matrix

$\therefore$ From (i) we get $|A|.|B| = 1$.

which implies that $|A| \neq 0$.

$\therefore$ The matrix A is non-singular.

The condition is sufficient: If A is an $n \times n$ non-singular matrix and there be another matrix B defined by

$$B = \frac{1}{|A|}(\text{Adj A})$$

Then $$AB = A\frac{1}{|A|}(\text{Adj A})$$

$$= \frac{1}{|A|}(\text{A.Adj A})$$

$$= \frac{1}{|A|}.|A|\ I_n$$

$$= I_n$$

$\therefore AB = BA = I_n$

$\therefore$ B is the inverse of A and it exists.

SOME IMPORTANT THEOREMS

Theorem 1:

The inverse of the transposed conjugate of a non-singular matrix A is the transposed conjugate of the inverse of A singular matrix A is the transposed conjugate of the inverse of A i.e. $(A^\theta)^{-1} = (A^{-1})^\theta$

Proof:

If A is a non-singular matrix, then A is invertible and we have

$$AA^{-1} = I = A^{-1}A$$

$\Rightarrow \quad (AA^{-1})^\theta = I^\theta = (A^{-1}A)^\theta$

$\Rightarrow \quad (A^{-1})^\theta A^\theta = I = A^\theta (A^{-1})^\theta$, since $(AB)^\theta = B^\theta A^\theta$, $I^\theta = I$.

$\therefore A^\theta$ is invertible and we have $(A^\theta)^{-1} = (A^{-1})^\theta$. **Hence proved.**

Theorem 2:

If a non-singular matrix A is symmetric, then A^{-1} *is also symmetric.*

Proof:

If A is symmetric, then $A = A'$ ...(i)

Also by definition if A is non-singular, then

$$A^{-1}A = I$$

$\Rightarrow \quad A^{-1}A = I'$, since $I' = I$

$= (AA^{-1})$, since $I = A^{-1}A = AA^{-1}$

$= (A^{-1})'A'$, since $(AB)' = B'A'$

i.e., $\quad A^{-1}A = (A^{-1})'A$, since $A = A'$, from (i)

$\Rightarrow \quad A^{-1} = (A^{-1})'$, by right cancellation law.

Hence A^{-1} is symmetric by definition. **Hence proved.**

Theorem 3:

The inverse of the inverse of a matrix is the matrix itself. i.e. $(A^{-1})^{-1} = A$, *where A is the inverse.*

Proof:

Let A be the given matrix. Then its inverse is A^{-1}.

Also by definition $AA^{-1} = I = A^{-1}A$.

$\therefore A^{-1}$ is invertible and we have $(A^{-1})^{-1} = A$

i.e., the inverse of the inverse of A is A itself. **Hence proved.**

Theorem 4:

If A, B are any two n × n matrices such that AB = O, where O is the null matrix, then at least one of them is singular.

Proof:

Since A, B are two n × n matrices.

so AB = O, where O is the null matrix

$\Rightarrow \quad |A| \cdot |B| = 0$ **(Note)**

$$\Rightarrow \begin{cases} \text{either } |A| = 0, \text{ which means A is singular} \\ \text{or } \quad |B| = 0, \text{ which means B is singular} \\ \text{or both } |A| \text{ and } |B| \text{ are zero which means both A and B are singular.} \end{cases}$$

Hence at least one of A and B is singular.

Theorem 5:

If A is a non-singular matrix of order n such that AX = AY, then X = Y.

Proof:

If A is a non-singular matrix, then A^{-1} exists

Given $\quad AX = AY$

$\Rightarrow \quad A^{-1}(AX) = A^{-1}(AY)$

$\Rightarrow \quad (A^{-1}A)X = (A^{-1}A)Y$

$\Rightarrow \quad IX = IY \qquad \because A^{-1}A = I$

$\Rightarrow \quad X = Y$, by left cancellation law. **Hence proved.**

Theorem 6:

If r be the rank of a matrix A of order m × n; A_r be the normal form of A, R be the product of elementary matrices of order m and S be the product of elementary matrices of order n, then $A_r = RAS$.

Proof:

Since R and S are non-singular (*i.e.,* their inverses exist), therefore

R^{-1} A, S^{-1} = A, where R^{-1} and S^{-1} are the inverses of R and S respectively.

$\Rightarrow$ $A = B\ A_r\ C$, where $B = R^{-1}$, $C = S^{-1}$

$\Rightarrow$ $A_r = B^{-1}\ A\ C^{-1}$ (Note)

Now if A is a non-singular matrix of order n, then r = n and

$A_r = I_n$

Hence $A = B\ I_n\ C$,

which is of the form A = B, since B and C are the product of elementary matrices.

Cor.:

If two matrices A and B are of the same order m × n and same rank, then there exists non-singular square matrices P, Q such that B = PAQ.

Proof:

From above theorem we find that

$A = CA_rD$, $B = C_1\ A_r\ D_1$

where C, C_1 are product of elementary matrices of order m and D, D_1 of order n.

From $A = C\ A_r\ D$, we get $A_r = C^{-1}\ A\ D^{-1}$

Substituting this is $B = C_1\ A_r\ D_1$, we get

$B = C_1\ (C^{-1}\ A\ D^{-1})\ D_1 = (C_1\ C^{-1})\ A\ (D^1\ D_1)$

which is of the form B = PAQ.

Example 1:

If two non-singular symmetric matrices A and B be such that AB = BA (i.e. commute under multiplication), then prove that A^{-1} B and A^{-1} B^{-2} are symmetric.

Solution:

Now we are given that AB = BA

$\therefore$ We have $A^{-1}\ AB = A^{-1}\ BA$ premultiplying by A^{-1}

$\Rightarrow$ $IB = A^{-1}\ BA$, $A^{-1}\ A = 1$

$\Rightarrow$ $B = A^{-1}\ BA$, $\because$ IB = B

$\Rightarrow \quad BA^{-1} = A^{-1}BAA^{-1}$, post multiplying by A^{-1}

$= A^{-1}BI = A^{-1}B,$ ...(i)

since $AA^{-1} = I$ and $BI = B$.

Again $(A^{-1}B)' = B'(A^{-1})'$, $\because (AB)' = B'A'$

$= B'(A')^{-1}$ $\because (A^{-1})' = (A')^{-1}$

$= BA^{-1}$, $\because A' = A$,

$B' = B$ as A and B are symmetric

i.e., $(A^{-1}B)' = A^{-1}B$, from (i).

Hence $A^{-1}B$ is symmetric.

Similarly $(A^{-1}B^{-1})' = (B^{-1})'(A^{-1})$, as $(CD)' = D'C'$

$\Rightarrow \quad (A^{-1}B^{-1}) = ((B')^{-1}(A')^{-1}$

$= B^{-1}A^{-1}$ $\because A' = A, B' = B$

$= (AB)^{-1}$, $\because (AB)^{-1} = B^{-1}A^{-1}$

$= (BA)^{-1}$, $\because AB = BA$ (given)

$\Rightarrow \quad (A^{-1}B^{-1})' = A^{-1}B^{-1}$.

Hence $A^{-1}B^{-1}$ is symmetric.

Example 2(a):

If $A = \begin{bmatrix} 3 & -3 & 4 \\ 2 & -3 & 4 \\ 0 & -1 & 1 \end{bmatrix}$, *show that* $A^3 = A^{-1}$.

Solution:

Here $|A| = \begin{vmatrix} 3 & -3 & 4 \\ 2 & -3 & 4 \\ 0 & -1 & 1 \end{vmatrix}$,

$= \begin{vmatrix} 1 & 0 & 0 \\ 2 & -3 & 4 \\ 0 & -1 & 1 \end{vmatrix}$, replacing R_1 by $R_1 - R_2$

$\Rightarrow \quad |A| = \begin{vmatrix} -3 & 4 \\ -1 & 1 \end{vmatrix} = -3 + 4 = 1$ ...(i)

Also for the matrix A, we have

$$C_{11} = \begin{vmatrix} -3 & 4 \\ -1 & 1 \end{vmatrix} = 1; \; C_{12} = -\begin{vmatrix} 2 & 4 \\ 0 & 1 \end{vmatrix} = -2; \; C_{13} = \begin{vmatrix} 2 & -3 \\ 0 & -1 \end{vmatrix} = -2;$$

$$C_{21} = -\begin{vmatrix} -3 & 4 \\ -1 & 1 \end{vmatrix} = -1; \; C_{22} = \begin{vmatrix} 3 & 4 \\ 0 & 1 \end{vmatrix} = 3; \; C_{23} = -\begin{vmatrix} 3 & -3 \\ 0 & -1 \end{vmatrix} = 3;$$

$$C_{31} = \begin{vmatrix} -3 & 4 \\ -3 & 4 \end{vmatrix} = 0; \; C_{32} = -\begin{vmatrix} 3 & 4 \\ 2 & 4 \end{vmatrix} = -4; \; C_{33} = \begin{vmatrix} 3 & -3 \\ 2 & -3 \end{vmatrix} = -3;$$

$$\therefore \quad C = \begin{bmatrix} 1 & -2 & -2 \\ -1 & 3 & 3 \\ 0 & -4 & -3 \end{bmatrix}$$

$$\therefore \quad \text{Adj. A} = C' = \begin{bmatrix} 1 & -1 & 0 \\ -2 & 3 & -4 \\ -2 & 3 & -3 \end{bmatrix} \qquad \ldots\text{(ii)}$$

$$\therefore \quad A^{-1} = \frac{\text{Adj. A}}{|A|} = \begin{bmatrix} 1 & -1 & 0 \\ -2 & 3 & -4 \\ -2 & 3 & -3 \end{bmatrix}, \text{ from (i), (ii)} \qquad \ldots\text{(iii)}$$

$$\text{Alos } A^2 = \begin{bmatrix} 3 & -3 & 4 \\ 2 & -3 & 4 \\ 0 & -1 & 1 \end{bmatrix} \times \begin{bmatrix} 3 & -3 & 4 \\ 2 & -3 & 4 \\ 0 & -1 & 1 \end{bmatrix}$$

$$= \begin{bmatrix} 9-6+0 & -9+9-4 & 12-12+4 \\ 6-6+0 & -6+9-4 & 8-12+4 \\ 0-2+0 & 0+3-1 & 0-4+1 \end{bmatrix} = \begin{bmatrix} 3 & -4 & 4 \\ 0 & -1 & 0 \\ -2 & 2 & -3 \end{bmatrix}$$

$$\therefore \quad A^2 = A^2.A = \begin{bmatrix} 3 & -4 & 4 \\ 0 & -1 & 0 \\ -2 & 2 & -3 \end{bmatrix} \times \begin{bmatrix} 3 & -3 & 4 \\ 2 & -3 & 4 \\ 0 & -1 & 1 \end{bmatrix}$$

$$= \begin{bmatrix} 9-8+0 & -9+12-4 & 12-16+4 \\ 0-2+0 & 0+3+0 & 0-4+0 \\ -6+4+0 & 6-6+3 & -8+8-3 \end{bmatrix} = \begin{bmatrix} 1 & -1 & 0 \\ -2 & 3 & -4 \\ -2 & 3 & -3 \end{bmatrix}$$

i.e., $A^3 = A^{-1}$, from (iii) **Hence proved.**

Example 2(b):

Show that the matrix $A = \begin{bmatrix} 1 & a & \alpha & a\alpha \\ 1 & b & \beta & b\beta \\ 1 & c & \gamma & c\gamma \end{bmatrix}$ *is of rank 3 provided no two of a, b, c are equal and no two of* $\alpha, \beta\ \gamma$ *are equal.*

Solution:

$$A \sim \begin{bmatrix} 1 & a & \alpha & a\alpha \\ 0 & b-a & \beta-\alpha & b\beta - a\alpha \\ 0 & c-a & \gamma-\alpha & c\gamma - a\alpha \end{bmatrix},$$

replacing R_2, R_3 by $R_2 - R_1, R_3 - R_1$ respectively.

$$\sim \begin{bmatrix} 1 & 0 & 0 & 0 \\ 0 & b-a & \beta-\alpha & b\beta - a\alpha \\ 0 & c-a & \gamma-\alpha & c\gamma - a\alpha \end{bmatrix},$$

replacing C_2, C_3, C_4 by $C_2 - aC_1, C_3 - \alpha C_1$ and $C_4 - a\alpha C_1$ respectively

$$\Rightarrow \quad A \sim \begin{bmatrix} 1 & 0 & 0 & 0 \\ 0 & b-a & \beta-\alpha & b\beta - a\alpha \\ 0 & c-a & \gamma-\alpha & c\gamma - a\alpha \end{bmatrix}, \text{ replacing } C_4 \text{ by } C_4 - \alpha C_2$$

$$\sim \begin{bmatrix} 1 & 0 & 0 & 0 \\ 0 & b-a & \beta-\alpha & 0 \\ 0 & c-a & \gamma-\alpha & \begin{matrix} c\gamma - a\alpha \\ -b\gamma + b\alpha \end{matrix} \end{bmatrix}, \text{ replacing } C_4 \text{ by } C_4 - bC_3$$

$$\sim \begin{bmatrix} 1 & 0 & 0 & 0 \\ 0 & b-a & \beta-\alpha & 0 \\ 0 & c-a & \gamma-\alpha & (c-b)(\gamma-\alpha) \end{bmatrix} = B \text{ (say)},$$

Now a minor of order 3 of B

$$= \begin{vmatrix} 1 & 0 & 0 \\ 0 & b-a & 0 \\ 0 & c-a & (c-b)(\gamma-\alpha) \end{vmatrix} = \begin{vmatrix} b-a & 0 \\ c-a & (c-b)(\gamma-\alpha) \end{vmatrix},$$

expanding with respect to R_1

$= (b - a)(c - b)(\gamma - \alpha) \neq 0$, as no two of a, b, c and no two of α, β, γ are equal (given).

$\therefore$ $\rho(B) \geq 3$ ...(i)

Also the matrix B does not posses any minor of order 4 *i.e.,*, of order 3 + 1, so $\rho(B) \leq 3$ ...(ii)

$\therefore$ From (i) and (ii) we get $\rho(B) = 3$

and therefore $\rho(A) = 3$, as $A \sim B$. **Hence proved.**

Example 2(c):

Find the rank of the matrix $A = \begin{bmatrix} 4 & 5 & 8 \\ 5 & 6 & 7 \\ 7 & 8 & 9 \end{bmatrix}$

Solution:

The determinant of order 3 formed by A

$= \begin{vmatrix} 4 & 5 & 8 \\ 5 & 6 & 7 \\ 7 & 8 & 9 \end{vmatrix} = \begin{vmatrix} 4 & 1 & 2 \\ 5 & 1 & 1 \\ 7 & 1 & 1 \end{vmatrix}$, replacing C_2, C_3 by $C_2 - C_1, C_3 - C_2$ respectively

$= \begin{vmatrix} 0 & 1 & 0 \\ 1 & 1 & -2 \\ 3 & 1 & -2 \end{vmatrix}$, replacing C_1, C_3 by $C_1 - 4C_2$ and $C_3 - 3C_2$ respectively

$= -\begin{vmatrix} 1 & -2 \\ 3 & -2 \end{vmatrix}$, expanding with respect to R_1

$= -[-2 + 6] = -4 \neq 0$

$\therefore$ $\rho(A) \geq 3$ ...(i)

Also the matrix A does not possess any minor of order 4 *i.e.,* 3 + 1 so

$\rho(A) \leq 3$ ...(ii)

$\therefore$ From (i) and (ii) we get $\rho(A) = 3$. **Ans.**

Example 3:

Find two non-singular matrices P and Q such that PAQ is in the normal form, where $A = \begin{bmatrix} 1 & 1 & 1 \\ 1 & -1 & -1 \\ 3 & 1 & 1 \end{bmatrix}$

Solution:

Here we find that A is a 3 × 3 matrix

$$\therefore \quad [A]_{3\times 3} = I_3\ A\ I_3$$

$$\Rightarrow \begin{bmatrix} 1 & 1 & 1 \\ 1 & -1 & -1 \\ 3 & 1 & 1 \end{bmatrix} = \begin{bmatrix} 1 & 0 & 0 \\ 0 & 1 & 0 \\ 0 & 0 & 1 \end{bmatrix} .A. \begin{bmatrix} 1 & 0 & 0 \\ 0 & 1 & 0 \\ 0 & 0 & 1 \end{bmatrix}$$

$$\Rightarrow \begin{bmatrix} 1 & 1 & 1 \\ 2 & 0 & 0 \\ 2 & 0 & 0 \end{bmatrix} = \begin{bmatrix} 1 & 0 & 0 \\ 1 & 1 & 0 \\ -1 & 0 & 1 \end{bmatrix} .A. \begin{bmatrix} 1 & 0 & 0 \\ 0 & 1 & 0 \\ 0 & 0 & 1 \end{bmatrix},$$

applying $R_2 + R_1$, $R_3 - R_1$

$$\Rightarrow \begin{bmatrix} 1 & 1 & 1 \\ 2 & 0 & 0 \\ 0 & 0 & 0 \end{bmatrix} = \begin{bmatrix} 1 & 0 & 0 \\ 1 & 1 & 0 \\ -2 & -1 & 1 \end{bmatrix} .A. \begin{bmatrix} 1 & 0 & 0 \\ 0 & 1 & 0 \\ 0 & 0 & 1 \end{bmatrix}, \text{ applying } R_3 - R_2$$

$$\Rightarrow \begin{bmatrix} 1 & 1 & 1 \\ 1 & 0 & 0 \\ 0 & 0 & 0 \end{bmatrix} = \begin{bmatrix} 1 & 0 & 0 \\ 1/2 & 1/2 & 0 \\ -2 & -1 & 1 \end{bmatrix} .A. \begin{bmatrix} 1 & 0 & 0 \\ 0 & 1 & 0 \\ 0 & 0 & 1 \end{bmatrix}, \text{ applying } R_2\left(\frac{1}{2}\right)$$

$$\Rightarrow \begin{bmatrix} 0 & 1 & 1 \\ 1 & 0 & 0 \\ 0 & 0 & 0 \end{bmatrix} = \begin{bmatrix} 1/2 & -1/2 & 0 \\ 1/2 & 1/2 & 0 \\ -2 & -1 & 1 \end{bmatrix} .A. \begin{bmatrix} 1 & 0 & 0 \\ 0 & 1 & 0 \\ 0 & 0 & 1 \end{bmatrix}, \text{ applying } R_1 - R_2$$

$$\Rightarrow \begin{bmatrix} 1 & 0 & 0 \\ 0 & 1 & 1 \\ 0 & 0 & 0 \end{bmatrix} = \begin{bmatrix} 1/2 & -1/2 & 0 \\ 1/2 & -1/2 & 0 \\ -2 & -1 & 1 \end{bmatrix} .A. \begin{bmatrix} 1 & 0 & 0 \\ 0 & 1 & 0 \\ 0 & 0 & 1 \end{bmatrix},$$

interchanging R_1 and R_2

$$\Rightarrow \begin{bmatrix} 1 & 0 & 0 \\ 0 & 1 & 0 \\ 0 & 0 & 0 \end{bmatrix} = \begin{bmatrix} 1/2 & 1/2 & 0 \\ 1/2 & -1/2 & 0 \\ -2 & -1 & 1 \end{bmatrix} .A. \begin{bmatrix} 1 & 0 & 0 \\ 0 & 1 & -1 \\ 0 & 0 & 1 \end{bmatrix},$$

applying $C_3 - C_2$ **(Note)**

Since L.H.S. is in the normal form, so we have

$$P = \begin{bmatrix} 1/2 & 1/2 & 0 \\ 1/2 & -1/2 & 0 \\ -2 & -1 & 1 \end{bmatrix}, Q = \begin{bmatrix} 1 & 0 & 0 \\ 0 & 1 & -1 \\ 0 & 0 & 1 \end{bmatrix},$$ **Ans.**

Example 4:

Find the non-singular matrices R and S, such that RAS is the normal form, where $A = \begin{bmatrix} 2 & 2 & -6 \\ -1 & 2 & 2 \end{bmatrix}$.

Solution:

Here we find that A is a 2 × 3 matrix.

$$\therefore \quad [A]_{2\times 3} = I_2\, A\, I_3$$

$$\Rightarrow \begin{bmatrix} 2 & 2 & -6 \\ -1 & 2 & 2 \end{bmatrix} = \begin{bmatrix} 1 & 0 \\ 0 & 1 \end{bmatrix} . A . \begin{bmatrix} 1 & 0 & 0 \\ 0 & 1 & 0 \\ 0 & 0 & 1 \end{bmatrix}$$

Now we are to bring L.H.S. to the normal form by applying elementary row and column operations.

$$\therefore \begin{bmatrix} 1 & 1 & -3 \\ -1 & 2 & 2 \end{bmatrix} = \begin{bmatrix} 1/2 & 0 \\ 0 & 1 \end{bmatrix} . A . \begin{bmatrix} 1 & 0 & 0 \\ 0 & 1 & 0 \\ 0 & 0 & 1 \end{bmatrix}, \text{ by } R_1\left(\frac{1}{2}\right)$$

$$\Rightarrow \begin{bmatrix} 1 & 1 & -3 \\ 0 & 3 & -1 \end{bmatrix} = \begin{bmatrix} 1/2 & 0 \\ 1/2 & 1 \end{bmatrix} . A . \begin{bmatrix} 1 & 0 & 0 \\ 0 & 1 & 0 \\ 0 & 0 & 1 \end{bmatrix}, \text{ by } R_2 + R_1$$

$$\Rightarrow \begin{bmatrix} 1 & 0 & 0 \\ 0 & 3 & -1 \end{bmatrix} = \begin{bmatrix} 1/2 & 0 \\ 1/2 & 1 \end{bmatrix} . A . \begin{bmatrix} 1 & -1 & 3 \\ 0 & 1 & 0 \\ 0 & 0 & 1 \end{bmatrix}, \text{ by } C_2 - C_1 \text{ and } C_3 + 3C_1$$

$$\Rightarrow \begin{bmatrix} 1 & 0 & 0 \\ 0 & 3 & -1 \end{bmatrix} = \begin{bmatrix} 1/2 & 0 \\ 1/2 & 1 \end{bmatrix} . A . \begin{bmatrix} 1 & -1/3 & 3 \\ 0 & 1/3 & 0 \\ 0 & 0 & 1 \end{bmatrix}, \text{ replacing } C_2 \text{ by } \frac{1}{3}C_2$$

$$\Rightarrow \begin{bmatrix} 1 & 0 & 0 \\ 0 & 1 & 0 \end{bmatrix} = \begin{bmatrix} 1/2 & 0 \\ 1/2 & 1 \end{bmatrix} . A . \begin{bmatrix} 1 & -1/3 & 8/3 \\ 0 & 1/3 & 1/3 \\ 0 & 0 & 1 \end{bmatrix}, \text{ by } C_3 + C_2$$

Since L.H.S. is in the normal form, so

$$R = \begin{bmatrix} 1/2 & 0 \\ 1/2 & 1 \end{bmatrix} \text{ and } S = \begin{bmatrix} 1 & -1/3 & 8/3 \\ 0 & 1/3 & 1/3 \\ 0 & 0 & 0 \end{bmatrix}$$

Ans.

Example 5:

Find the rank of an m × n matrix, every element of which is unity.

Solution:

Let an m × n matrix be $A = \begin{bmatrix} 1 & 1 & 1 & \dots & 1 \\ 1 & 1 & 1 & \dots & 1 \\ \dots & \dots & \dots & \dots & \dots \\ 1 & 1 & 1 & \dots & 1 \end{bmatrix}$

Then we find that every square submatrix of A higher than 1 × 1 will be a matrix each element of which is unity and therefore the value of the determinant will be always zero, since its rows and columns are identical. But the square sub-matrices of order 1 × 1 are [1] and the determinants of these are $| A | \neq 0$.

Hence the rank of A is 1. **Ans.**

Example 6:

Prove that the rank of a matrix remains unaltered by the application of elementary row and column operations.

Or

Prove that two equivalent matrices have the same rank.

Solution:

Let an m × n matrix A be given by

$$A = \begin{bmatrix} a_{11} & a_{12} & \dots & a_{1n} \\ a_{21} & a_{22} & \dots & a_{2n} \\ \dots & \dots & \dots & \dots \\ a_{m1} & a_{m2} & \dots & a_{mn} \end{bmatrix}$$

Let M be any minor of order r belonging to the first r rows of | A |.

Now firstly if we interchange any two rows or columns of A, then the minor M either remains unaltered or changes sign.

Secondly if we multiply one row or column of A by a number λ, then either the minor M remains unaltered or changes into λ M.

Thirdly if we replace any row R_i (or column C_i) by $R_i + \lambda R_j$ (or $C_i + \lambda C_j$), then either the minor M remains unaltered or change into a sum or difference of two of the original minors.

Let B the matrix obtained from A by the application of any one of the above three elementary row or column operations.

Thus if all the minors of order r in | A | are zero, then all the minors of order r in | B | are also zero.

$\therefore$ rank of B $\leq$ rank of A ...(i)

Similarly if all the minors of order r in | B | are zero, then all the minors of order r in | A | are also zero.

$\therefore$ rank of A $\leq$ rank of B ...(ii)

$\therefore$ From (i) and (ii) we get

rank of A = rank of B. **Hence proved.**

Example 7(a):

If A is of order m × n, R is a non-singular matrix of order m, show that

Rank of RA = Rank of A.

Solution:

Let $A = E\,A_r\,F$ and $R = E_1$

Then $RA = E_1\,(E\,A_r\,F) = E_1\,E\,A_r\,F$

i.e., RA has been expressed as the result of elementary operations on A_r.

Thus Rank of (RA) = Rank A_r = Rank A.

Example 7(b):

Find the reciprocal (or inverse) of the matrix

$$S = \begin{bmatrix} 0 & 1 & 1 \\ 1 & 0 & 1 \\ 1 & 1 & 0 \end{bmatrix}$$ *and show that the transform of*

the matrix $$A = \frac{1}{2}\begin{bmatrix} b+c & c-a & b-a \\ c-b & c+a & a-b \\ b-c & a-c & a+b \end{bmatrix}$$ *by S i.e.*

Solution:

In the usual way we can show that

$$S^{-1} = \text{inverse of } S = \frac{1}{2}\begin{bmatrix} -1 & 1 & 1 \\ 1 & -1 & 1 \\ 1 & 1 & -1 \end{bmatrix}$$

$$\therefore\ SA = \begin{bmatrix} 0 & 1 & 1 \\ 1 & 0 & 1 \\ 1 & 1 & 0 \end{bmatrix} \times \frac{1}{2}\begin{bmatrix} b+c & c-a & b-a \\ c-b & c+a & a-b \\ b-c & a-c & a+b \end{bmatrix}$$

$$= \frac{1}{2}\begin{bmatrix} 0 & 2a & 2a \\ 2b & 0 & 2b \\ 2c & 2c & 0 \end{bmatrix},$$ multiplying the matrices in the usual way.

$$\Rightarrow\quad SA = \begin{bmatrix} 0 & a & a \\ b & 0 & b \\ c & c & 0 \end{bmatrix}$$

$$\therefore\ SAS^{-1} = \begin{bmatrix} 0 & a & a \\ b & 0 & b \\ c & c & 0 \end{bmatrix} \times \frac{1}{2}\begin{bmatrix} -1 & 1 & 1 \\ 1 & -1 & 1 \\ 1 & 1 & -1 \end{bmatrix}$$

$$= \frac{1}{2}\begin{bmatrix} 2a & 0 & 0 \\ 0 & 2b & 0 \\ 0 & 0 & 2c \end{bmatrix},$$

multiplying the two matrices in the usual way.

$$= \begin{bmatrix} a & 0 & 0 \\ 0 & b & 0 \\ 0 & 0 & c \end{bmatrix},$$ which is a diagonal matrix.

Example 8:

Find the rank of the matrix $A = \begin{bmatrix} 4 & 5 & 8 \\ 5 & 6 & 7 \\ 7 & 8 & 9 \end{bmatrix}$

Solution:

The determinant of order 3 formed by A

$$= \begin{vmatrix} 4 & 5 & 8 \\ 5 & 6 & 7 \\ 7 & 8 & 9 \end{vmatrix} = \begin{vmatrix} 4 & 1 & 2 \\ 5 & 1 & 1 \\ 7 & 1 & 1 \end{vmatrix}$$, replacing C_2, C_3 by $C_2 - C_1, C_3 - C_2$ respectively

$$= \begin{vmatrix} 0 & 1 & 0 \\ 1 & 1 & -2 \\ 3 & 1 & -2 \end{vmatrix}$$, replacing C_1, C_3 by $C_1 - 4C_2$ and $C_3 - 3C_2$ respectively

$$= -\begin{vmatrix} 1 & -2 \\ 3 & -2 \end{vmatrix}$$, expanding with respect to R_1

$= -[-2 + 6] = -4 \neq 0$

$\therefore \quad p(A) \geq 3$ ...(i)

Also the matrix A does not possess any minor of order 4 *i.e.*, 3 + 1 so

$p(A) \leq 3$ ...(ii)

$\therefore$ From (i) and (ii) we get $p(A) = 3$. **Ans.**

Example 9:

Find the rank of the matrix

$$A = \begin{bmatrix} 1^2 & 2^2 & 3^2 & 4^2 \\ 2^2 & 3^2 & 4^2 & 5^2 \\ 3^2 & 4^2 & 5^2 & 6^2 \\ 4^2 & 5^2 & 6^2 & 7^2 \end{bmatrix}$$

Solution:

$$\text{Given } A = \begin{bmatrix} 1 & 4 & 9 & 16 \\ 4 & 9 & 16 & 25 \\ 9 & 16 & 25 & 36 \\ 16 & 25 & 36 & 49 \end{bmatrix}$$

$$\therefore A \sim \begin{bmatrix} 1 & 0 & 0 & 0 \\ 4 & -7 & -20 & -39 \\ 9 & -20 & -56 & -108 \\ 16 & -39 & -108 & -107 \end{bmatrix}$$, replacing C_2, C_3, C_4 by $C_2 - 4C_1, C_2 - 9C_1$ and $C_4 - 16C_1$ respectively.

$$\sim \begin{bmatrix} 1 & 0 & 0 & 0 \\ 0 & -7 & -20 & -39 \\ 1 & -6 & -16 & -30 \\ 0 & -11 & -28 & -51 \end{bmatrix},$$ replacing R_2, R_3, R_4 by $R_2 - 4R_1, R_3 - 2R_2$ and $R_4 - 4R_2$ respectively

$$\sim \begin{bmatrix} 1 & 0 & 0 & 0 \\ 0 & -7 & -20 & -39 \\ 0 & -6 & -16 & -30 \\ 0 & -4 & -8 & -12 \end{bmatrix},$$ replacing R_3, R_4 by $R_3 - R_1$ and $R_4 - R_2$ respectively

$$\sim \begin{bmatrix} 1 & 0 & 0 & 0 \\ 0 & 1 & 4 & 9 \\ 0 & 2 & 8 & 18 \\ 0 & 4 & 8 & 12 \end{bmatrix},$$ replacing R_2, R_3, R_4 by $-(R_2 - R_3) - (R_3 - R_4)$ and $-R_4$ respectively.

$$\sim \begin{bmatrix} 1 & 0 & 0 & 0 \\ 0 & 1 & 4 & 9 \\ 0 & 0 & 0 & 0 \\ 0 & 0 & -8 & -24 \end{bmatrix},$$ replacing R_3, R_4 by $R_3 - 2R_2$ $R_4 - 4R_2$ respectively.

$$\sim \begin{bmatrix} 1 & 0 & 0 & 0 \\ 0 & 1 & 0 & 0 \\ 0 & 0 & 0 & 0 \\ 0 & 0 & -8 & -24 \end{bmatrix},$$ replacing C_3, C_4 by $C_3 - 4C_2$, $C_4 - 9C_2$ respectively.

$$\sim \begin{bmatrix} 1 & 0 & 0 & 0 \\ 0 & 1 & 0 & 0 \\ 0 & 0 & 0 & 0 \\ 0 & 0 & 1 & 3 \end{bmatrix},$$ replacing R_4 by $-\frac{1}{8}R_4$

$$\sim \begin{bmatrix} 1 & 0 & 0 & 0 \\ 0 & 1 & 0 & 0 \\ 0 & 0 & 1 & 3 \\ 0 & 0 & 0 & 0 \end{bmatrix},$$ int erchanging R_3 and R_4

$$\sim \begin{bmatrix} 1 & 0 & 0 & 0 \\ 0 & 1 & 0 & 0 \\ 0 & 0 & 1 & 0 \\ 0 & 0 & 0 & 0 \end{bmatrix}, \text{ replacing } C_4 \text{ by } C_4 - 3C_2$$

$$\sim \begin{bmatrix} I_2 & 0 \\ 0 & 0 \end{bmatrix}$$

∴ The rank of the matrix A is 3. **Ans.**

Example 10:

If $A = \begin{bmatrix} a_1 & 0 & 0 \\ 0 & a_2 & 0 \\ 0 & 0 & a_3 \end{bmatrix}$, *where none of a's is zero, then show that A is invertible. Also evaluate* A^{-}.

Solution:

$$|A| = \begin{vmatrix} a_1 & 0 & 0 \\ 0 & a_2 & 0 \\ 0 & 0 & a_3 \end{vmatrix} = a_1\ a_2\ a_3 \text{ on evaluating}$$

i.e., $|A| \neq 0$. Hence A is invertible.

Also

$$C_{11} = \begin{vmatrix} a_2 & 0 \\ 0 & a_3 \end{vmatrix} = a_2\, a_3;\ C_{12} = -\begin{vmatrix} 0 & 0 \\ 0 & a_3 \end{vmatrix} = 0;\ C_{13} = \begin{vmatrix} 0 & a_2 \\ 0 & 0 \end{vmatrix} = 0;$$

$$C_{21} = -\begin{vmatrix} 0 & 0 \\ 0 & a_3 \end{vmatrix} = 0;\ C_{22} = \begin{vmatrix} a_1 & 0 \\ 0 & a_3 \end{vmatrix} = a_1\, a_3;\ C_{23} = -\begin{vmatrix} a_1 & 0 \\ 0 & 0 \end{vmatrix} = 0;$$

$$C_{31} = \begin{vmatrix} 0 & 0 \\ a_2 & 0 \end{vmatrix} = 0;\ C_{32} = -\begin{vmatrix} a_1 & 0 \\ 0 & 0 \end{vmatrix} = 0;\ C_{33} = \begin{vmatrix} a_1 & 0 \\ 0 & a_2 \end{vmatrix} = a_1\, a_3;$$

$$\therefore \quad C = \begin{bmatrix} a_2a_3 & 0 & 0 \\ 0 & a_3a_1 & 0 \\ 0 & 0 & a_1a_2 \end{bmatrix}$$

$$\therefore \quad \text{Adj. } A = C' = \begin{bmatrix} a_2a_3 & 0 & 0 \\ 0 & a_3a_1 & 0 \\ 0 & 0 & a_1a_2 \end{bmatrix}$$

$$A^{-1} = \frac{\text{Adj. } A}{|A|} = \frac{1}{a_1\, a_2\, a_3}\begin{bmatrix} a_2a_3 & 0 & 0 \\ 0 & a_3a_1 & 0 \\ 0 & 0 & a_1a_2 \end{bmatrix}$$

$$= \begin{bmatrix} 1/a_1 & 0 & 0 \\ 0 & 1/a_2 & 0 \\ 0 & 0 & 1/a_3 \end{bmatrix}$$

Ans.

Example 11:

If A is invertible show that $\overline{A}$ is invertible

Solution:

If A is invertible, then we know that

$$AA^{-1} = I = A^{-1}A$$

$$\Rightarrow \qquad \overline{(AA^{-1})} = \overline{I} = \overline{(A^{-1}A)}$$

$$\Rightarrow \qquad \overline{A}\,\overline{(A^{-1})} = I = \overline{(A^{-1})}\,\overline{A}. \quad \because \overline{AB} = \overline{A}.\overline{B}.$$

Hence $\overline{A}$ is invertible and we have $(\overline{A})^{-1} = \overline{(A^{-1})}$. **Hence proved.**

Example 12:

Find the rank of $A = \begin{bmatrix} 1 & 2 & 3 & 1 \\ 2 & 4 & 6 & 2 \\ 1 & 2 & 3 & 2 \end{bmatrix}$

Solution:

$$A \sim \begin{bmatrix} 1 & 2 & 3 & 1 \\ 0 & 0 & 0 & 0 \\ 0 & 0 & 0 & 1 \end{bmatrix}$$, replacing R_2, R_3 by $R_2 - 2R_1$ and $R_3 - R_1$ respecitvely

$$\sim \begin{bmatrix} 1 & 2 & 3 & 0 \\ 0 & 0 & 0 & 0 \\ 0 & 0 & 0 & 1 \end{bmatrix}$$, replacing R_1 by $R_1 - R_2$

$$\sim \begin{bmatrix} 1 & 0 & 0 & 0 \\ 0 & 0 & 0 & 0 \\ 0 & 0 & 0 & 1 \end{bmatrix}, \text{ replacing } C_2, C_3 \text{ by } C_2 - 2C_1 \text{ and } C_3 - 3C_1 \text{ respectively}$$

$$\sim \begin{bmatrix} 1 & 0 & 0 & 0 \\ 0 & 0 & 0 & 0 \\ 0 & 1 & 0 & 0 \end{bmatrix}, \text{ interchaing } C_2 \text{ and } C_3$$

$$\sim \begin{bmatrix} 1 & 0 & 0 & 0 \\ 0 & 1 & 0 & 0 \\ 0 & 0 & 0 & 0 \end{bmatrix}, \text{ interchaing } R_2 \text{ and } R_3$$

$$\sim \begin{bmatrix} I_2 & 0 \\ 0 & 0 \end{bmatrix}$$

∴ The rank of matrix A is 2. **Ans.**

ECHELON FORM OF A MATRIX

Definition: *If in a matrix,*

(i) all the non-zero rows, if any, precede the zero rows,

(ii) the number of zero proceeding the first non-zero element in a row is less than the number of such zero in the succeeding row,

(iii) the first non-zero elemen[illegible] a row is unity, then it is in the Echelon form.

Note: The number of non-zero rows of a matrix given in the Echelon form is its rank.

Examples of a matrix in the Echelon Form :-

$$\begin{bmatrix} 1 & 3 & 4 & 5 & 6 \\ 0 & 1 & 2 & 3 & 4 \\ 0 & 0 & 1 & 2 & 3 \\ 0 & 0 & 0 & 0 & 0 \end{bmatrix}$$

In this matrix we observe that

(i) the first three non-zero rows precede the fourth row which is a zero row,

(ii) the number of zero is R_1, R_2 and R_3 are 5, 2 and 1 respectively which are in descending order,

(iii) the first non-zero term in each row is unity.

Hence all the three conditions of the Echelon form are satisfied.

Also there being three non-zero rows in this matrix, its rank is 3. This fact can be proved by actually finding the rank of this matrix.

In this matrix, a minor of order 4.

$$= \begin{vmatrix} 1 & 3 & 4 & 5 \\ 0 & 1 & 2 & 3 \\ 0 & 0 & 1 & 2 \\ 0 & 0 & 0 & 0 \end{vmatrix} = 0, \text{ one row being of zero.}$$

In a similar way we can show that all minors of order 4 are zero.

Now a minor of order 3 $= \begin{vmatrix} 1 & 3 & 4 \\ 0 & 1 & 2 \\ 0 & 1 & 0 \end{vmatrix}$

$$= \begin{vmatrix} 1 & 2 \\ 0 & 1 \end{vmatrix}, \text{ expanding w.r. to } C_1$$

$$= 1 \neq 0$$

Hence the rank of this matrix = 3.

Example:

Find the rank of the matrix $A = \begin{bmatrix} 0 & 1 & 2 & 3 \\ 0 & 0 & 1 & -1 \\ 0 & 0 & 0 & 0 \end{bmatrix}$

Solution:

In the given matrix we observe that

(i) the first two non-zero rows precede the third row which is a zero row,

(ii) the number of zero in R_2, R_3 and R_1 are 4, 2 and 1 respectively which are in descending order, and

(iii) the first non-zero term in each row is unity.

Hence all three conditions of the Echelon form are satisfied.

Also there being two non-zero rows in this matrix, its rank is 2. **Ans.**

INVARIANCE OF RANK UNDER ELEMENTARY OPERATIONS

Theorem:

All equivalent matrices have the same ranks i.e. the rank of a matrix remains unaltered by the application of elementary row and column operations.

Proof:

Let r be the rank of an m × n matrix $A = [a_{ij}]$.

Case I: *If ith and jth rows are interchanged (which may be written symbolically as R_{ij} or $R_i \leftrightarrow R_j$) then it does not effect the rank.*

Let B denote the matrix obtained from the matrix A by the elementary operation $R_i \leftrightarrow R_j$ and let p be the rank of B.

Also if D be any (r + 1) rowed square sub-matrix of B, then | D | = ± | C |, where C is a particular (r + 1) rowed submatrix of A.

As r is the rank of the matrix A so every (r + 1) rowed minor of A vanishes and therefore p, the rank osw *are multiplied by a non-zero number λ (which may be written symbolically, as $R_i \to \lambda R_j$, $\lambda \neq 0$) it does not effect the rank.*

Let B denote the matrix obtained from the matrix A by the elementary operation $R_i \to \lambda R_j$ and let p be the rank of B.

Let D be any (r + 1) rowed square submatrix of B and let C be the sub-matrix of A having the same position as D. Then either | D | = | C | or | D | = λ | C |.

[Here | D | = | C | happens if the ith row of A is one of those rows which are removed to obtain D from B and | D | = λ | C | happens when the ith row is not removed while obtaining C from A].

Also as r is the rank of the matrix A so every (r + 1) rowed minor of A vanishes and therefore in particular | C | = 0 and consequently in both the above cases | D | = 0.

∴ p, the rank of B, cannot exceed r, the rank of A.

i.e., $p \leq r$

Also we can obtain A from B by the elementary operation $R_i \to \lambda^{-1} R_j$, therefore in that case interchanging the roles of A and B we shall gets $r \leq p$.

Hence r = p.

Case III: *If to the elements of the ith row are added the products by any non-zero numbers λ of the corresponding elements of jth row (which may be written symbolically as $R_i \rightarrow R_j + \lambda R_j$, $\lambda \neq 0$)*

Let B denote the matrix obtained from the matrix A by the elementary operation $R_i \rightarrow R_j + \lambda R_i$ and let p be the rank of B.

Let D be any (r + 1) rowed square submatrix of B and let C be the submatrix of A having the same position as D.

Now three sub-cases arise:-

(i) If A and B differ only in the ith row *i.e.*, if ith row of B is one of those rows which have been removed while obtaining C.

In this case D = C and therefore | D | = | C |.

$\therefore$ The rank of A is r, so | C | = 0 and consequently | D | = 0

(ii) If ith row of B has not been removed but jth row has been removed while obtaining D.

In this case $|D| = |C| + \lambda |C_0|$, where C_0 is an (r + 1) rowed matrix which is obtained from C by replacing A_{ik} by a_{jk} *i.e.*, C_0 is obtained from C by performing the elementary operation R_{ij} or $R_i \leftrightarrow R_j$ and then removing those rows and columns of the new matrix which are removed to obtain D from B.

$\therefore$ $|C_0|$ is negative of some (r + 1) rowed minor of A and as the rank of A is r so every (r + 1) rowed minor of A is zero *i.e.*, | C | = 0, $|C_0| = 0$ and consequently | D | = 0.

(iii) If neither the ith row nor the jth row of B has been removed while obtained D.

Here | D | = | C | and so as before | D | = 0.

$\therefore$ Every (r + 1) rowed minor of B vanishes so p, the rank of B, cannot exceed r, the rank of A *i.e.*, $p \leq r$.

Also we can obtain A from B by the elementary operation $R_i \rightarrow R_j - \lambda R_j$, therefore in that case interchanging the roles of A and B we shall get $r \leq p$.

Hence r = p.

Thus we have observed that the rank of a matrix remains invariant under elementary row operations. Similarly it can be shown that the rank of a matrix remains invariant under elementary column operations too.

Note: By the application of the above theorem we can easily obtain the rank of a matrix for if we can obtain a matrix B by elementary operations on a matrix A and of the rank of B can be easily determined by inspection or simple calculations as given in previous articles in this chapter, then we can determine the rank of A.

RANK OF A MATRIX

Definition: *If in an m × n matrix A, at least one of its r × r minors is different from zero while all the minors of order (r + 1) are zero, then r is defined as the rank of the matrix A.*

Or

A number r is defined as the rank of an m × n matrix A provide (i) A has at least one minor of order r which does not vanish and (ii) there is no minor of order (r + 1) which is nor equal to zero.

Note 1: The rank of a matrix remains unaltered by the application of elementary row or column operations *i.e.,* all equivalent matrices have the same rank.

Note 2: From the definition of the rank of a matrix we conclude that :

(a) If a matrix A does not posses any minor of order (r+1) then p(A)< r.

(b) If at least one minor of order r of the matrix A is not equal to zero, then p (A) < r.

Note 3: If every minor of order p of a matrix A is zero, the every minor of order higher than p is definitely zero.

Note 4: The rank of a matrix A is also denoted by p (A).

Note 5: The rank of a zero matrix by definition is 0 *i.e.,* p (O) = 0.

Consider the matrix $A = \begin{bmatrix} 1 & 2 & 3 \\ 2 & 3 & 4 \\ 3 & 5 & 7 \end{bmatrix}$

This matrix A has only one three rowed minor *i.e.,* minor of order 3,

viz. $\begin{vmatrix} 1 & 2 & 3 \\ 2 & 3 & 4 \\ 3 & 5 & 7 \end{vmatrix}$ and its value can easily be calculated to zero, by expanding

with respect to first row.

The matrix A has 9 minors of order 2 (or two-rowed minors) and one of them is $\begin{vmatrix} 3 & 4 \\ 5 & 7 \end{vmatrix}$ which has the value

$$(3 \times 7) - (5 \times 4) = 21 - 20 = 1 \neq 0.$$

This fact that A is a matrix whose every minor of order 3 is zero and there is at least one minor of order 2 which is not equal to zero is also expressed as 'the rank of the matrix A is 2'.

Example 1:

Find the rank of the matrix

$$A = \begin{bmatrix} 1 & 1 & 1 \\ a & b & c \\ a^3 & b^3 & c^3 \end{bmatrix}, \textit{ where a, b, c are all real.}$$

Solution:

$$|A| = \begin{vmatrix} 1 & 1 & 1 \\ a & b & c \\ a^3 & b^3 & c^3 \end{vmatrix} = \begin{vmatrix} 1 & 0 & 0 \\ a & b-a & c-a \\ a^3 & b^3-a^3 & c^3-a^3 \end{vmatrix}, \text{ replacing } C_2, C_3 \text{ by } C_2 - C_1, C_3 - C_1$$

$$= \begin{vmatrix} b-a & c-a \\ b^3-a^3 & c^3-a^3 \end{vmatrix}, \text{ expanding with respect to } R_1$$

$$= (b-a)(c-a) \begin{vmatrix} 1 & 1 \\ b^2+ab+a^2 & c^2+ca+a^2 \end{vmatrix},$$

taking (b – a), (c – a) common from C_1 and C_2

$$= (b-a)(c-a) \begin{vmatrix} 1 & 1 \\ b^2+ab+a^2 & c^2+ca-b^2-ab \end{vmatrix},$$

replacing C_2 by $C_2 - C_1$

$$= (b-a)(c-a)[(c^2+ca-b^2-ab) - 0]$$

$$= (b-a)(c-a)[(c^2-b^2) + a(c-b)] \quad \textbf{(Note)}$$

$$= (b-a)(c-a)[(c-b)(c+b+a)]$$

$$\Rightarrow |A| = (a-b)(b-c)(c-a)(a+b+c) \quad \ldots(i)$$

Now following cases arise :

Case I: $a = b = c$.

If a = b = c, then $A = \begin{bmatrix} 1 & 1 & 1 \\ a & a & a \\ a^3 & a^3 & a^3 \end{bmatrix}$

Therefore all minors of order 3 and 2 of A are zero.

Also as no element of A is zero, so A has non zero minors of order 1.

Hence in this case, the rank of A is 1. **Ans.**

Case II: *Two of the numbers a, b, c are equal but are different from the third.*

let a = b ≠ c

Then $= |A| = \begin{vmatrix} 1 & 1 & 1 \\ a & a & a \\ a^3 & a^3 & a^3 \end{vmatrix} = 0$, as C_1 and C_2 are identical.

Also A has a minor of order 2, viz. $\begin{vmatrix} 1 & 1 \\ a & c \end{vmatrix} = c - a \neq 0$. $\because a \neq c$

Hence in this case the rank of A is 2. **Ans.**

Similarly we can discuss the cases b = c ≠ a, c = a ≠ b.

Case III: *a, b, c are all different but a + b + c = 0.*

In this case from (i), it is evident that | A | = 0. **(Note)**

Also A has a minor of order 2, viz. $\begin{vmatrix} 1 & 1 \\ a & b \end{vmatrix} = b - a \neq 0$, $\because a \neq b$

Hence in this case the rank of A is 2. **Ans.**

Case IV: *a, b, c are all different but a + b + c ≠ 0.*

In this case from (i), it is evident that | A | ≠ 0. **(Note)**

i.e., A has a non-zero minor of order 3.

Also A has no minor of order greater than 3.

Hence in this case the rank of A is 3. **Ans.**

Example 2:

Find the rank of the matrix $A = \begin{bmatrix} 6 & 1 & 3 & 8 \\ 4 & 2 & 6 & -1 \\ 10 & 3 & 9 & 7 \\ 16 & 4 & 12 & 15 \end{bmatrix}$

Solution:

The det. of order 4 formed by this matrix

$$= \begin{vmatrix} 6 & 1 & 3 & 8 \\ 4 & 2 & 6 & -1 \\ 10 & 3 & 9 & 7 \\ 16 & 4 & 12 & 15 \end{vmatrix}$$

$$= \begin{vmatrix} 6 & 1 & 3 & 8 \\ 4 & 2 & 6 & -1 \\ 6 & 1 & 3 & 8 \\ 6 & 1 & 3 & 5 \end{vmatrix}$$, replacing R_3 and R_4 by $R_3 - R_2$ and $R_4 - R_3$ respectively

= 0, as its three rows are identical

A minor of order 3

$$= \begin{vmatrix} 6 & 1 & 3 \\ 4 & 2 & 6 \\ 10 & 3 & 9 \end{vmatrix} = \begin{vmatrix} 6 & 1 & 3 \\ 4 & 2 & 6 \\ 6 & 1 & 2 \end{vmatrix}$$, repacing R_3 by $R_3 - R_2$

= 0, two rows being identical.

In a similar way we can prove that all the minors of order 3 are zero.

Now a minor of order 2 = $\begin{vmatrix} 6 & 1 \\ 4 & 2 \end{vmatrix} = 12 - 4 = 8 \neq 0$

Hence, the rank of the given matrix = 2. **Ans.**

Example 3:

Find the rank of the matrix $\begin{bmatrix} 3 & 4 & 5 & 6 & 7 \\ 4 & 5 & 6 & 7 & 8 \\ 5 & 6 & 7 & 8 & 9 \\ 15 & 16 & 17 & 18 & 19 \end{bmatrix}$

Solution:

One minor of order 3 of A

$$= \begin{vmatrix} 5 & 7 & 8 \\ 6 & 8 & 9 \\ 6 & 18 & 19 \end{vmatrix} = \begin{vmatrix} 5 & 7 & 8 \\ 1 & 1 & 1 \\ 11 & 11 & 11 \end{vmatrix}$$, replacing R_2 and R_3 by $R_2 - R_1$ and $R_3 - R_1$ respectively.

$$= \begin{vmatrix} 5 & 7 & 8 \\ 1 & 1 & 1 \\ 0 & 0 & 0 \end{vmatrix}, \text{ replacing } R_3 \text{ by } R_3 - 11R_2$$

$= 0.$

In a similar way we can prove that all the minors of order 3 of A are zero.

This shows that all minors of order 4 of A are automatically zero.

Now one minor of order 2 of A

$$= \begin{vmatrix} 7 & 8 \\ 8 & 9 \end{vmatrix} = (7 \times 9) - (8 \times 8) = 63 - 64 == -1 \neq 0.$$

Hence the rank of A is 2. **Ans.**

Example 4:

Find the rank of A $= \begin{bmatrix} 1 & 1 & 1 \\ b+c & c+a & a+b \\ bc & ca & ab \end{bmatrix}$

Solution:

$$|A| = \begin{vmatrix} 1 & 1 & 1 \\ b+c & c+a & a+b \\ bc & ca & ab \end{vmatrix}$$

$= -(a-b)(b-c)(c-a)$, on evaluating. ...(i)

Now following cases arises :

Case I: $a = b = c.$

If a = b = c, then A = $\begin{vmatrix} 1 & 1 & 1 \\ 2a & 2a & 2a \\ a^2 & a^2 & a^2 \end{vmatrix}$

Therefore all minors of order 2 and 3 of A vanish.

Also A has non-zero minor of order 1, since no element of A is zero.

Hence the rank of A in this case is 1. **Ans.**

Case II: *Two of numbers a, b, c are equal but are different from the third.*

Let $a = b \neq c$.

Then $|A| = \begin{vmatrix} 1 & 1 & 1 \\ a+c & c+a & 2a \\ ac & ca & a^2 \end{vmatrix} = 0$, as C_1, C_2 are identical.

Also A has a minor order 2 viz. $\begin{vmatrix} 1 & 1 \\ a+c & 2a \end{vmatrix}$

$= 2a - (a + c) = a - c \neq 0, \quad \because a \neq c,$

Hence the rank of A in this case is 2.

Similarly we can discuss the cases $b = c \neq a$, $c = a \neq b$. **Ans.**

Case III: a, b, c are all different.

In this case $|A| \neq 0$, as is evident from (i) above.

i.e., A has a non-zero minor of order 3 and there exists no minor of order greater than 3.

Example 5(a):

Find the rank of the matrix $A = \begin{bmatrix} 1 & a & b & 0 \\ 0 & c & d & 1 \\ 1 & a & b & 0 \\ 0 & c & d & 1 \end{bmatrix}$

Solution:

$|A| = \begin{vmatrix} 1 & a & b & 0 \\ 0 & c & d & 1 \\ 1 & a & b & 0 \\ 0 & c & d & 1 \end{vmatrix}$

$= 0$, as R_1, R_3 are identical.

A minor of order 3 of A

$= \begin{vmatrix} a & b & 0 \\ c & d & 1 \\ a & b & 0 \end{vmatrix} = 0$, as R_1, R_3 are identical.

In a similar way we can show that all the minors of order 3 are zero in value.

A minor order 2 of A

$$= \begin{vmatrix} a & b \\ c & d \end{vmatrix} = ad - bc \neq 0$$

Hence the rank of the matrix A is 2. **Ans.**

Example 5(b):

Under what condition the rank of the following matrix A is 3? Is it possible for the rank to be 1? Why?

$$A = \begin{bmatrix} 2 & 4 & 3 \\ 3 & 1 & 2 \\ 1 & 0 & x \end{bmatrix}$$

Solution:

If the rank of the matrix A is 3, then the minor of order 3 of A should be non-zero

i.e.,, $\begin{vmatrix} 2 & 4 & 3 \\ 3 & 1 & 2 \\ 1 & 0 & x \end{vmatrix} \neq 0$, which is the required condition.

Also the rank of A can not be 1 as at least one minor of order 1 of A *i.e.*, one element of A is zero.

[If we are to find the condition under which the rank of A is 2, then the same is | A | = 0 *i.e.*, minor of order 3 of A must be zero.]

i.e., $\begin{vmatrix} 2 & 4 & 2 \\ 3 & 1 & 2 \\ 1 & 0 & x \end{vmatrix} = 0$ *i.e.*, $\begin{vmatrix} 2 & 4 & 2 \\ 1 & -3 & 0 \\ 1 & 0 & x \end{vmatrix} = 0$,

replacing R_2 by $R_2 - R_1$

i.e., $\begin{vmatrix} 0 & 10 & 2 \\ 1 & -3 & 0 \\ 1 & 0 & x \end{vmatrix} = 0$, replacing R_1 by $R_1 - 2R_2$

i.e., $\begin{vmatrix} 10 & 2 \\ -3 & 0 \end{vmatrix} - 0 + x \begin{vmatrix} 0 & 10 \\ 1 & -3 \end{vmatrix} = 0$, expanding with respect to R_2

i.e., $6 - 10x = 0$ *i.e.*, $x = 6/10 = 3/5$. **Ans.**

Example 5(c):

Prove that the points (x_1, y_1), (x_2, y_2), (x_3, y_3) are collinear if the rank of the matrix $\begin{bmatrix} x_1 & y_1 & 1 \\ x_2 & y_2 & 1 \\ x_3 & y_3 & 1 \end{bmatrix}$ *is less than 3.*

Solution:

If the rank of the given matrix is less than 3, then the minor of order 3 of this matrix must be zero.

i.e., $$\begin{vmatrix} x_1 & y_1 & 1 \\ x_2 & y_2 & 1 \\ x_3 & y_3 & 1 \end{vmatrix} = 0 \qquad ...(i)$$

Now the area of the triangle whose vertices are (x_1, y_1), (x_2, y_2) and

$$(x_3, y_3) = \frac{1}{2}\begin{vmatrix} x_1 & y_1 & 1 \\ x_2 & y_2 & 1 \\ x_3 & y_3 & 1 \end{vmatrix}$$

$= 0$, from (i)

Since the area of this triangle is zero, so its vertices (x_1, y_1) (x_2, y_2) and (x_3, y_3) are collinear. **Hence proved.**

Example 6:

Are the matrices $A = \begin{bmatrix} 1 & 2 & 3 \\ 2 & 5 & 4 \\ 3 & 7 & 9 \end{bmatrix}$ *and*

$B = \begin{bmatrix} 1 & 0 & -5 & 6 \\ 3 & -2 & 1 & 2 \\ 5 & -2 & -9 & 14 \\ 4 & -2 & -4 & 8 \end{bmatrix}$ *equivalent?*

Example 7:

Find the rank of the matrix $A = \begin{bmatrix} 1 & 3 & 2 \\ 1 & 2 & 3 \\ 1 & 5 & 4 \end{bmatrix}$

Solution:

The determinant of order 3 formed by A

$$= \begin{vmatrix} 1 & 3 & 2 \\ 1 & 2 & 3 \\ 1 & 5 & 4 \end{vmatrix} = \begin{vmatrix} 1 & 3 & 2 \\ 0 & -1 & 1 \\ 0 & 2 & 2 \end{vmatrix},$$ replacing R_2, R_3 by $R_2 - R_1$, $R_3 - R_1$ respectively.

$$= \begin{vmatrix} -1 & 1 \\ 2 & 2 \end{vmatrix} = -2 - 2 = -4 \neq 0.$$

$\therefore$ p (A) $\geq$ 3 ...(i)

Also the matrix A does not possess any matrix of order 4 *i.e.*, 3 + 1, so

p(A) $\leq$ 3. ...(ii)

$\therefore$ From (i) and (ii) we get p (A) = 3. **Ans.**

Example 8:

Find the rank of the matrix A = $\begin{bmatrix} 0 & 1 & 2 \\ 1 & 2 & 3 \\ 3 & 1 & 1 \end{bmatrix}$

Solution:

The determinant of order 3 formed by A

$$= \begin{vmatrix} 0 & 1 & 2 \\ 1 & 2 & 3 \\ 3 & 1 & 1 \end{vmatrix} = \begin{vmatrix} 0 & 1 & 0 \\ 1 & 2 & -1 \\ 3 & 1 & -1 \end{vmatrix},$$ replacing C_3 by $C_3 - 2C_3$

$$= - \begin{vmatrix} 1 & -1 \\ 3 & -1 \end{vmatrix} = -[-1 + 3] = -2 \neq 0.$$

$\therefore$ p (A) $\geq$ 3 ...(i)

Also the matrix A does not possesses any matrix of order 4 *i.e.*, 3 + 1, so p (A) $\leq$ 3. ...(ii)

$\therefore$ From (i) and (ii) we get p (A) = 3. **Ans.**

Example 9:

Find the rank of the matrix A = $\begin{bmatrix} 1 & 2 & 3 \\ 2 & 5 & 8 \\ 4 & 10 & 18 \end{bmatrix}$

Solution:

The determinant of order 2 formed by A

$$= \begin{vmatrix} 1 & 2 & 3 \\ 2 & 5 & 8 \\ 4 & 10 & 18 \end{vmatrix}$$

$$= \begin{vmatrix} 1 & 0 & 0 \\ 2 & 1 & 2 \\ 4 & 2 & 6 \end{vmatrix}$$,replacing C_2, C_3 by $C_2 - 2C_1$, $C_3 - 3C_1$ respectively

$$= \begin{vmatrix} 1 & 2 \\ 2 & 6 \end{vmatrix} = (1 \times 6) - (2 \times 2) = 6 - 4 = 2 \neq 0$$

$\therefore$ $\rho(A) \geq 3$ ...(i)

Also matrix A does not possesses any matrix of order 4 *i.e.*, 3 + 1, so

$\rho(A) \leq 3$. ...(ii)

$\therefore$ From (i) and (ii) we get $\rho(A) = 3$. **Ans.**

Example 10:

Find the rank of the matrix $A = \begin{bmatrix} 1 & 2 & 3 \\ 2 & 3 & 4 \\ 4 & 10 & 18 \end{bmatrix}$

Solution:

The determinant of order 3 formed by A

$$= \begin{vmatrix} 1 & 2 & 3 \\ 2 & 3 & 4 \\ 4 & 10 & 18 \end{vmatrix} = \begin{vmatrix} 1 & 0 & 0 \\ 2 & -1 & -2 \\ 4 & 2 & 6 \end{vmatrix}$$, replacing C_2, C_3 by $C_2 - 2C_1$, $C_3 - 3C_1$ respectively.

$$= \begin{vmatrix} -1 & -2 \\ 2 & 6 \end{vmatrix} = -6 + 4 = -2 \neq 0$$

$\therefore$ $\rho(A) \geq 3$ ...(i)

Also the matrix A does not possesses any matrix of order 4 *i.e.*, 3 + 1, so

$\rho(A) \leq 3$. ...(ii)

$\therefore$ From (i) and (ii) we get $\rho(A) = 3$. **Ans.**

Example 11(a):

Find the rank of the matrix $A = \begin{bmatrix} 1 & -3 & 2 \\ 3 & -9 & 6 \\ -2 & 6 & -4 \end{bmatrix}$

Solution:

The determinant of order 3 formed by this matrix A

$$= \begin{vmatrix} 1 & -3 & 2 \\ 3 & -9 & 6 \\ -2 & 6 & -4 \end{vmatrix} = \begin{vmatrix} 1 & 0 & 0 \\ 3 & 0 & 0 \\ -2 & 0 & 0 \end{vmatrix}$$, replacing C_2, C_3 by $C_2 + 3C_1$ and $C_3 - 2C_1$ respectively

$$= 0$$

Also there exists no minor of order 2 of A which is not equal to zero. (Students can verify for themselves).

Finally all minors of order 1 of the matrix A are non-zero, as no element of the matrix A is zero.

Hence the rank of A is 1. **Ans.**

Example 11(b):

Find the rank of the matrix $\begin{bmatrix} 1 & 2 & 3 & 1 \\ 2 & 4 & 6 & 2 \\ 1 & 2 & 3 & 2 \end{bmatrix}$

Solution:

In this matrix, a minor of order

$$= \begin{vmatrix} 1 & 2 & 3 \\ 2 & 4 & 6 \\ 1 & 2 & 3 \end{vmatrix} = 0, \because R_1, R_2 \text{ are identical}$$

In a similar way we prove that all the minors of order 3 are zero.

Now a minor order 2 = $\begin{vmatrix} 1 & 2 \\ 2 & 4 \end{vmatrix} = 0.$

But another minor order 2 = $\begin{vmatrix} 3 & 1 \\ 3 & 2 \end{vmatrix} \neq 0$

Hence rank of the given matrix is 2. **Ans.**

Example 11(c):

Find the rank of the matrix $\begin{bmatrix} 13 & 16 & 19 \\ 14 & 17 & 20 \\ 15 & 18 & 21 \end{bmatrix}$

Solution:

The determinant of order 3 formed by this matrix

$$= \begin{vmatrix} 13 & 16 & 19 \\ 14 & 17 & 20 \\ 15 & 18 & 21 \end{vmatrix}$$

$$= \begin{vmatrix} 13 & 3 & 3 \\ 14 & 3 & 3 \\ 15 & 3 & 3 \end{vmatrix},$$ replacing C_2 and C_3 by $C_2 - C_1$ and $C_3 - C_2$ respectively.

= 0, since two columns viz. C_2 and C_3 are identical.

A minor of order 2 = $\begin{vmatrix} 13 & 16 \\ 14 & 17 \end{vmatrix} \neq 0.$

Hence the rank of the given matrix is 2. **Ans.**

Example 12:

Find the rank of the matrix A = $\begin{bmatrix} 1 & 2 & 3 \\ 2 & 3 & 1 \\ -2 & -3 & -1 \end{bmatrix}$

Solution:

The determinant of order 3 formed by this matrix A

$$= \begin{vmatrix} 1 & 2 & 3 \\ 2 & 3 & 1 \\ -2 & -3 & -1 \end{vmatrix} = \begin{vmatrix} 1 & 2 & 3 \\ 2 & 3 & 1 \\ 0 & 0 & 0 \end{vmatrix},$$ replacing R_3 by $R_3 + R_2$

= 0

Also there exists a minor of order 2 of A

viz. $\begin{vmatrix} 1 & 2 \\ 2 & 3 \end{vmatrix} = 3 - 4 = -1 \neq 0$

Hence the rank of the given matrix A is 2. **Ans.**

Example 13:

Find the rank of the matrix $A = \begin{bmatrix} 1 & 2 & 3 \\ 2 & 3 & 4 \\ 4 & 5 & 6 \end{bmatrix}$

Solution:

The determinant of order 3 formed by A

$$= \begin{vmatrix} 1 & 2 & 3 \\ 2 & 3 & 4 \\ 4 & 5 & 6 \end{vmatrix} = \begin{vmatrix} 1 & 0 & 0 \\ 2 & -1 & -2 \\ 4 & -3 & -6 \end{vmatrix},$$ replacing C_2, C_3 by $C_2 - 2C_1, C_3 - 3C_1$ respectively.

$$= \begin{vmatrix} -1 & -2 \\ - & -6 \end{vmatrix} = 6 - 6 = 0$$

Also there exists a minor of order 2 of A

viz. $\begin{vmatrix} 3 & 4 \\ 5 & 6 \end{vmatrix} = 18 - 20 = -2 \neq 0.$

Hence the rank of the given matrix A is 2. **Ans.**

Example 14:

Find the rank of the matrix $\begin{bmatrix} 1 & -1 & -2 & -4 \\ 3 & 1 & 3 & -2 \\ 6 & 3 & 0 & -7 \\ 2 & 3 & -1 & -1 \end{bmatrix}$

Solution:

The determinant of order 4 formed by the given matrix

$$= \begin{vmatrix} 1 & -1 & -2 & -4 \\ 3 & 1 & 3 & -2 \\ 6 & 3 & 0 & -7 \\ 2 & 3 & -1 & -1 \end{vmatrix}$$

$$= \begin{vmatrix} 1 & 0 & 0 & 0 \\ 3 & 4 & 9 & 10 \\ 6 & 9 & 12 & 17 \\ 2 & 5 & 3 & 7 \end{vmatrix},$$ replacing C_2, C_3, C_4 by $C_2 + C_1, C_3 + 2C_1, C_4 + 4C_1$ respectively.

$$= \begin{vmatrix} 4 & 9 & 10 \\ 9 & 12 & 17 \\ 5 & 3 & 7 \end{vmatrix} = \begin{vmatrix} 4 & 9 & 10 \\ 5 & 3 & 7 \\ 5 & 3 & 7 \end{vmatrix}, \text{ replacing } R_2 \text{ by } R_2 - R_1$$

= 0, as its two row and identical.

A minor of order 3

$$= \begin{vmatrix} 1 & -1 & -2 \\ 3 & 1 & 3 \\ 6 & 3 & 0 \end{vmatrix} = \begin{vmatrix} 1 & 0 & 0 \\ 3 & 4 & 9 \\ 6 & 9 & 12 \end{vmatrix}, \text{ replacing } C_2, C_3 \text{ by } C_2 + C_1, C_3 + 2C_1 \text{ respectively.}$$

$$= \begin{vmatrix} 4 & 9 \\ 9 & 12 \end{vmatrix} = 48 - 81 = -33 \neq 0.$$

Hence the rank of the given matrix is 3. **Ans.**

Example 15:

Find the rank of the matrix A $= \begin{bmatrix} 1 & 1 & 1 & -1 \\ 1 & 2 & 3 & 4 \\ 3 & 4 & 5 & 2 \end{bmatrix}$

Solution:

In this matrix there are four minors of order 3 of a Viz.

$$\begin{vmatrix} 1 & 1 & 1 \\ 1 & 2 & 3 \\ 3 & 4 & 5 \end{vmatrix}; \begin{vmatrix} 1 & 1 & -1 \\ 1 & 2 & 4 \\ 3 & 4 & 2 \end{vmatrix}; \begin{vmatrix} 1 & 1 & -1 \\ 1 & 3 & 4 \\ 3 & 5 & 2 \end{vmatrix}; \begin{vmatrix} 1 & 1 & -1 \\ 2 & 3 & 4 \\ 4 & 5 & 2 \end{vmatrix}$$

and each one of them is zero in value (prove it).

Also therexsts a minor of or 2 of A viz.

$$\begin{vmatrix} 2 & 3 \\ 4 & 5 \end{vmatrix} = 10 - 12 = -2 \neq 0$$

Hence the rank of the given matrix A is 2. **Ans.**

Example 16:

Find the rank of the matrix A $= \begin{bmatrix} 6 & 1 & 3 & 8 \\ 16 & 4 & 12 & 15 \\ 5 & 3 & 3 & 4 \\ 4 & 2 & 6 & -1 \end{bmatrix}$

Solution:

The determinant of order 4 formed by A

$$= \begin{vmatrix} 6 & 1 & 3 & 8 \\ 16 & 4 & 12 & 15 \\ 5 & 3 & 3 & 4 \\ 4 & 2 & 6 & -1 \end{vmatrix} = \begin{vmatrix} 0 & 1 & 0 & 0 \\ -8 & 4 & 0 & -17 \\ -13 & 3 & -6 & -20 \\ -8 & 2 & 0 & -17 \end{vmatrix},$$

replacing C_1, C_3, C_4 by $C_1 - 6C_2$, $C_3 - 3C_2$ and $C_4 - 8C_2$ respectively.

$$= \begin{vmatrix} -8 & 0 & -17 \\ -13 & -6 & -20 \\ -8 & 0 & -17 \end{vmatrix} = 0, \text{ as } R_1\ R_2 \text{ are identical.}$$

Also one minor of order 3 viz.

$$\begin{vmatrix} 1 & 3 & 8 \\ 3 & 3 & 4 \\ 2 & 6 & -1 \end{vmatrix} = \begin{vmatrix} 1 & 0 & 0 \\ 3 & -6 & -20 \\ 2 & 0 & -17 \end{vmatrix} = \begin{vmatrix} -6 & -20 \\ 0 & -17 \end{vmatrix} \neq 0.$$

Hence the rank of given matrix A is 3. **Ans.**

Example 17:

Fine the rank of the matrix $\begin{bmatrix} 1 & 3 & 4 & 3 \\ 3 & 9 & 12 & 9 \\ -1 & -3 & -4 & -3 \end{bmatrix}$

Solution:

In this matrix, a minor of order 3

$$= \begin{vmatrix} 1 & 3 & 4 \\ 3 & 9 & 12 \\ -1 & -3 & -4 \end{vmatrix} = 3 \begin{vmatrix} 1 & 3 & 4 \\ 1 & 3 & 4 \\ -1 & -3 & -4 \end{vmatrix}, \text{ taking common from } R_3$$

= 0, as R_1 and R_2 are identical.

In a similar way we can prove that all minors of order 3 are zero.

Now a minor of order 2.

$$= \begin{vmatrix} 1 & 3 \\ 3 & 9 \end{vmatrix} = 3 \begin{vmatrix} 1 & 3 \\ 1 & 3 \end{vmatrix}, \text{ taking out 3 common from } R_3$$

= 0, as rows are identical.

Similarly, all the minors of order 2 are zero.

Hence, we are left with minors of order unity, viz. the elements of the given matrix, which are not equal to zero.

Hence rank of the given matrix = 1. **Ans.**

Example 18(a):

Find the rank of the matrix $A = \begin{bmatrix} 1 & -1 & 3 & 6 \\ 1 & 3 & -3 & -4 \\ 5 & 3 & 3 & 11 \end{bmatrix}$

Solution:

The given matrix A possesses a minor of order 3 viz.

$$\begin{vmatrix} 1 & 3 & 6 \\ 1 & -3 & -4 \\ 5 & 3 & 11 \end{vmatrix} = \begin{vmatrix} 2 & 0 & 2 \\ 1 & -3 & -4 \\ 6 & 0 & 7 \end{vmatrix}, \text{ replacing } R_1, R_2 \text{ by } R_2 + R_1, R_3 + R_2$$

$$= -3 \begin{vmatrix} 2 & 2 \\ 6 & 7 \end{vmatrix} = -3\,(14 - 12) = -6 \neq 0$$

$\therefore$ p (A) ≥ 3 ...(i)

Also A does not posses any minor of order 4 *i.e.*,, 3 + 1, so

p (A) ≤ 3 ...(ii)

$\therefore$ From (i) and (ii) we get p (A) = 3. **Ans.**

Example 18(b):

Find the rank of the matrix $A = \begin{bmatrix} 1 & 3 & 4 & 5 \\ 1 & 2 & 6 & 7 \\ 1 & 5 & 0 & 1 \end{bmatrix}$

Solution:

One minor of order three of A

$$= \begin{vmatrix} 1 & 4 & 5 \\ 1 & 6 & 7 \\ 1 & 0 & 1 \end{vmatrix} = \begin{vmatrix} 1 & 0 & 0 \\ 1 & 2 & 2 \\ 1 & -4 & -4 \end{vmatrix}, \text{ replacing } C_2, C_3 \text{ by } C_2 - 4C_1 \text{ and } C_3 - 5\,C_1 \text{ respectively.}$$

$$= \begin{vmatrix} 2 & 2 \\ -4 & -4 \end{vmatrix}, \text{ expanding with respect to } R_1$$

$= 2\,(-4) - 2\,(-4) = 0.$

In a similar way we can prove that all the four minors of order three are zero.

Now a minor or order 2 is $\begin{vmatrix} 2 & 6 \\ 5 & 0 \end{vmatrix} = 2.0 - 6.5 \neq 0.$

Hence the rank of A is 2.

Example 18(c):

Find the rank of the matrix $A = \begin{bmatrix} 1 & 3 & 5 & 1 \\ 2 & 4 & 8 & 0 \\ 3 & 1 & 7 & 5 \end{bmatrix}$

Solution:

The given matrix A possesses a minor of order 3 viz.

$$\begin{vmatrix} 1 & 3 & 1 \\ 2 & 4 & 0 \\ 3 & 1 & 5 \end{vmatrix} = \begin{vmatrix} 0 & 0 & 1 \\ 2 & 4 & 0 \\ -2 & -14 & 5 \end{vmatrix}, \text{ repacing } C_1 \text{ and } C_2 \text{ by } C_1 - C_3 \text{ and } C_2 - 3C_2$$

$$= \begin{vmatrix} 2 & 4 \\ -2 & -14 \end{vmatrix}, \text{expanding with respect to } R_1$$

$$= 2\,(-14) - (4)\,(-2) = -28 + 8 \neq 0$$

$\therefore \qquad p\,(A) \geq 3$ ...(i)

Also A does not posses any minor of order 4 *i.e.*,, 3 + 1, so

$p\,(A) \leq 3$ ...(ii)

From (i) and (ii) we get p (A) = 3 *i.e.*, the rank of A is 3. **Ans.**

Example 19:

Determinant the rank of $A = \begin{bmatrix} 6 & 1 & 8 & 3 \\ 2 & 1 & 0 & 2 \\ 4 & -1 & -8 & -3 \end{bmatrix}$

Solution:

The given matrix A possesses a minor of order 3 viz.

$$\begin{vmatrix} 6 & 1 & 8 \\ 2 & 1 & 0 \\ 4 & -1 & -8 \end{vmatrix} = \begin{vmatrix} 10 & 0 & 0 \\ 6 & 0 & -8 \\ 4 & -1 & -8 \end{vmatrix}, \text{ replacing } R_1, R_2 \text{ by } R_1 + R_3, R_2 + R_3$$

$$= 10 \begin{vmatrix} 0 & -8 \\ -1 & -8 \end{vmatrix} = 10\,(0 - 8) = -\,80 \neq 0$$

$\therefore$ $\quad p\,(A) \geq 3$ ...(i)

Also A does not posses any minor of order 4 *i.e.*,, 3 + 1, so

$p\,(A) \leq 3$ ...(ii)

$\therefore$ From (i) and (ii) we get p (A) = 3. **Ans.**

SWEEP OUT METHOD OF FINDING THE RANK OF A MATRIX

In the process of evaluation of the rank of a matrix by means of elementary row and column transformations, if certain rows of columns are zero-rows or zero-columns (*i.e.*, each element of these rows or columns are zero) then we can remove these rows or columns without any effect on the rank of the matrix. This method is generally called the *Sweep out* method.

Example 1:

Find the rank of the matrix $\begin{bmatrix} 0 & 1 & -3 & -1 \\ 1 & 0 & 1 & 1 \\ 3 & 1 & 0 & 2 \\ 1 & 1 & -2 & 0 \end{bmatrix}$

Solution:

Let $A = \begin{bmatrix} 0 & 1 & -3 & -1 \\ 1 & 0 & 1 & 1 \\ 3 & 1 & 0 & 2 \\ 1 & 1 & -2 & 0 \end{bmatrix}$

Now $A \sim \begin{bmatrix} 0 & 1 & -3 & -1 \\ 1 & 1 & -2 & 0 \\ 3 & 3 & -6 & 0 \\ 1 & 1 & -2 & 0 \end{bmatrix}$, replacing R_2 and R_3 by $R_2 + R_1$ and $R_3 + 2R_1$ respectively

$\sim \begin{bmatrix} 0 & 1 & -3 & -1 \\ 1 & 1 & -2 & 0 \\ 1 & 1 & -2 & 0 \\ 0 & 0 & 0 & 0 \end{bmatrix}$, replacing R_3 and R_4 by $\frac{1}{3}R_3$ and $R_4 - R_2$ respectively.

$$\sim \begin{bmatrix} 0 & 1 & -3 & -1 \\ 1 & 0 & 0 & 0 \\ 0 & 0 & 0 & 0 \\ 0 & 0 & 0 & 0 \end{bmatrix},$$ replacing R_3 by $R_3 - R_2$ and then C_2 and C_3 by $C_2 - C_1$ and $C_3 + 2C_2$ respectively

$$\sim \begin{bmatrix} 0 & 1 & -3 & -1 \\ 1 & 0 & 0 & 0 \end{bmatrix}$$

Now a minor of order 2 is $\begin{vmatrix} 0 & 1 \\ 1 & 0 \end{vmatrix} = -1 \neq 0.$

Hence its rank is 2. **Ans.**

Example 2:

Find the rank of the matrix $A = \begin{bmatrix} 6 & 1 & 3 & 8 \\ 4 & 2 & 6 & -1 \\ 10 & 3 & 9 & 7 \\ 16 & 4 & 12 & 15 \end{bmatrix}$

Solution:

$$A \sim \begin{bmatrix} 6 & 1 & 3 & 8 \\ 4 & 2 & 6 & -1 \\ 4 & 2 & 6 & -1 \\ 4 & 2 & 6 & -1 \end{bmatrix},$$ replacing R_3 and R_4 by $R_3 - R_1$ and $R_4 - 2R_1$ respectively

$$\therefore\ A \sim \begin{bmatrix} 6 & 1 & 3 & 8 \\ 4 & 2 & 6 & -1 \end{bmatrix}$$

Now a minor of order 2 is

$$\begin{vmatrix} 1 & 8 \\ 2 & -1 \end{vmatrix} = -1 - 16 = -17 \neq 0.$$

Hence its rank is 2. **Ans.**

SOME IMPORTANT THEOREMS

Theorem 1:

The rank of the product matrix AB of two matrices A and B is less than the rank of either of the matrices A and B.

Proof:

Let r_1 and r_2 be the rank of the matrices A and B.

$\because$ r_1 is the rank of A, therefore $A \sim \begin{bmatrix} M \\ O \end{bmatrix}$, where M is a submatrix of ran r_1 and contains r_1 rows.

Post-Multiplying it by B, we get

$$AB \sim \begin{bmatrix} M \\ O \end{bmatrix} B.$$

But $\begin{bmatrix} M \\ O \end{bmatrix}$ B can have r_1 non-zero rows at the most which are obtained on multiplying r_1 non-zero rows of M with columns of B.

$\therefore$ Rank of AB = Rank of $\begin{bmatrix} M \\ O \end{bmatrix} B \leq r_1$

i.e., Rank of AB £ rank of A ...(i)

In a similar way we get B ~ [N O], where N is a submatrix of ran r_2 and contains r_2 columns.

Pre-multiplying it by A, we get

$$AB \sim A\,[N\ O].$$

But [N O] can have r_2 non-zero columns at the most which are obtained on multiplying the rows of A with r_2 non-zero columns of [N O].

$\therefore$ Rank of AB = Rank of A[N O] $\leq r_2$

i.e., Rank of AB $\leq$ rank of B ...(ii)

Hence the theorem from (i) and (ii).

Theorem 2:

The rank of a matrix is equal to the rank of the transposed matrix.

Proof:

Let $A = [a_{ij}]$ be any m × n matrix.

The transposed matrix $A' = [A_{ij}]$ is an n × m matrix.

Let the rank of A be r and let B be the r × r sub-matrix of A such that $|B| \neq 0$.

Also we know that the value of a determinant remains unaltered if its rows and columns are interchanged.

i.e., | B' | = | B | ≠ 0, where B' is evidently a r × r sub-matrix of A'

∴ The rank of A' ≥ r.

Again if C be a (r + 1) × (r + 1) submatrix of A, then by definition of rank we must e all | C | = 0.

Also C' is a (r + 1) × (r + 1) submatrix of A' so we have

| C' | = | C | = 0, as explained above.

∴ We conclude that there cannot be any (r + 1) × (r + 1) submatrix of A' with non-zero determinant.

∴ The rank of A' ≥ r and it cannot be greater than r as above.

∴ The rank of A' is r which is also the rank of A. **Hence proved.**

NORMAL FORM OF A MATRIX

Every non-zero matrix A of order m × n can be reduced by application of elementary row and column operations into equivalent matrix of one of the following forms:

(i) $\begin{bmatrix} I_r & 0 \\ 0 & 0 \end{bmatrix}$,

(ii) $\begin{bmatrix} I_r \\ 0 \end{bmatrix}$,

(iii) $[I_r\ 0]$,

(iv) $[I_r]$,

where I_r is r × r identity matrix and 0 is null matrix of any order.

These four forms are called *Normal or canonical form* of **A.**

Important Theorems (Without Proof)

Theorem 1:

If m × n matrix A is reduced to the canonical form or normal form $\begin{bmatrix} I_r & 0 \\ 0 & 0 \end{bmatrix}$ *by application of elementary row or column operations, then r, the order to the identity sub-matrix I_r is the rank of the matrix A.*

Theorem 2:

If a non-singular matrix of order n × n is reduce to the identity matrix I_n (which is its canonical or normal form), then the rank of the matrix is ~

MISCELLANEOUS EXAMPLES

Example 1:

Find the rank of the matrix $A = \begin{bmatrix} 1 & 2 & -1 & 3 \\ 2 & 4 & -4 & 7 \\ -1 & -2 & -1 & 1 \end{bmatrix}$

Solution:

$$A \sim \begin{bmatrix} 1 & 0 & 0 & 0 \\ 2 & 0 & -2 & 1 \\ -1 & 0 & -2 & 5 \end{bmatrix}, \text{ replacing } C_2, C_3, C_4 \text{ by } C_2 - 2C_1, C_3 + C_1 \text{ and } C_4 - 3C_1 \text{ respectively}$$

$$\sim \begin{bmatrix} 1 & 0 & 0 & 0 \\ 2 & 0 & -2 & 1 \\ -3 & 0 & 0 & 4 \end{bmatrix}, \text{ replacing } R_3 \text{ by } R_3 - R_2$$

$$\sim \begin{bmatrix} 1 & 0 & 0 & 0 \\ 0 & 0 & -2 & 1 \\ 0 & 0 & 0 & 4 \end{bmatrix}, \text{ replacing } R_2, R_3 \text{ by } R_2 - 2R_1 \text{ and } R_3 + 3R_1 \text{ respectively}$$

$$\sim \begin{bmatrix} 1 & 0 & 0 & 0 \\ 0 & 0 & 1 & 1 \\ 0 & 0 & 0 & 1 \end{bmatrix}, \text{ replacing } C_3 \text{ by } -\frac{1}{2}C_3 \text{ and } R_3 \text{ by } \frac{1}{4}R_3$$

$$\sim \begin{bmatrix} 1 & 0 & 0 & 0 \\ 0 & 0 & 1 & 0 \\ 0 & 0 & 0 & 1 \end{bmatrix}, \text{ replacing } C_4 \text{ by } C_4 - C_3$$

$$\sim \begin{bmatrix} 1 & 0 & 0 & 0 \\ 0 & 1 & 0 & 0 \\ 0 & 0 & 0 & 1 \end{bmatrix}, \text{ interchanging } C_2 \text{ and } C_3$$

$$\sim \begin{bmatrix} 1 & 0 & 0 & 0 \\ 0 & 1 & 0 & 0 \\ 0 & 0 & 1 & 0 \end{bmatrix}, \text{ interchanging } C_3 \text{ and } C_4$$

$\sim [\, I_3 \; 0 \,]$ **(Note)**

Hence the rank of A is . **Ans.**

Example 2:

Determine by reducing to normal form the rank of the matrix

$$A = \begin{bmatrix} 8 & 1 & 3 & 6 \\ 0 & 3 & 2 & 2 \\ -8 & -1 & -3 & 4 \end{bmatrix}$$

Solution:

$$A \sim \begin{bmatrix} 1 & 1 & 3 & 3 \\ 0 & 3 & 2 & 1 \\ -1 & -1 & -3 & 2 \end{bmatrix}, \text{ replacing } C_1 \text{ by } \frac{1}{8} C_1 \text{ and } C_4 \text{ by } \frac{1}{2} C_4$$

$$\sim \begin{bmatrix} 1 & 0 & 0 & 0 \\ 0 & 3 & 2 & 1 \\ -1 & 0 & 0 & 5 \end{bmatrix}, \text{ replacing } C_2, C_3 \text{ and } C_4 \text{ by } C_2 - C_1, C_3 - 3C_1, C_4 - 3C_1 \text{ respectively.}$$

$$\sim \begin{bmatrix} 1 & 0 & 0 & 0 \\ 0 & 3 & 2 & 1 \\ 0 & 0 & 0 & 5 \end{bmatrix}, \text{ replacing } R_3 \text{ by } R_3 + R_1$$

$$\sim \begin{bmatrix} 1 & 0 & 0 & 0 \\ 0 & 1 & 1 & 1 \\ 0 & 0 & 0 & 5 \end{bmatrix}, \text{ repacing } C_2 \text{ by } \frac{1}{3} C_2 \text{ and } C_2 \text{ by } \frac{1}{2} C_3$$

$$\sim \begin{bmatrix} 1 & 0 & 0 & 0 \\ 0 & 1 & 0 & 0 \\ 0 & 0 & 0 & 5 \end{bmatrix}, \text{ repacing } C_3 \text{ and } C_4 \text{ by } C_3 - C_2 \text{ and } C_4 - C_2 \text{ respectively.}$$

$$\sim \begin{bmatrix} 1 & 0 & 0 & 0 \\ 0 & 1 & 0 & 0 \\ 0 & 0 & 5 & 0 \end{bmatrix}, \text{ int erchanging } C_3 \text{ and } C_4$$

$$\Rightarrow \quad A \sim \begin{bmatrix} 1 & 0 & 0 & 0 \\ 0 & 1 & 0 & 0 \\ 0 & 0 & 1 & 0 \end{bmatrix}, \text{ replacing } C_3 \text{ by } \frac{1}{5} C_3$$

$$\sim [\, I_3 \; 0 \,]$$ **(Note)**

Hence the rank of A is . **Ans.**

Example 3(a):

Find the rank of the matrix $A = \begin{bmatrix} 1 & 1 & 1 \\ 2 & 2 & 2 \\ 3 & 3 & 3 \end{bmatrix}$

Solution:

$A \sim \begin{bmatrix} 1 & 0 & 0 \\ 2 & 0 & 0 \\ 3 & 0 & 0 \end{bmatrix}$, replacing C_2, C_3 by $C_3 - C_1$, $C_3 - C_1$ respectively

$\sim \begin{bmatrix} 1 & 0 & 0 \\ 0 & 0 & 0 \\ 0 & 0 & 0 \end{bmatrix}$, replacing R_2, R_3 by $R_2 - 2R_1$ $R_3 - 3R_1$ respectively

$\sim \begin{bmatrix} I_1 & 0 \\ 0 & 0 \end{bmatrix}$

Hence the rank of the matrix A is 1. **Ans.**

Example 3(b):

Find the rank of the matrix $A = \begin{bmatrix} 0 & 2 & 3 \\ 0 & 4 & 6 \\ 0 & 6 & 9 \end{bmatrix}$

Solution:

$A \sim \begin{bmatrix} 0 & 2 & 1 \\ 0 & 4 & 2 \\ 0 & 6 & 3 \end{bmatrix}$, replacing C_3 by $C_3 - C_2$

$\sim \begin{bmatrix} 0 & 0 & 1 \\ 0 & 0 & 2 \\ 0 & 0 & 3 \end{bmatrix}$, replacing C_2 by $C_2 - 2C_3$

$\sim \begin{bmatrix} 0 & 0 & 1 \\ 0 & 0 & 0 \\ 0 & 0 & 0 \end{bmatrix}$, replacing R_2, R_3 by $R_2 - 2R_1, R_3 - 3R_1$

$\sim \begin{bmatrix} 1 & 0 & 0 \\ 0 & 0 & 0 \\ 0 & 0 & 0 \end{bmatrix}$, int erchanging C_1 and C_2

$$\sim \begin{bmatrix} I_1 & 0 \\ 0 & 0 \end{bmatrix}$$

Hence the rank of A is 1. **Ans.**

Example 3(c):

Reduce the matrix A to its normal form where

$A = \begin{bmatrix} 0 & 1 & 2 & -2 \\ 4 & 0 & 2 & 6 \\ 2 & 1 & 3 & 1 \end{bmatrix}$ *and hence determine its rank.*

Solution:

$$A \sim \begin{bmatrix} 0 & 1 & 0 & 0 \\ 4 & 0 & 2 & 8 \\ 2 & 1 & 1 & 4 \end{bmatrix}, \text{ replacing } C_3, C_4 \text{ by } C_3 - 2C_2 \text{ and } C_4 + C_3 \text{ respectively}$$

$$\sim \begin{bmatrix} 0 & 1 & 0 & 0 \\ 4 & 0 & 2 & 8 \\ 2 & 1 & 1 & 4 \end{bmatrix}, \text{ replacing } R_2 \text{ by } R_2 - 2R_3$$

$$\sim \begin{bmatrix} 0 & 1 & 0 & 0 \\ 0 & 0 & 0 & 0 \\ 2 & 0 & 1 & 4 \end{bmatrix}, \text{ replacing } R_2, R_3 \text{ by } R_2 + 2R_1 \text{ and } R_3 - R_1 \text{ respectively.}$$

$$\sim \begin{bmatrix} 0 & 1 & 0 & 0 \\ 0 & 0 & 0 & 0 \\ 0 & 0 & 1 & 0 \end{bmatrix}, \text{ replacing } C_1, C_4 \text{ by } C_1 - 3C_3, C_4 - 4C_3 \text{ respectively.}$$

$$\sim \begin{bmatrix} 1 & 0 & 0 & 0 \\ 0 & 0 & 0 & 0 \\ 0 & 0 & 1 & 0 \end{bmatrix}, \text{ interchanging } C_1 \text{ and } C_2$$

$$\sim \begin{bmatrix} 1 & 0 & 0 & 0 \\ 0 & 0 & 0 & 0 \\ 0 & 1 & 0 & 0 \end{bmatrix}, \text{ interchanging } C_2, C_3$$

$$\sim \begin{bmatrix} 1 & 0 & 0 & 0 \\ 0 & 1 & 0 & 0 \\ 0 & 0 & 0 & 0 \end{bmatrix}, \text{ interchanging } R_2, R_3$$

$$\sim \begin{bmatrix} I_2 & 0 \\ 0 & 0 \end{bmatrix}$$

Hence the rank of A is 2. **Ans.**

Example 4:

Find the rank of the matrix $A = \begin{bmatrix} 1 & 2 & 3 \\ 2 & 3 & 4 \\ 3 & 5 & 7 \end{bmatrix}$ *after reducing it to the normal form.*

Solution:

$$A \sim \begin{bmatrix} 1 & 1 & 1 \\ 2 & 1 & 1 \\ 3 & 2 & 2 \end{bmatrix}, \text{ replacing } C_2 \text{ and } C_3 \text{ by } C_2 - C_1 \text{ and } C_3 - C_2$$

$$\sim \begin{bmatrix} 1 & 1 & 1 \\ 1 & 0 & 0 \\ 1 & 1 & 1 \end{bmatrix}, \text{ replacing } R_2 \text{ and } R_3 \text{ by } R_2 - R_1 \text{ and } R_3 - R_2$$

$$\sim \begin{bmatrix} 1 & 1 & 1 \\ 1 & 0 & 0 \\ 0 & 0 & 0 \end{bmatrix}, \text{ replacing } R_3 \text{ by } R_3 - R_1$$

$$\sim \begin{bmatrix} 1 & 1 & 0 \\ 1 & 0 & 0 \\ 0 & 0 & 0 \end{bmatrix}, \text{ replacing } C_3 \text{ by } C_3 - C_2$$

$$\sim \begin{bmatrix} 0 & 1 & 0 \\ 1 & 0 & 0 \\ 0 & 0 & 0 \end{bmatrix}, \text{ replacing } C_1 \text{ by } C_1 - C_2$$

$$\sim \begin{bmatrix} 1 & 0 & 0 \\ 0 & 1 & 0 \\ 0 & 0 & 0 \end{bmatrix}, \text{ interchanging } C_1 \text{ and } C_2$$

$$\sim \begin{bmatrix} I_2 & 0 \\ 0 & 0 \end{bmatrix}$$

Hence the rank of A is 2. **Ans.**

Example 5:

Find the rank of the matrix $A = \begin{bmatrix} -1 & -2 & -1 \\ 6 & 12 & 6 \\ 5 & 10 & 5 \end{bmatrix}$

Solution:

$$A \sim \begin{bmatrix} -1 & 0 & 0 \\ 6 & 0 & 0 \\ 5 & 0 & 0 \end{bmatrix}, \text{ replcaing } C_2, C_3 \text{ by } C_2 - 2C_1, C_3 - C_1 \text{ respectively.}$$

$$\sim \begin{bmatrix} -1 & 0 & 0 \\ 0 & 0 & 0 \\ 0 & 0 & 0 \end{bmatrix}, \text{ replacing } R_2, R_3 \text{ by } R_2 + 6R_1, R_3 + 5R_1 \text{ respectively}$$

$$\sim \begin{bmatrix} 1 & 0 & 0 \\ 0 & 0 & 0 \\ 0 & 0 & 0 \end{bmatrix}, \text{ replacing } R_1 \text{ by } - R_1$$

$$\sim \begin{bmatrix} I_1 & 0 \\ 0 & 0 \end{bmatrix}$$

Hence the rank of the matrix A is 2. **Ans.**

Example 6:

Show that AA', has the same rank as A, where A' is the transpose of A.

Solution:

Let B = AA', then rank of B $\leq$ rank of A ...(i)

Also A^{-1} B = A', and so we have

rank A = rank A' $\leq$ rank of B ...(ii)

$\therefore$ From (i) and (ii), rank of A = rank of B.

Example 7:

Prove that if A is a matrix of order m × n and if B is a non singular matrix of order n, then the product P = AB has the same rank as A.

Solution:

Here A ~ (m × n), B ~ (n × n)

P = AB ~ (m × n)

If m < n, rank of A $\leq$ m but rank of B = n

$\therefore$ rank of A < rank of B.

Now rank (P) = rank (AB $\leq$ rank A ...(i)

But we can write A = PB^{-1}

$\therefore$ rank of A = rank of (PB^{-1}) $\leq$ rank of P ...(ii)

$\therefore$ From (i) and (ii) we get rank of P = rank of A.

Example 8:

Show that the rank of a matrix (A does not alter by pre or post) multiplying it with any non-singular matrix R.

Solution:

Let B = RA

Then rank of B = rank of RA $\leq$ rank A ...(i)

Also A = R^{-1} B, where R^{-1} is the inverse matrix of R.

$\therefore$ rank of A = rank of $(R^{-1} B)$ $\leq$ rank of B. ...(ii)

$\therefore$ From (i) and (ii) we conclude that

rank of A = rank of B. **Hence proved.**

Example 9:

Find the rank of the matrix $A = \begin{bmatrix} 1 & 2 & 3 \\ 4 & 5 & 6 \\ 2 & 1 & 2 \end{bmatrix}$

Solution:

$A \sim \begin{bmatrix} 1 & 0 & 0 \\ 4 & -3 & -6 \\ 2 & -3 & -4 \end{bmatrix}$, replacing C_2, C_3 by $C_2 - 2C_1$, $C_3 - 3C_1$ respectively.

$$\sim \begin{bmatrix} 1 & 0 & 0 \\ 4 & -3 & -6 \\ 2 & -3 & -4 \end{bmatrix}, \text{ replacing } R_2, R_3 \text{ by } R_2 - 4R_1, \; R_3 - 2R_1 \text{ respectively}$$

$$\sim \begin{bmatrix} 1 & 0 & 0 \\ 0 & 0 & -2 \\ 0 & -3 & 4 \end{bmatrix}, \text{ replacing } R_3 \text{ by } R_2 - R_3$$

$$\sim \begin{bmatrix} 1 & 0 & 0 \\ 0 & 0 & 1 \\ 0 & 1 & 2 \end{bmatrix}, \text{ replacing } C_2, C_3 \text{ by } -\frac{1}{3}C_2, -\frac{1}{2}C_3 \text{ respectively}$$

$$\sim \begin{bmatrix} 1 & 0 & 0 \\ 0 & 0 & 1 \\ 0 & 1 & 0 \end{bmatrix}, \text{ replacing } R_3 \text{ by } R_3 - 2R_2$$

$$\sim \begin{bmatrix} 1 & 0 & 0 \\ 0 & 1 & 0 \\ 0 & 0 & 1 \end{bmatrix}, \text{ interchanging } C_2 \text{ and } C_3$$

$$\sim [\, I_3 \,]$$

Hence the rank of A is 3. **Ans.**

Example 10:

Find the rank of the matrix $A = \begin{bmatrix} -2 & -1 & -3 & -1 \\ 1 & 2 & 3 & -1 \\ 1 & 0 & 1 & 1 \\ 0 & 1 & 1 & -1 \end{bmatrix}$

Solution:

$$A \sim \begin{bmatrix} 0 & -1 & -1 & 1 \\ 0 & 2 & 2 & -2 \\ 1 & 0 & 1 & 1 \\ 0 & 1 & 1 & -1 \end{bmatrix}, \text{ replacing } R_1, R_2 \text{ by } R_1 + 2R_3, R_2 - R_3 \text{ respectively}$$

$$\sim \begin{bmatrix} 0 & 0 & 0 & 1 \\ 0 & 0 & 0 & -2 \\ 1 & 1 & 2 & 1 \\ 0 & 0 & 0 & -1 \end{bmatrix}, \text{ replacing } C_2, C_3 \text{ by } C_2 + C_4, C_3 + C_4 \text{ respectively}$$

$$\Rightarrow \quad A \sim \begin{bmatrix} 0 & 0 & 0 & 1 \\ 0 & 0 & 0 & 0 \\ 1 & 1 & 2 & 0 \\ 0 & 0 & 0 & 0 \end{bmatrix}, \text{ replacing } R_2, R_3, R_4 \text{ by } R_2 + 2R_1, R_3 - R_1, R_4 + R_1 \text{ respectively}$$

$$\sim \begin{bmatrix} 0 & 0 & 0 & 1 \\ 0 & 0 & 0 & 0 \\ 1 & 0 & 0 & 0 \\ 0 & 0 & 0 & 0 \end{bmatrix}, \text{ replacing } C_2, C_3 \text{ by } C_2 - C_1, C_3 - 2C_1 \text{ respectively.}$$

$$\sim \begin{bmatrix} 1 & 0 & 0 & 0 \\ 0 & 0 & 0 & 0 \\ 0 & 0 & 0 & 1 \\ 0 & 0 & 0 & 0 \end{bmatrix}, \text{ interchanging } R_1 \text{ and } R_4$$

$$\sim \begin{bmatrix} 1 & 0 & 0 & 0 \\ 0 & 0 & 0 & 0 \\ 0 & 1 & 0 & 0 \\ 0 & 0 & 0 & 0 \end{bmatrix}, \text{ interchanging } C_2 \text{ and } C_4$$

$$\sim \begin{bmatrix} 1 & 0 & 0 & 0 \\ 0 & 1 & 0 & 0 \\ 0 & 0 & 0 & 0 \\ 0 & 0 & 0 & 0 \end{bmatrix}, \text{ interchanging } R_2 \text{ and } R_3$$

$$\sim \begin{bmatrix} I_2 & 0 \\ 0 & 0 \end{bmatrix}$$

Hence the rank of A is 2. **Ans.**

Example 11:

Find the rank of the matrix $A = \begin{bmatrix} 1 & -1 & 2 & -3 \\ 4 & 1 & 0 & 2 \\ 0 & 3 & 0 & 4 \\ 0 & 1 & 0 & 2 \end{bmatrix}$

Solution:

$$A \sim \begin{bmatrix} 1 & 0 & 2 & 0 \\ 4 & 5 & 0 & 14 \\ 0 & 3 & 0 & 4 \\ 0 & 1 & 0 & 2 \end{bmatrix}, \text{ replacing } C_2, C_4 \text{ by } C_2 + C_1, C_4 + 3C_1 \text{ respectively}$$

$$\sim \begin{bmatrix} 1 & 0 & 1 & 0 \\ 4 & 5 & 0 & 7 \\ 0 & 3 & 0 & 2 \\ 0 & 1 & 0 & 1 \end{bmatrix}, \text{ replacing } C_3, C_4 \text{ by } \frac{1}{2}C_3, \frac{1}{2}C_4 \text{ respectively}$$

$$\Rightarrow \quad A \sim \begin{bmatrix} 0 & 0 & 1 & 0 \\ 4 & 5 & 0 & 2 \\ 0 & 3 & 0 & -1 \\ 0 & 1 & 0 & 0 \end{bmatrix}, \text{ replacing } C_1, C_4 \text{ by } C_1 - C_3, C_4 - C_2 \text{ respectively}$$

$$\sim \begin{bmatrix} 0 & 0 & 1 & 0 \\ 1 & 5 & 0 & 2 \\ 0 & 3 & 0 & -1 \\ 0 & 1 & 0 & 0 \end{bmatrix}, \text{ replacing } C_1 \text{ by } \frac{1}{2}C_1$$

$$\sim \begin{bmatrix} 0 & 0 & 1 & 0 \\ 1 & 0 & 0 & 0 \\ 0 & 3 & 0 & -1 \\ 0 & 1 & 0 & 0 \end{bmatrix}, \text{ replacing } C_2, C_4 \text{ by } C_2 - 5C_1, C_4 - 2C_1 \text{ respectively}$$

$$\sim \begin{bmatrix} 0 & 0 & 1 & 0 \\ 1 & 0 & 0 & 0 \\ 0 & 0 & 0 & 1 \\ 0 & 1 & 0 & 0 \end{bmatrix}, \text{ replacing } C_2, C_4 \text{ by } C_2 + 3C_4, -C_4 \text{ respectively}$$

$$\sim \begin{bmatrix} 1 & 0 & 0 & 0 \\ 0 & 1 & 0 & 0 \\ 0 & 0 & 1 & 0 \\ 0 & 0 & 0 & 1 \end{bmatrix}, \text{ rearranging columns}$$

$$\sim [\, I_4 \,].$$

$\therefore$ Rank of A is 4.

Ans.

Example 12:

Find the rank of the matrix $A = \begin{bmatrix} 1 & -1 & 2 & -3 \\ 4 & 1 & 0 & 2 \\ 0 & 3 & 0 & 4 \\ 0 & 1 & 0 & 2 \end{bmatrix}$.

Solution:

$$A \sim \begin{bmatrix} 1 & 0 & 1 & 0 \\ 4 & 5 & 0 & 14 \\ 0 & 3 & 0 & 4 \\ 0 & 1 & 0 & 2 \end{bmatrix}, \text{ replacing } C_2, C_3, C_4 \text{ by } C_2 + C_1, \frac{1}{2}C_3, C_4 + 3C_1 \text{ respectively.}$$

$$\sim \begin{bmatrix} 0 & 0 & 1 & 0 \\ 4 & 5 & 0 & 7 \\ 0 & 3 & 0 & 2 \\ 0 & 1 & 0 & 1 \end{bmatrix}, \text{ replacing } C_1, C_4 \text{ by } C_1 - C_3, \frac{1}{2}C_4 \text{ respectively.}$$

$$\sim \begin{bmatrix} 0 & 0 & 1 & 0 \\ 1 & 5 & 0 & 2 \\ 0 & 3 & 0 & -1 \\ 0 & 1 & 0 & 0 \end{bmatrix}, \text{ replacing } C_1, C_4 \text{ by } \frac{1}{4}C_1, C_4 - C_2 \text{ respectively.}$$

$$\sim \begin{bmatrix} 0 & 0 & 1 & 0 \\ 1 & 0 & 0 & 0 \\ 0 & 3 & 0 & -1 \\ 0 & 1 & 0 & 0 \end{bmatrix}, \text{ replacing } C_2, C_4 \text{ by } C_2 - 5C_1, C_4 - 2C_1 \text{ respectively.}$$

$$\Rightarrow A \sim \begin{bmatrix} 1 & 0 & 0 & 0 \\ 0 & 0 & 1 & 0 \\ 0 & 0 & 0 & 1 \\ 0 & 1 & 0 & 0 \end{bmatrix}, \text{ replacing } C_2, C_4 \text{ by } C_2 + 3C_4, -C_4 \text{ respectively and interchanging } C_1, C_2$$

$$\sim \begin{bmatrix} 1 & 0 & 0 & 0 \\ 0 & 1 & 0 & 0 \\ 0 & 0 & 1 & 0 \\ 0 & 0 & 0 & 1 \end{bmatrix}, \text{ rearranging columns}$$

$\sim [\, I_4 \,]$

Hence the rank of A is 4. **Ans.**

Example 13:

Find the rank of the matrix $A = \begin{bmatrix} 1 & 3 & 4 & 7 \\ 2 & 4 & 5 & 8 \\ 3 & 1 & 2 & 4 \end{bmatrix}$

Solution:

$$A \sim \begin{bmatrix} 1 & 3 & 4 & 7 \\ 1 & 1 & 1 & 1 \\ 3 & 1 & 2 & 4 \end{bmatrix}, \text{ replacing } R_2 \text{ by } R_2 - R_1$$

$$\sim \begin{bmatrix} 1 & 2 & 3 & 6 \\ 1 & 0 & 0 & 0 \\ 3 & -2 & -1 & 1 \end{bmatrix}, \text{ replacing } C_2, C_3, C_4 \text{ by } C_2 - C_1, C_3 - C_1, C_4 - C_1 \text{ respectively.}$$

$$\sim \begin{bmatrix} 0 & 2 & 3 & 6 \\ 1 & 0 & 0 & 0 \\ 3 & -2 & -1 & 1 \end{bmatrix}, \text{ replacing } R_1, R_3 \text{ by } R_1 - R_2, R_3 - 3R_2$$

$$\sim \begin{bmatrix} 0 & 1 & 3 & 0 \\ 1 & 0 & 0 & 0 \\ 3 & -1 & -1 & 3 \end{bmatrix}, \text{ replacing } C_2, C_4 \text{ by } \frac{1}{2}C_2, C_4 - 2C_3 \text{ respectively.}$$

$$\Rightarrow \quad A \sim \begin{bmatrix} 0 & 1 & 0 & 0 \\ 1 & 0 & 0 & 0 \\ 0 & -1 & 2 & 3 \end{bmatrix}, \text{ replacing } C_3 \text{ by } C_3 - 3C_2$$

$$\sim \begin{bmatrix} 0 & 1 & 0 & 0 \\ 1 & 0 & 0 & 0 \\ 0 & 0 & 2 & 3 \end{bmatrix}, \text{ replacing } R_3 \text{ by } R_3 + R_1$$

$$\sim \begin{bmatrix} 0 & 1 & 0 & 0 \\ 1 & 0 & 0 & 0 \\ 0 & 0 & 1 & 1 \end{bmatrix}, \text{ replacing } C_3, C_4 \text{ by } \frac{1}{2}C_3 \text{ and } \frac{1}{3}C_4 \text{ respectively.}$$

$$\sim \begin{bmatrix} 0 & 1 & 0 & 0 \\ 1 & 0 & 0 & 0 \\ 0 & 0 & 1 & 0 \end{bmatrix}, \text{ replacing } C_4 \text{ by } C_4 - C_3$$

$$\sim \begin{bmatrix} 1 & 0 & 0 & 0 \\ 0 & 1 & 0 & 0 \\ 0 & 0 & 1 & 0 \end{bmatrix}, \text{ interchanging } R_1 \text{ and } R_2$$

$$\sim [\, I_3 \; 0 \,]$$

Hence the rank of A is 3. **Ans.**

Example 14:

Find the rank of $A = \begin{bmatrix} 0 & 1 & -3 & -1 \\ 1 & 0 & 1 & 1 \\ 3 & 1 & 0 & 2 \\ 1 & 1 & -2 & 0 \end{bmatrix}$

Solution:

$$A \sim \begin{bmatrix} 0 & 1 & -3 & -1 \\ 1 & 0 & 0 & 0 \\ 3 & 1 & -3 & -1 \\ 1 & 1 & -3 & -1 \end{bmatrix}, \text{ replacing } C_4, C_4 \text{ by } C_3 - C_1 \text{ and } C_4 - C_1 \text{ respectively}$$

$$\sim \begin{bmatrix} 0 & 1 & -3 & -1 \\ 1 & 0 & 0 & 0 \\ 3 & 0 & 0 & 0 \\ 1 & 0 & 0 & 0 \end{bmatrix}, \text{ replacing } R_3, R_4 \text{ by } R_3 - R_1 \text{ and } R_4 - R_1 \text{ respectively.}$$

$$\sim \begin{bmatrix} 0 & 1 & 0 & 0 \\ 1 & 0 & 0 & 0 \\ 3 & 0 & 0 & 0 \\ 1 & 0 & 0 & 0 \end{bmatrix}, \text{ replacing } C_3, C_4 \text{ by } C_3 + 3C_2, C_4 + C_2 \text{ respectively}$$

$$\sim \begin{bmatrix} 0 & 1 & 0 & 0 \\ 1 & 0 & 0 & 0 \\ 0 & 0 & 0 & 0 \\ 0 & 0 & 0 & 0 \end{bmatrix}, \text{ replacing } R_3, R_4 \text{ by } R_3 - 3R_2, R_4 - R_2 \text{ respectively.}$$

$$\sim \begin{bmatrix} 1 & 0 & 0 & 0 \\ 0 & 1 & 0 & 0 \\ 0 & 0 & 0 & 0 \\ 0 & 0 & 0 & 0 \end{bmatrix}, \text{ interchanging } C_1, C_2$$

$$\sim \begin{bmatrix} I_2 & 0 \\ 0 & 0 \end{bmatrix}$$

Hence the rank of A is 2. **Ans.**

Example 15:

Find the rank of the matrix $A = \begin{bmatrix} 1 & 0 & 2 & 1 \\ 0 & 1 & -2 & 1 \\ 1 & -1 & 4 & 0 \\ -2 & 2 & 8 & 0 \end{bmatrix}$

Solution:

$$A \sim \begin{bmatrix} 1 & 0 & 2 & 1 \\ 0 & 1 & -2 & 1 \\ 0 & -1 & 2 & -1 \\ 0 & 2 & 12 & 2 \end{bmatrix}, \text{ replacing } R_3, R_4 \text{ by } R_3 - R_2 \text{ and } R_4 + 2R_1 \text{ respectively}$$

$$\sim \begin{bmatrix} 1 & 0 & 2 & 1 \\ 0 & 1 & -2 & 1 \\ 0 & 0 & 0 & 0 \\ 0 & 0 & 16 & 0 \end{bmatrix}, \text{ replacing } R_3 \text{ and } R_4 \text{ by } R_3 + R_2 \text{ and } R_4 - 2R_2 \text{ respectively.}$$

$$\sim \begin{bmatrix} 1 & 0 & 2 & 1 \\ 0 & 1 & 0 & 1 \\ 0 & 0 & 0 & 0 \\ 0 & 0 & 16 & 0 \end{bmatrix}, \text{ replacing } C_2 \text{ by } C_3 + 2C_2$$

$$\sim \begin{bmatrix} 1 & 0 & 0 & 0 \\ 0 & 1 & 0 & 0 \\ 0 & 0 & 0 & 0 \\ 0 & 0 & 16 & 0 \end{bmatrix}, \text{ replacing } C_3, C_4 \text{ by } C_3 - 2C_1, C_4 - C_1 - C_2 \text{ respectively}$$

$$\sim \begin{bmatrix} 1 & 0 & 0 & 0 \\ 0 & 1 & 0 & 0 \\ 0 & 0 & 0 & 0 \\ 0 & 0 & 1 & 0 \end{bmatrix}, \text{ replacing } C_3 \text{ by } \frac{1}{16} C_2$$

$$\sim \begin{bmatrix} 1 & 0 & 0 & 0 \\ 0 & 1 & 0 & 0 \\ 0 & 0 & 1 & 0 \\ 0 & 0 & 0 & 0 \end{bmatrix}, \text{ interchanging } R_3 \text{ and } R_4$$

$$\sim \begin{bmatrix} I_3 & 0 \\ 0 & 0 \end{bmatrix}$$

Hence the rank of the given matrix = 3. **Ans.**

Example 16:

Use elementary transformation to reduce the following matrix A to triangular form and hence find the rank of A.

$$A = \begin{bmatrix} 5 & 3 & 14 & 4 \\ 0 & 1 & 2 & 1 \\ 1 & -1 & 2 & 0 \end{bmatrix}$$

Solution:

$$A \sim \begin{bmatrix} 5 & 3 & 8 & 1 \\ 0 & 1 & 0 & 0 \\ 1 & -1 & 4 & 1 \end{bmatrix}, \text{ replacing } C_3, C_4 \text{ by } C_3 - 2C_2, C_4 - C_2 \text{ respectively}$$

$$\Rightarrow \quad A \sim \begin{bmatrix} 5 & 8 & -12 & -4 \\ 0 & 1 & 0 & 0 \\ 1 & 0 & 0 & 0 \end{bmatrix}, \text{ replacing } C_2, C_3, C_4 \text{ by } C_2 + C_1, C_3 - 4C_1, C_4 - C_1$$

$$\sim \begin{bmatrix} 5 & 0 & 0 & -4 \\ 0 & 1 & 0 & 0 \\ 1 & 0 & 0 & 0 \end{bmatrix}, \text{ replacing } C_2, C_3 \text{ by } C_2 + 2C_4, C_3 - 3C_4 \text{ respectively}$$

$$\sim \begin{bmatrix} 5 & 0 & 0 & 1 \\ 0 & 1 & 0 & 0 \\ 1 & 0 & 0 & 0 \end{bmatrix}, \text{ replacing } C_4 \text{ by } -\frac{1}{4} C_4$$

$$\sim \begin{bmatrix} 0 & 0 & 0 & 1 \\ 0 & 1 & 0 & 0 \\ 1 & 0 & 0 & 0 \end{bmatrix}, \text{ replacing } C_1 \text{ by } C_1 - 5C_4$$

$$\sim \begin{bmatrix} 0 & 1 & 0 & 0 \\ 0 & 0 & 0 & 1 \\ 1 & 0 & 0 & 0 \end{bmatrix}, \text{ interchanging } R_1, R_2$$

$$\sim \begin{bmatrix} 0 & 1 & 0 & 0 \\ 1 & 0 & 0 & 0 \\ 0 & 0 & 0 & 1 \end{bmatrix}, \text{ interchanging } R_2, R_3$$

$$\sim \begin{bmatrix} 1 & 0 & 0 & 0 \\ 0 & 1 & 0 & 0 \\ 0 & 0 & 0 & 1 \end{bmatrix}, \text{ interchanging } R_1, R_2$$

$$\sim \begin{bmatrix} 1 & 0 & 0 & 0 \\ 0 & 1 & 0 & 0 \\ 0 & 0 & 1 & 0 \end{bmatrix}, \text{ interchanging } C_3, C_4$$

$$\sim [\, I_3 \; 0 \,]$$

Hence the rank of A is 3. **Ans.**

EXERCISES

1. *Show that the inverse of*

$$\begin{bmatrix} 3 & -2 & -1 \\ -4 & 1 & -1 \\ 2 & 0 & 1 \end{bmatrix} \text{ is } \begin{bmatrix} 1 & 2 & 3 \\ 2 & 5 & 7 \\ -2 & -4 & -5 \end{bmatrix}$$

2. *If* $A = \begin{bmatrix} a_1 & b_1 & c_1 \\ a_2 & b_2 & c_2 \\ a_3 & b_3 & c_3 \end{bmatrix}$, *prove that* $AA^{-1} = I_3$.

3. *Prove that* $Adj. (A') = (Adj. A)'$.

4. *Compute rank of the matrix* $\begin{bmatrix} 1 & 2 & 3 \\ 4 & 5 & 6 \\ 7 & 8 & 9 \end{bmatrix}$

5. $\begin{bmatrix} 1 & 2 & 3 \\ 2 & 4 & 9 \\ 3 & 6 & 10 \end{bmatrix}$

6. If A and B are symmetric (or skew-symmetric) matrices, then so is A + B.

7. $\begin{bmatrix} 0 & 1 & 1 \\ 1 & 0 & 1 \\ 1 & 1 & 1 \end{bmatrix}$

8. If A is any matrix, then show that A' A is a symmetric matrix.

9. If A is symmetric matrix, then show that AA' = A'A and A^2 is symmetic.

10. Show that the matrix $A = \frac{1}{\sqrt{2}}\begin{bmatrix} 1 & i \\ -i & -1 \end{bmatrix}$ is unitary.

11. *Show that the inverse of*

$$\begin{bmatrix} 2 & -1 & 1 \\ -15 & 6 & -5 \\ 5 & -2 & 2 \end{bmatrix} \text{is} \begin{bmatrix} -2 & 0 & 1 \\ -5 & 1 & 5 \\ 0 & 0 & 3 \end{bmatrix}$$

12. *Show that the adjoint of*

$$\begin{bmatrix} 1 & 1 & 3 \\ 0 & 1 & -1 \\ 2 & 0 & -4 \end{bmatrix} \text{is} \begin{bmatrix} -4 & 4 & -4 \\ -2 & -10 & 1 \\ -2 & 2 & 1 \end{bmatrix}$$

12. If A and B are symmetric matrices, then prove that AB + BA is symmertic and AB – BA is Skew Symmettic,

14. Show that all positive integal powers of s symmetric matrix are symmertic.

15. If B is any square matrix, show that B' AB is symmetric provided B' AB is defined.

16. Show that $\begin{bmatrix} 1 & 1 & 1 \\ 0 & 1 & 1 \\ 0 & 0 & 1 \end{bmatrix}^{n} = \begin{bmatrix} 1 & n & \frac{1}{2}n\,(n+1) \\ 0 & 1 & n \\ 0 & 0 & 1 \end{bmatrix}$ for all natural numbers n

17. $\begin{bmatrix} 1 & 1 & 1 \\ a & b & c \\ a^2 & b^2 & c^2 \end{bmatrix}$, *where a, b, c are all real.*

18. $\begin{bmatrix} 2 & 1 & 3 & 5 \\ 4 & 2 & 1 & 3 \\ 8 & 4 & 7 & 13 \\ 8 & 4 & -3 & -1 \end{bmatrix}$

19. A fruit seller has in stock 20 dozen mangoes, 16 dozen apples and 32 dozen bananas. Suppose the selling prices are Rs. 0.35, Rs. 0.75 and Rs. 0.08 per mango, apple and banana respectively. Find the total amount the fruit seller will get by selling his whole stock.

20. If A = $\begin{bmatrix} 2 & 3 & 1 \\ 0 & 2 & -1 \end{bmatrix}$ and B = $\begin{bmatrix} 1 & -2 \\ 3 & 0 \\ -1 & 2 \end{bmatrix}$ then find Ab and BA.

21. Show that

$\begin{bmatrix} 1 & 0 & 0 & 0 \\ 2 & 1 & 0 & 0 \\ 4 & 2 & 1 & 0 \\ -2 & 3 & 1 & 1 \end{bmatrix}$ *is the inverse of* $\begin{bmatrix} 1 & 0 & 0 & 0 \\ -2 & 1 & 0 & 0 \\ 0 & -2 & 1 & 0 \\ 8 & -1 & -1 & 1 \end{bmatrix}$

22. If A be any square matrix, then show that

$A + A^{\Theta}$ is Hermitian.

23. For any two orthogonal matrices A and B, show that BA is an orthogonal matrix.

24. For any two unitary matrices A and B, show that BA is an unitary matrix.

25. If A and B are symmetric and they commute, then $A^{-1}B$ and $A^{-1}B^{1}$ are symmetic

26. Show that every square matrix can be expressed is one and only one way as P + iQ, where P and Q are Hermitian.

27. If $A_\alpha = \begin{bmatrix} \cos\alpha & \sin\alpha \\ -\sin\alpha & \cos\alpha \end{bmatrix}$, then show that $A_\alpha^{\,n} = \begin{bmatrix} \cos n\alpha & \sin n\alpha \\ -\sin n\alpha & \cos n\alpha \end{bmatrix}$,

where n is any positive integer.

Also prove that A_α and A_β commute and $A_\alpha . A_\beta = A_{\alpha+\beta}$.

28. Show that the matrix $A = \begin{bmatrix} 1 & 2 \\ 3 & 1 \end{bmatrix}$ satisfies the equation $A^2 - 2A - 5I = O$, where O is the 2×2 null matrix.

29. Evaluate $A^2 - 3A - 13I$, where I is the 2×2 unit matrix and

$$A = \begin{bmatrix} 2 & 5 \\ 3 & 1 \end{bmatrix}$$

30. Show that matrix $A = \begin{bmatrix} 1 & 0 & 0 \\ 2 & 1 & 0 \\ 3 & 2 & 1 \end{bmatrix}$

satisfies the equation $A^3 - 3A^2 + 3A - I = O$, where I is the unit matrix and O the null matrix of order 3.

5

Binomial Theorem

INTRODUCTION

Any expression of the type $x \pm y$ is called a binomial expression, x is called first term and y, second term. by elementary algebra, we know that (x + y)z = x2 + 2xy + y2; (x + y)3 = x3 + 3x2y + 3xy2 +y3. in this section, we decelop a formula for the nth power of x + y. n being a positive integer. We shall make use of Principle of Mathematical Induction in proving the expansion of (x + y)n.

Binomial Theorem: *If n is a positive integer, then*

$$(x + y)^n = x^n + {}^nC_1x^{n-1}y + {}^nC_2x^{n-2}y2 \ldots + {}^nC_ny^n.$$

***Proof*:** Clearly for $n = 1$, LHS $= x = y$.

and RHS $= x + {}^1C_1\ y = x + y$, so that result is true for $n = 1$.

Let $n + 1 > 1$ and the result be true for n.

i.e. $\quad (x + y)^n = x^n + {}^nC_1\,x^{n-1}\,y + \ldots + {}^nC_n\,y^n.$

Consider $\quad (x + y)^{n+1} = (x + y)^n\,(x + y)$

$$= (x^n + {}^nC_1\,x^{n-1}y + \ldots + {}^nC_ny^n)\,(x + y)$$

$$= x^{n+1} + ({}^nC_1\,x^n\ y + x^n\ y) + ({}^nC_2\,x^{n-1}y^2 + {}^nC_1\,x^{n-1}y^2)$$

$$+ \ldots + ({}^nC_{n-1}x\ y^n + {}^nC_nx\ y^n) + {}^nC_ny^{n+1}$$

$$= x^{n+1} + ({}^nC_0 + {}^nC_1)\,x^n y + ({}^nC_1 + {}^nC_2)\,x^{n-1}y^2$$

$$+ ({}^nC_2 + {}^nC_3)\,x^{n-2}y^3 + \ldots +$$

$$+ \ldots + ({}^nC_{n-1} + {}^nC_n)\,x\,y^2 + {}^nC_ny^{n+1}$$

But $\quad {}_nC_r + {}^nC_{r-1} = {}^{n+1}C_r$ for all $1 < r < n$.

Hence we get that

$$(x + y)^{n+1} = x^{n+1} + {}^{n+1}C_1\,x^n y + {}^{n+1}C_2\,x^{n-1}y^2 + \ldots +$$

$$+ \ldots + {}^{n+1}C_2\,x\,y + {}^{n+1}C_{n+1}\ y^{n+1}.$$

Consequently Binomial Theorem holds for $n + 1$.

Thus by Mathematical Induction, it is true for all positive integers n.

Remark: The expansion of $(x + y)^n$ is given by

$(x + y)^n = x^n + {}^nC_1\, x^{n-1}(-y) + {}^nC_2\, x^{n-2}(-y)^2 + \ldots + {}^nC_n(-y)^n$

$= x^n - {}^nC_1\, x^{n-1} y + {}^nC_2\, x^{n-2} y^2 - {}^nC_3\, x^{n-3} y^3 + \ldots + (-1)^n\, {}^nC_n\, y^n.$

Example 1:

Expand $(x + y)^n$.

Solution:

$$(x + y)^5 = x^5 + {}^5C_1\, x^4 y + {}^5C_2\, x^3 y^2 + {}^5C_2\, x^2 y^3 + {}^5C_4\, x\, y^4 + {}^5C_5\, x^0 y^5.$$

Now $\quad {}^5C_1 = {}^5C_4 = {}^5C_2 = {}^5C_3 = \dfrac{5.4}{2.1} = 10.$

Hence $\quad (x + y)^5 = x^5 + 5x^4y + 10x^3y^2 + 10x^2y^3 + 5xy^4 + y^5$.

Example 2:

Expand $(x - 1)^7$.

Solution:

By Binomial Theorem

$(x - 1)^7 = x^7 + {}^7C_1\, x^6(-1) + {}^7C_2 x^5(-1)^2 + {}^7C_3\, x^4(-1)^3 + {}^7C_4 x^3(-1)^4$
$\quad + {}^7C_5\, x^2(-1)^5 + {}^7C_6\, x(-1)^6 + {}^7C_7(-1)$

$= x^7 - 7x^6 + 21x^5 - 35x^4 + 35x^3 - 21x^2 + 7x - 1.$

Example 3:

Expand $\left(3x - \dfrac{1}{2y}\right)^4$ *with the help of Binomisl theorem. By giving suitable values to x and y obtain the value of* $(29.5)^4$ *correct to fourth decimal place*

Solution:

We have

$$\left(3x - \frac{y}{2}\right)^4 = (3x)^4 + {}^4C_1(3x)^3\left(-\frac{y}{2}\right) + {}^4C_2(3x)^2\left(-\frac{y}{2}\right)^2$$

$$+ 4\sqrt{}\, C_3(3x)\left(-\frac{y}{2}\right)^3 + {}^4C_4\left(-\frac{y}{2}\right)^4$$

$$=(3x)^4 - 4(3x)^3\frac{y}{2} + 6(3x)^2\frac{y^2}{4} - 12x\frac{y^3}{8} + \frac{y^4}{16}$$

$$= 81x^4 - 54x^3y + \frac{27}{2}x^2y^2 - \frac{3}{2}xy^3 + \frac{y^4}{16}$$

Now we can write

$(29.5)^4 = (30 - 0.5)^4\ (3 \times 10 - 1/2 \times 1)^4$

Thus if we put $x = 10$ and $y = 1$ in the above expansion, we get

$$(29.5)^4 = 81\times(10)^4 - 54(10)^3 + \frac{27}{2}(10)^2 - \frac{3}{2}.10 + \frac{1}{16}$$

$$= 810000 - 54000 + \frac{27}{2}\times 100 - 15\frac{1}{16}$$

$$= 810000 - 54000 + 1350 - 15 + \frac{1}{16}$$

$= 757335.0625$

Example 4:

Show that $(\sqrt{3} + \sqrt{3})^3 + (\sqrt{3} + \sqrt{3})^3 = 18\sqrt{3}$.

Solution:

We have

$(\sqrt{3} + \sqrt{2})^3 = (\sqrt{3})^3 + 3(\sqrt{3})^2\sqrt{2} + 3(\sqrt{3})(\sqrt{2})^2 + (\sqrt{2})^3$

$(\sqrt{3} + \sqrt{2})^3 = (\sqrt{3})^3 - 3(\sqrt{3})^2\sqrt{2} + 3(\sqrt{3})(\sqrt{2})^2 - (\sqrt{2})^3$

Addition gives

$(\sqrt{3} + \sqrt{2})^3 + (\sqrt{3} - \sqrt{3})^3 = 2(\sqrt{3})^3 + 6(\sqrt{3})(\sqrt{2})^2$

$= 6\sqrt{3} + 12\sqrt{3}$

$= 18\sqrt{3}$.

TO FIND THE GENERAL TERM IN THE EXPANSION OF $(x + y)^n$

Let T_{r+1} be the $(r + 1)$th term in the expansion of $(x + y)^n$. Now lst term $= x^n = {}^nC_0x^ny^o = {}^nC_0x^{n-0}y^0$ second term $= {}^nC_1x^{n-1}y$; third term $= {}^nC_2x^{n-2}y^2$. This gives that $(r+1)$th term $= {}^nC_rx^{n-r}y^r$.

Thus $\quad T_{r+1} = {}^nC_rx^{n-r}y^r$.

Also, it can be easily seen that number of terms in the expansion of $(x + y)^n$ is $n + 1$.

Example 5:

Find the 9th term in the expansion of $(\sqrt{a} - 2\sqrt{b})^{12}$.

Solution:

The general term in the expansion of $(x + y)^n$ is

$T_{r+1} = {}^nC_r x^{n-r} y^r$.

We want to find T_9 *i.e.,* $r = 8$

Now $T_9 = {}^{12}C_8 x^{12-8} y^8$.

$\Rightarrow T_9 = {}^{12}C_8 (\sqrt{a})^4 (-2\sqrt{b})^8$

$= \dfrac{12.11.10.9}{4.3.2.1} a^2.256.b^4$

$= 126720\ a^2b^4$.

Example 6:

Find the coeffcient of x^6 *in* $(x + 2)^9$.

Solution:

Let x^6 occcur in T_{r+1}.

Then $T_{r+1} = {}^9C_r x^{6-r} 2r$ .Since x^6 is to occur in this, $9 - r = 6 \Rightarrow r = 3$.
Hence the term containing x^6 is given by ${}^9C_3 x^6\, 2^3$. So the coeffcient of x^6 is ${}^9C_3.2^3 = \dfrac{9.8.7}{3.2.1} \times 8 = 672.$

Example 7:

Find the term independent of x in the expnsion of $\left(x - \dfrac{1}{x}\right)^{10}$

Solution:

Suppose the required tern is $(r - 1)$th term T_{r+1} .

Then $T_{r+1} = {}^{10}C_r x^{10-r} \left(-\dfrac{1}{x}\right)^r$.

The power of x present in T_{r+1} . is $10 - 2r$.

Since there must be no x in T_{r+1} $10 - 2r = 0 \Rightarrow r = 5$. Hence T_6 is the required term.

Now $T_6 = {}^{10}C_5 (-1)^5$

$= \dfrac{10.9.8.7.6}{5.4.3.2.1}(-1) = 9.4.7(-1) = 252.$

Example 8:

The first three terms in the expansion of a binomial are 729, 7290, 30375. Find it.

Solution:

Let the binomial be $(x + a)^n$.

Then it is given that

$x^n = 729$

${}^nC_1 x^{n-1} a = 7290$

${}^nC_2 x^{n-2} a^2 = 30375$

i.e. $x^n = 729$ ...(*i*)

$nx^{n-1} a = 7290$ ...(*ii*)

$$\frac{n(n-1)}{2} x^{n-2} a^2 = 30375 \quad ...(iii)$$

(*i*) and (*iii*) on multiplication give

$$\frac{n(n-1)}{2} x^{2n-2}.a^2 = 729 \times 30375$$

Also square of (*ii*) is given by

$n^2 x^{2n-2}.a^2 = 7290 \times 7290$

Dividing these two results, we get

$$\frac{n(n-1)}{2n^2} = \frac{729 \times 30375}{7290 \times 7290}$$

$\Rightarrow n = 6$ (on simplification)

Hence, from (*i*), $x^n = x^6 = 792$

$\Rightarrow x = 3$

Again (*ii*) gives $6.(3)^5.a = 7290$

$\Rightarrow a = 5.$

The required binomial expansion is, therefore,

$(3 + 5)^6$

Example 9:

If the 21st and 22nd terms in the expansion of $(1 + x)^{44}$ are equal, find the value of x.

Solution:

21st and 22nd terms in the expansion of $(1 + x)^{44}$ are given by

$T_{21} = {}^{44}C_{20}\, x^{20}$

$T_{21} = {}^{44}C_{21}\, x^{21}$

By the given condition

$${}^{44}C_{20}\, x^{20} = {}^{44}C_{21}\, x^{21}$$

$$\Rightarrow x = \frac{{}^{44}C_{20}}{{}^{44}C_{21}} = \frac{44!}{24!20!} \times \frac{23!21!}{44!}$$

$$= \frac{23!21!}{24!20!} = \frac{23!20!.21}{23!20!.24}$$

$$= \frac{21}{24} = \frac{7}{8}$$

which is the required value of x.

Example 10:

If the absolute term in the expansion of $\left(\sqrt{x} - \frac{k}{x^2}\right)^{10}$ *is 405, find the value of k.*

Solution:

By the absolute term, we mean the term independent of x (also called the constant term).

So suppose T_{r+1} is the absolute term in the expansion of

$$\left(\sqrt{x} - \frac{k}{x^2}\right)^{10}$$

Now $\quad T_{r+1} = {}^{10}C_r\, (\sqrt{x})^{10-r} \left(-\frac{k}{x^2}\right)^r$

$= {}^{10}C_r\, x^{(10-r)/2}\, x^{-2r}.\, k^r\, (-1)^r$

$= (-1)^r\, {}^{10}C_r\, k^r x^{(10-5r)/2}$

T_{r+1} will be insependent of x if $\frac{10-5r}{2} = 0$

i.e.., $\quad$ if $r = 2$.

Thus $\quad T_3$ is the absolute term.

Now $\quad T_3 = (-1)2\ {}^{10}C_2\, k^2$.

Also it is given that $T_3 = 405$

$\Rightarrow (-1)2\ {}^{10}C_2\, k^2 = 405$

$$\Rightarrow k^2 = \frac{405}{45} = 9$$

$\Rightarrow k = \pm 3$

These are the required values of k.

EXERCISES

1. Expand the following:

(*i*) $(2x - 3y)^5$ (*ii*) $(1 - 2x)^{10}$

(*iii*) $\left(\frac{2x}{3} - \frac{3}{2x}\right)^4$ (*iv*) $\left(x + 1 - \frac{1}{x}\right)^6$

(*v*) $(x + \sqrt{x^2 - 1})^7 + (x - \sqrt{x^2 - 1})^7$.

2. Find the 8th term in the expansion of $(x^{3/2} y^{1/2} - x^{1/2} y^{3/2})^{10}$.

3. Find the 9th term in the expansion of $(2x^{1/2} - y^{1/2})^{20}$.

4. (*i*) Determine the coefficient of x^{15} in the expansion of $(x - x^2)^{10}$.

(*ii*) Find the coefficient of x^{-6} in the xpansion of $\left(x^2 + \frac{1}{x^3}\right)^{12}$.

5. What is the term independent of x in the expansion of $\left(\frac{3}{2}x^2 - \frac{1}{3x}\right)^6$?

6. Find the term independent of x in the expansion of $\left(x^2 + \frac{1}{x}\right)^{12}$

7. If x^{4r} is present in the expansion of $\left(x - \frac{1}{x^2}\right)^{4n}$, then show that its coefficient is $\dfrac{4n!}{\left[\frac{4}{3}(n-r)\right]!\left[\frac{4}{3}(2n+r)\right]!}$.

8. If in the expansion of $(x + y)^n$, sum of all odd terms is a and sum of all even terms is b, prove that $(x^2 + y^2)^n = a^2 + b^2$.

9. Use Binomial theorem to evaluate

(*i*) $(98)^4$ (*ii*) $(999)^4$

10. The first three terms in the expansion of a binomial are 32, 240, 720 respectively. Find it.

11. The first three terms in the expansion of a binomial are 1, 10, 40. Find it.

12. If the second, third and fourth terms in the expansion of $(x + a)^n$ are 240, 720 and 1080 respectively. find x, a, n.

13. In the expansion of $(1+x)^{11}$, the fifth term is 24 times the third term. Find the value of x.

14. If three successive terms in the expansion of $(1+x)^n$ are 8064, 13440, 15360, find the values of n and x.

15. If in the expansion of $(1+x)^n$, the fifth term is four times the fourth term and the fourth term is six times the third term find the values of n and x.

16. If the absolute term in the expansion 0 $\left(x^2+\frac{k}{x^2}\right)^{12}$ is 495, find the value of k.

17. Find the coefficient of x^4 in the expansion of $\left(x^4+\frac{k}{x^2}\right)^{15}$

ANSWERS

1. (*i*) $32x^5 - 240x^5y + 720x^3y^2 - 1080x^2y^3 + 810xy^4 - 243y^5$.

   (*ii*) $1 - 20x + 180x^2 - 960x^3 + 3360x^4 - 8064x^5\ 13440x^6 - 15360x^7 + 11520x^8 - 5120x^9 + 1024x^{10}$.

   (*iii*) $\frac{64}{729}x^6 - \frac{32}{27}x^1 + \frac{20}{3}x^2 - 20 + \frac{135}{4x^2} - \frac{243}{8x^4} + \frac{729}{64x^6}$.

   (*iv*) $x^6 + 6x^5 + 9x^4 - 10x^3 - 30x^2 + 6x + 41 - \frac{6}{x} - \frac{30}{x^2} + \frac{10}{x^2} + \frac{9}{x^4} - \frac{6}{x^5} + \frac{1}{x^6}$.

   (*v*) $2x\,(64x^6 - 112x^4 + 56x^2 - 7)$

2. $-120\,x^8y^{12}$.
3. $760xy^6$.
4. (*i*) -252. (*ii*) 924.
5. $\frac{7}{18}$
6. 495.
9. (*i*) 9236816. (*ii*) 996, 005, 996, 001.
10. $(2+3)^5$
11. $(1+2)^5$.
12. $x = 2$, $a = 3$, $n = 5$.
13. ± 12.
14. $n = 10$, $x = 2$.
15. $x = 2$, $n = 11$.
16. 1.
17. ${}^{15}C_8$

TO FIND OUT THE MIDDLE TERM IN THE EXPANSILN OF $(x + y)$

Case I. n is even.

Number of terms in the expansion of $(x+y)^n$ is $n+1$ *i.e.*, an odd term.

So, the middle term will be $\left(\frac{n}{2}+1\right)$th term *i.e.*, $T_{(n/2+1)}$.

Hence middle term $= {}^nC_{n/2}\, x^{n-(n/2)}\, y^{n/2}$

$= {}^nC_{n/2}\, x^{n/2}\, y^{n/2}$

$$= \frac{n!}{\left[\left(\frac{n}{2}\right)!\right]^2} x^{n/2}\, y^{n/2}$$

***Case* II.** n is odd.

Number of terms in the expansion of $(x + y)^n$ is $n + 1$ *i.e.*, an even number.

In this case there are two middle terms, namely, $\frac{n+1}{2}$th and $\frac{n+3}{2}$th terms.

$\frac{n+1}{2}$th term $= T_{(n+1)/2} = T_{(n-1)/2+1} = {}^nC_{(n-1)/2}$
$x^{n-(n-1)/2}\, y^{(n-1)/2} = {}^nC_{(n-1)/2}\, x^{(n+1)/2}\, y^{(n-1)/2}$

and $\quad \frac{n+3}{2}$th term $= T_{(n+3)/2} = T_{(n+1)/2+1} = {}^nC_{(n+1)/2}$
$x^{n-(n+1)/2}\, y^{(n+1)/2} = {}^nC_{(n+1)/2}\, x^{(n-1)/2}\, y^{(n+1)/2}$

note that coefficient of both the middle terms are same !

TO FIND THE GREATEST TERM (NUMERICALLY) IN THE EXPANSION OF $(x + y)^n$

Let rth term be T_r and $(r + 1)$th term be T_{r+1}

Then $\quad T_{r+1} = {}^nC_r\, x^{n-r}\, y^r$

and $\quad T_r = {}^nC_{r-1}\, x^{n-r+1}\, y^{r-1}$

So $\quad \frac{T_{r+1}}{Tr} = \frac{{}^nC_r\, y}{{}^nC_{r-1}}\frac{y}{x} = \frac{n!}{r!(n-r)!}\quad \frac{(n-r+1)!(r-1)!}{n!}\frac{y}{x}$

$= \frac{(n-r+1)}{r}\frac{x}{y}$.

Thus $\quad T_{r+1} \gtrless T_r \Leftrightarrow (n-r+1)y \gtrless rx$

$\Leftrightarrow \frac{n+1}{r} - 1 \gtrless \frac{x}{y}$

$\Leftrightarrow \frac{n+1}{r} \gtrless \left(\frac{x}{y}+1\right)$

$$\Leftrightarrow \frac{n+1}{\frac{x}{y}+1} \gtrless r$$

$$\Leftrightarrow r \gtrless \frac{n+1y}{x+y}.$$

(Note that $y > 0$ as we are/interested only in numerically geratest term)

Case I. If $\frac{(n+1)v}{x+y}$ is a whole integer, say m.

Then for all r varying from 1 to $m-1$, $T_{r+1} > T_r$ i.e. each of $T_1, T_2, \ldots, T_{m-1}, T_m$ is greater than its preceding term. For $r = m$. $T_{r+1} = T_r$ and for $r >$ m $T_{r+1} < T_r$ i.e, each of $T_{m+2}, T_{m+3}, \ldots, T_n, T_{n+1}$. is lesser than the preceding one.

Hence T_m and T_{m+1} are equal and are of greatesst value among all terms of $(x+y)^n$. in other words mth and $(m+1)$th tenn are of numerically greatest value.

Case II. If $\frac{(n+1)v}{x+y}$ is not a whole integer, let m be the integral part of $\frac{(n+1)y}{x+y}$. For all $r = 1, \ldots, m$; $T_{r+1} > T_r$ and for all $r = m+1, \ldots, n+1$, $T_{r+1} < T_r$.

Hence T_{m+1} is of greatest value, *i.e.* $(m+1)$th term is of greatest numerical value.

Example 11:

Find the middle term in the expansion of $(x+y)^{13}$

Solution:

in this case there are two middle terms *viz.*, T_7, T_8.

$$\text{Now } T_7 = {}^{13}C^6 x^7 y^6 = \frac{13.12.11.10.9.8}{6.5.4.3.2.1} x^7 y^6$$

$$= 1716x^6y^7$$

$$T_8 = {}^{13}C_7 x^6 y^7 = {}^{13}C_6 x^6 y^7 = 1716x^6y^7.$$

Example 12:

Find out the term of greatest value in the expansion of $(4+3x)^{25}$ *when* $x = 4$.

Solution:

$$\frac{T_{r+1}}{T_r} = \frac{25-r+1}{r}\left(\frac{3x}{4}\right)$$

$$= \frac{25-r+1}{r}\left(\frac{34}{4}\right) \text{ as } x = 4$$

$$= \frac{78-3r}{r}$$

So $\quad T_{r+1} \gtrless T_r \quad \Leftrightarrow \quad \frac{78-3r}{r} \gtrless 1$

$\Leftrightarrow \; 78 - 3r \gtrless r$

$\Leftrightarrow \; 78 \gtrless 4r$

$\Leftrightarrow \; 39 \gtrless 2r \qquad \Leftrightarrow \; r \gtrless 19 \; 1/2.$

Thus T_{20} is of greatest value.

Example 13:

Show that the middle term in the expansion of $(1 + x)^{2n}$ *is* $\frac{1.3.5.\ldots(2n-1)}{n\,!}2^n x^n$.

Solution:

In the expansion of $(1 + x)^{2n}$ there will be $2n + 1$ (*i.e.*, odd number) of terms. The middle trm will thus be the $(n + 1)$th term.

Now $\quad T_{n+1} = {}^{2n}C_n x^n$

$$= \frac{(2n)\,!}{n\,!\,n\,!}.x^n$$

$$= \frac{2n(2n-2)(2n-3)\ldots 4.3.2.1}{n\,!\,n\,!}x^n$$

$$= \frac{[2n(2n-2)(2n-4)\ldots 4.2][(2n-1)(2n-3)\ldots 3.1]}{n\,!\,n\,!}x^n.$$

$$= \frac{2^n[n(n-1)(n-2)\ldots 2.1][(2n-1)(2n-3)\ldots 3.1]}{n\,!\,n\,!}x^n.$$

$$= \frac{2^n[(2n-1)(2n-3)\ldots 3.1]}{n\,!}x^n.$$

$$= \frac{1.3.5\ldots(2n-1)}{n\,!}2^n \; .x^n$$

Example 14:

Prove that the middle term in the expansion of

$$\left(x+\frac{1}{2x}\right)^{2n} \text{ is } \frac{1.3.5\ldots(2n-1)}{n!}$$

Solution:

Here again the middle term is the $(n + 1)$th term. we have

$$T_{n+1} = {}^{2n}C_n x^n \left(\frac{1}{2x}\right)^n = {}^{2n}C_n \frac{1}{2_n}$$

$$= \frac{(2n)!}{n!\,n!} \cdot \frac{1}{2_n}$$

$$= \frac{1.3.5\ldots(2n-1)^n}{n!} \cdot \frac{1}{2_n} \quad \text{(As in the previous example)}$$

$$= \frac{1.3.5\ldots(2n-1)}{n!}.$$

BINOMIAL COEFFICIENTS

The coffrcients ${}^nC_0, {}^nC_1, {}^nC_2 \ldots$ in the expansion of a binomial are called *Binomial coefficients.*

We state and prove a few properties of the binomial coefficients.

(1) Sum of the binomial coefficients in the expansion of $(1 + x)^n$ is $2n$

i.e., we want to show that

$${}^nC_0, + {}^nC_1 + \ldots + {}^nC_n = 2^n$$

We know, $(1 + x)^n = {}^nC_0 + {}^nC_1x + {}^nC_2\, x^2 + \ldots + {}^nC_n\, x^n$

Put $x = 1$ to obtain

$$2^n = {}^nC_0 + {}^nC_1 + {}^nC_2 + \ldots + {}^nC_n.$$

which proves the result.

(2) The coefficients of the terms in the expansion of $(1 + x)^n$, equidistant from the beginning and end are equal.

The general term T_{r+1} from the beginning is ${}^nC_r\, x^r \Rightarrow$ its coefficient is ${}^nC_r$

Since total number of terms in $n + 1$, the $(r + 1)$th term from the end has $(n + 1) - (r + 1) = n - r$ terms before it.

i.e., $(r + 1)$th term from the end is $(n - r + 1)$th term the beginning and

$T_{n-r+1} = {}^nC_{n-r} x^{n-r} \Rightarrow$ its coefficient $= {}^nC_{n-r}$

But ${}^nC_r = {}^nC_{n-r}$

Hence the result follows.

(3) In the expansion of $(1+x)^n$, the sum of the even coefficients is equal to the sum of the odd coefficients.

Put $x = -1$ in the expansion

$(1+x)^n = {}^nC_0 + {}^nC_1 x + {}^nC_2 x^2 + \ldots + {}^nC_n x^n$

We have, $0 = {}^nC_0 - {}^nC_1 + {}^nC_2 - {}^nC_3 + \ldots (-1)^n\, {}^nC_n$

$\Rightarrow {}^nC_0 + {}^nC_2 + {}^nC_4 + \ldots = {}^nC_1 + {}^nC_3 + {}^nC_5 + \ldots$

which proves our result.

Note that the sum of each $= 1/2.\ 2^n = 2^{n-1}$

Example 15:

If $(1+x)^n = C_0 + C_1 x + C_2 x^2 + \ldots + C_n x^n$,

prove that $C_0C_2 + C_1C_3 + C_2C_4 + \ldots + C_{n-2} C^n = \dfrac{\underline{|2n}}{\underline{|n2} - \underline{|n+2}}$

Solution:

since $(1+x)^n = C_0 + C_1 x + C_2 x^2 + \ldots + C_n x^n$

we get $\left(1+\dfrac{1}{x}\right)^n = C_0 + \dfrac{C_1}{x} + \dfrac{C_2}{x_2} + \ldots + \dfrac{C_n}{x_n}$

Multiplying these two expansions, we have

$$(1+x)^n \left(1+\frac{1}{x}\right)^n = (C_0 + C_1 x + C_2 x^2 + \ldots + \ldots + C_n x^n)$$

$$\left(C_0 + \frac{C_1}{x} + \frac{C_2}{x_2} + \ldots + \frac{C_n}{x_n}\right).$$

Clearly $C_0C_2 + C_1C_3 + \ldots + C_{n-2} C^n =$ coefficient of $\dfrac{1}{x^2}$ in the expansion of RHS.

Thus

$C_0C_2 + C_1C_3 + \ldots + C_{n-2} C^n =$ coefficient of $\dfrac{1}{x^2}$ in $(1+x)^n \left(1+\dfrac{1}{x^2}\right)^n$

=coefficient of x^{n-2} in $(1+x)^{2n}$

Suppose x^{n-2} occurs in $T_{n+1} = {}^{2n}C_r x^r$

Then $r = n-2$

Hence coefficient of x^{n-2} in $(1+x)^{2n}$ is ${}^{2n}C_r = {}^{2n}C_r = {}^{2n}C_{n-2}$

$$\frac{\underline{|2n}}{\underline{|2n-(n-2)}\underline{|n-2}} = \frac{\underline{|2n}}{\underline{|n-2}\underline{|n+2}}.$$

Consequently $C_0C_2 + C_1C_3 + \ldots + C_{n-2}\, C_n = \dfrac{\underline{|2n}}{\underline{|n-2}\underline{|n+2}}.$

Example 16:

If $(1 + x)^n = C_0 + C_1x + C_2x^2 + \ldots + C_nx^n$,

Prove that $C_1^2 + 2C_2^2 + 3_3^2 + \ldots + nC_n^2 = \dfrac{\underline{|2n-1}}{(\underline{|n-1})^2}.$

Solution:

Now $\quad C_1x + 2C_2x^2 + 3C_3x^3 + \ldots + n\,C_nx^n$,

$$= nx\left\{1+(n-1)x+\frac{(n-1)(n-2)}{1.2}x^2+\ldots x^{n-1}\right\} = nx\,(1 + x)^{n-1}$$

Changing x into $\dfrac{1}{x}$, we get

$$\frac{C_1}{x}+\frac{2C_2}{x^2}+\frac{3C_3}{x^3}+\ldots+\frac{nC_n}{x^n} = \frac{n}{x}\left(1+\frac{1}{x}\right)^{n-1} = \frac{n}{x^n}(1+x)^{n-1} \quad \ldots(1)$$

Also $\quad C_0 + C_1x + C_2x^2 + \ldots + C_nx^n = (1 + x)^n \quad \ldots(2)$

Multiplying the expansion (1) and (2), we note that $C_1^2 + 2C_2^2 + 3C_3^2 + \ldots + nC_n^2$ is term independent of x in LHS of the product of (1) and (2).

Hence $C_1^2 + 2C_2^2 + 3C_3^2 + \ldots + nC_n^2$ = term independent of x in

$$\frac{n}{x^n}(1+x)^{n-1}(1+x)^n$$

= coeffcient of x^n in $n\,(1 + x)^{2n-1}$

= n times the coefficient of x^n in $(1 + x)^{2n-1}$

Suppose x^n *occurs in* $T_{r+1} = {}^{2n-1}C_rx^r$, then $r = n$.

As a consequence, coefficient of x^n in $(1 + x)^{2n-1}$ is ${}^{2n-1}C_n$

Thus $C_1^2 + 2C_2^2 + 3C_3^2 + \ldots + nC_n^2 = n.\dfrac{\underline{|2n-1}}{\underline{|2n-1-n}\underline{|n}}$

$$= \frac{\underline{|2n-1}}{\underline{|n-1}\underline{|n-1}}$$

$$= \frac{\underline{|2n-1}}{(\underline{|n-1})^2}$$

Example 17:

If a_1, a_2, a_3, a_4 *are the coefficients of the seond, third, fourth and fifthe terms respectively in binomial expansion of* $(1 + x)^n$, *prove that*

$$\frac{a_1}{a_1+a_2}+\frac{a_3}{a_3+a_4}=\frac{2a_2}{a_2+a_3}$$

Solution:

We know $(1 + x)^n = {}^nC_0 + {}^nC_1x + {}^nC_2x^2 + \ldots + {}^nC_nx^2$

Thus $a_1 = {}^nC_1$

$a_2 = {}^nC_2$

$a_3 = {}^nC_3$

$a_4 = {}^nC_4$.

So $$\frac{a_1}{a_1+a_2}+\frac{a_3}{a_3+a_4}=\frac{{}^nC_1}{{}^nC_1+{}^nC_2}+\frac{{}^nC_3}{{}^nC_3+{}^nC_4}$$

$$=\frac{{}^nC_1}{{}^{n+1}C_2}+\frac{{}^nC_3}{{}^{n+1}C_4}$$

$$=\frac{n}{\frac{(n+1)n}{2}}+\frac{\frac{n(n-1)(n-2)}{6}}{\frac{(n+1)n(n-1)(n-2)}{24}}$$

$$=\frac{2}{n+1}+\frac{4}{n+1}=\frac{6}{n+1} \quad \text{...(i)}$$

$$=\frac{1846}{1850}$$

$$\left(1+\frac{1}{n}\right)^{nx}=1+nx\frac{1}{n}+\frac{nx(nx-1)}{2\ !}\frac{1}{n^2}+\ldots\infty$$

$$=1+x+\frac{x(x-1/n)}{2\ !}+\frac{x(x-1/n)(x-2/n)}{3\ !}+\ldots\infty$$

$$e^x=1+x+$$

Again $$\lim_{x\to\infty}\left(\frac{1+x}{1-x}\right)^{nx}=\left[\lim_{x\to\infty}\left(1+\frac{1}{n}\right)^n\right]^x=e^x.1+x+\frac{x^2}{2\ !}+\frac{x^3}{3\ !}+\ldots\quad\infty$$

$$x-\frac{[y\log(x+1)]^2}{2\ !}+\ldots$$

$$2\left(x+\frac{x^3}{3}+\frac{x^5}{5}+\ldots.\right).+\ldots$$

$$= \frac{n(n-1)}{\frac{(n+1)n(n-1)}{6}} = \frac{6}{n+1} \quad \text{...(ii)}$$

(i) and (ii) give that result.

Example 18:

If the coefficients of x^2 and x^3 in the expansion of $(3 + kx)^9$ are equal, find the value of k.

Solution:

Coefficient of x^2 is the expansion of $(3 + kx)^9$ is

$^9C_2\ (3)^7k^2 = (3)^6\ 36k^2.$

Whereas coefficient of x^3 in the expansion of $(3 + kx)^9$ is

$^9C_2\ (3)^6k^3 = (3)^6\ 84k^3.$

By hypothesis $(3)^7\ 36k^2 = (3)^6\ 84k^3.$

$\Rightarrow k = 0$

or $k = \frac{3.36}{84} = \frac{9}{7}$

$k = 0$ will imply that $3 + kx$ is no more a binomial expression, so we reject this value of k

Hence $k = 9/7$.

EXERCISES

1. Find out the numerically greatest term in the expansion of

(*i*) $\left(\frac{x}{2} - \frac{y}{3}\right)^{15}$ when $x = 8$, $y = 9$.

(*ii*) $(2 + 3x)^{12}$ when $x = \frac{5}{6}$,

(*iii*) $(3 - 5y)^{11}$ when $y = \frac{1}{6}$.

2. Find the middle term in the expansion of $\left(1 - \frac{x^2}{2}\right)^{14}$

3. If $(1 + x)^n = C_0 + C_1x + C_2x^2 + \ldots + C_nx^n$, find the values of $C_0 + 2C_1 + 3C_3 + \ldots + (n + 1)C_n$ and $C_0C_1 + C_1C_2 + C_2C_3 + C_{n-1}C_n$.

4. If $C_0, C_1, C_2 \ldots C_n$ denoted the coefficid\ents in the expansion of $(1 + x)^n$, prove that

(i) $C_1 + 2C_2 + 3C_2 + \ldots + nC_n = n.\ 2^{n-1}$.

(ii) $\frac{C_0}{1} + \frac{C_1}{2} + \ldots + \frac{C_n}{n+1} = \frac{2^{n+1}-1}{n+1}$.

5. In the expansion of $(1 + x)^{43}$, the coefficients of the $(2r + 1)$th term and the $(r + 2)$th term are equal find r.

6. In the expansion of $(1 + x)^{20}$, the coefficient of the rth term is to that of the $(r + 1)$th term in the ratio 1 : 2. Find r.

7. In the expansion of $(1 + x)^{10}$, the coefficient of $(2r + 1)$th term is equal to the coefficient of the $(4r - 5)$th term, Find r.

8. If three successive coefficient in the expansion of $(1 + x)^n$ are 210, 120 and 45 respectively. Find n.

9. Prove that if x lies between $\frac{n}{n+1}$ and $\frac{n+1}{n}$, the greatest term in the expansion of $(1 + x)^{2n}$ has the greatest coefficient.
[Hint. Greatest coefficient in the expansion of $(1+x)n$ is ${}^nC_{n/2}$, when n is even and it is ${}^nC_r$ where $r = \frac{n-1}{2}$ or $\frac{n+1}{2}$, when n is odd].

10. Find the middle term / terms in the expansion of

(*i*) $\left(2x^2 - \frac{1}{3x^2}\right)^{10}$ (*ii*) $\left(\frac{3}{4}x + \frac{4}{3}y\right)^{11}$

(*iii*) $\left(3a - \frac{a^3}{b}\right)^9$ (*iv*) $\left(2x + \frac{3}{x}\right)^{20}$

ANSWERS

1. (*i*) 7th term (*ii*) 8th term
(*iii*) 3rd and 4th terms.

2. $\frac{429}{16}x^{14}$.

3. $2^{n-1}(n+2); \frac{\lfloor 2n}{\lfloor n-1 \lfloor n+1}$

5. 14.

6. 7

7. 1

8. 10

10. (*i*) $\frac{-896}{27}$ (*ii*) $\frac{693}{2}x^6y^5,\ 616x^5y^6$

(*iii*) ${}^9C_4 3^5.\frac{a^{17}}{b^4}$ (*iv*) $19.17.13.11.3^{10}.2^{12}$.

BINOMIAL THEOREM (ANY INDEX)

We have already found out the expansion of $(x + y)^n$ when n is a +ve integer. We now give the expansion of $(1 + y)^n$, where n can be negative integer or fraction.

Binomial theorem for any index:

If $|x| < 1$, *then*

$$(1 + x)^n = 1 + nx + \frac{n(n-1)}{\underline{|2}}x^2 + \frac{n(n-1)(n-2)}{\underline{|3}}x^3 + \ldots\infty$$

where n may be negative integer or fraction.

The following examples will illustrate the use of the above expansion.

Example 19:

Expand $\left(1-\frac{x}{2}\right)^{-\frac{1}{2}}$ *giving the first four terms.*

Solution:

We have

$$\left(1-\frac{x}{2}\right)^{-\frac{1}{2}} = 1+(-\frac{1}{2})\left(-\frac{x}{2}\right)+\frac{(-\frac{1}{2})(-\frac{1}{2}-1)}{\underline{|2}}\left(-\frac{x}{2}\right)^2$$

$$+\frac{(-\frac{1}{2})(-\frac{1}{2}-1)(-\frac{1}{2}-2)}{\underline{|3}}\left(-\frac{x}{2}\right)^3 + \ldots\infty$$

$$= 1+\frac{x}{4}+\frac{3x^2}{32}+\frac{5}{128}x+\ldots$$

Example 20:

Expand $(5-x)^{-\frac{1}{2}}$ *up to terms containing* x^3, *when* $x < 5$.

Solution:

We have $(5 - x)^{-1/2} = (5)^{-1/2}\left(1-\frac{x}{5}\right)^{-1/2}$

Since $|x| < |x/5| < 1$.

Thus by Binomial Theorem for anu index

$$\left(1-\frac{x}{2}\right)^{-1/2} = 1+(-\frac{1}{2})\left(-\frac{x}{5}\right)+\frac{(-\frac{1}{2})(-\frac{1}{2}-1)}{\underline{|2}}\left(-\frac{x}{5}\right)^2$$

$$+\frac{(-\frac{1}{2})(-\frac{1}{2}-1)(-\frac{1}{2}-2)}{\underline{|3}}\left(-\frac{x}{5}\right)^3+\ldots$$

$$=1+\frac{1}{2}+\frac{1.3}{2.4}\cdot\frac{x^2}{25}+\frac{1.3.5}{2.4.6}\cdot\frac{x^3}{125}+\ldots$$

and thus

$$(5-x)^{-1/2}=\frac{1}{\sqrt{5}}\left[1+\frac{1}{2}\cdot\frac{x}{5}+\frac{1.3}{2.4}\cdot\frac{x^2}{25}+\frac{1.3.5}{2.4.6}\cdot\frac{x^3}{125}+\ldots\right]$$

Example 21:

Using Binomial Theorem, find the value of $3\sqrt{128}$ *to four decimal places.*

Solution:

We have

$$3\sqrt{128}=(128)^{1/3}=(125+3)^{1/3}$$

$$=(125)^{1/3}\left[1+\frac{3}{125}\right]^{1/3}$$

$$=5\,[1+.024]^{1/3}$$

$$=5\left[1+\frac{1}{3}\times.024+\frac{\frac{1}{3}(\frac{1}{3}-1)}{2}\times(.024)^2+\ldots\right]$$

$$=5\,[1+.008-.000064+\ldots]$$

$= 5.\dot{}0397$ correct up to four places of decimal.

Example 22:

Find the first negative term in the expnsion of $(1 + 3/4x)^{13/3}$ *where* $0 < x < 4/3$.

Solution:

Suppose

$$T_{r+1}=\frac{n(n-1)(n-2)\ldots(n-r+1).}{\underline{|r}}\left(\frac{3}{4}x\right)^r$$

is the first negative term. As $(3/4x)^r$ is positive for $0 < x < 4/3$.

T^{r+1} will be the first negative term if r is the least integer satisfying $n - r + 1 < 0$.

i.e. $13/3 - r + 1 < 0$

$\Rightarrow$ $r > 16/3$

Least r greater than 16/3 is 6.

Hence value of r is 6.

Thus T_7 is the first negative term.

Now

$$T_7 = \frac{\frac{13}{3}\left(\frac{13}{3}-1\right)\left(\frac{13}{3}-2\right)\left(\frac{13}{3}-3\right)\left(\frac{13}{3}-4\right)\left(\frac{13}{3}-5\right)}{\underline{|6}} \cdot \left(\frac{3}{4}x\right)^6$$

$$= \frac{-91x^6}{36864}.$$

Example 23:

Find the first three terms of

$$= \frac{1}{(1+x)^2 + \sqrt{1+4x}}$$

essumig the validity of the expansion in ascending powers of x.

Solution:

We have

$$= \frac{1}{(1+x)^2 + \sqrt{1+4x}} = (1+x)^{-2}(1+4x)^{-1/2}$$

Now $(1+x)^{-2} = 1 + (-2)x + \frac{(-2)(-2)}{\underline{|2}}x^2 + \frac{(-3)(-2-1)(-2-2)}{\underline{|3}} + x^2...$

and $(1+4x)^{-1/2} = 1 + (-1/2)4x + \frac{\left(-\frac{1}{2}\right)\left(-\frac{1}{2}-1\right)}{\underline{|2}}(4x^2 + ...$

Hence

$(1 + x)^{-2}\ (1 + 4x)^{-1/2} = (- - 2x + 3x^2 + \ldots)(1 - 2x + 6x^2 + \ldots)$

Hence first three terms in the expansion are

$1 - 4x + 13x^2$.

Example 24:

Product in thousand kilograms of a certain firm in the first second, third, etc, weeks is the same as the coefficient of the first, second, third etc., powers of x in the expansion of $(1 + x)(1 - x)^{-2}$. *Find the production in the sixth week.*

Solution:

The expansion of $(1 + x)(1 - x)^{-2}$ is given by

$$(1+x)\left[1+(-2)(-x)+\frac{(-2)(-2-1)}{2}(-x)^2+\frac{(-2)(-2-1)(-2-2)}{\lfloor 3}(-x)^3+\frac{(-2)(-2-1)(-2-2)(-2-3)}{\lfloor 4}(-x)^4+\ldots\right]$$

$= (1 + 2x)[1 + 2x + 3x^2 + 4x^3 + 5x^4 + 6x^5 + 7x^6 + \ldots]$

$= (1 + 2x + 3x^2 + 4x^3 + 5x^4 + 6x^5 + 7x^6 + \ldots)$

$\quad + (x + 2x^2 + 3x^3 + 4x^4 + 5x^5 + 6x^6 + \ldots)$

$= 1 + 3x + 5x^2 + 7x^3 + 9x^4 + 11x^5 + 13x^6 + \ldots$

As the coefficient of the sixth power of x is 13, the required production in the sixth week is 13 thousand kilograms.

EXERCISES

1. Expand the following

   (*i*) $(4 - 3x)^{3/2}$, $\quad |x| < 4/3$

   (*ii*) $(1 + x)^{-4}$, $\quad |x| <$

   (*iii*) $(1 + x)^{-2}$, $\quad |x| <$

   (*iv*) $(1 + x)^{-1}$, $\quad |x| <$

   (*v*) $(1 + 3x)1/2 \quad (1 - 2x)^{-1/3}$, $|x| <$ Min (1/3, 1/2).

2. Find

   (*i*) the 8th term of $(1 - 3a^2)^{16/3}$, $\quad |a| < 1/\sqrt{3}$

   (*ii*) the 5th term of $(1 - 2x^{-3})^{11/2}$, $\quad |x| < 1/3\sqrt{2}$

   (*iii*) general term of $(2 - 3x)^{3/2}$, $\quad |x| < 2/3$

   (*iv*) general term of $(1 + 2x)^{-1/2}$, $\quad |x| < 1/2$

3. Find the first negative term in the expansion of

   (*i*) $\left(1+\frac{4}{3}x\right)^{27/3}$ where $0 < x < \frac{3}{4}$.

   (*ii*) $(1+3x)^{16/2}$ where$0 < x < \frac{1}{3}$

   (*iii*) $\left(1+\frac{3}{4}x\right)^{19/4}$ where$0 < x < \frac{3}{4}$

4. Find $\frac{1}{\sqrt[3]{62}}$, correct to four decimal places.

5. Find $\frac{1}{\sqrt[3]{630}}$, correct to five decimal places.

6. Find the coefficient of x^n in the expansion of

$$\frac{(1+x)^2}{(1+x)^2}, \quad |x| < 1.$$

7. Find the coefficient of x^n in the expansion of

$$\frac{x-2}{(1+2)x-1)^2}, \quad |x| < 1.$$

8. What is the value of n for which the coefficient of x^2 in the expansion of $(1 + x^2)(1 + x)^n$ is six times the coefficient of x? $(|x| < 1)$.

9. Use Binomial Theorem to expand $(1 + x + x^2)^{-3}$.

[Hint. $1 + x + x^2 = \frac{1-x^2}{1-x}$].

10. Expand $(x + y)^n$ when n is not a positive integer.

11. Find the coefficient of x^n in the expansion of $(1 + x + x^2 - x^2)^{-1}$

12. Show that

$$x^n = 1 - n\left(1-\frac{1}{x}\right) + \frac{n(n-1)}{2!}\left(1-\frac{1}{x}\right)^2 + \ldots$$

13. Find the value of $(126)^{1/3}$ correct to 4 decimal places, using Binomial Theorem.

ANSWERS

1. (*i*) $8 + 9x + \frac{27}{16}x^2 + \ldots$

(*ii*) $1 - 4x + 10x^2 - 10x^3 + \ldots$

(*iii*) $1 - 2x + 3x^2 - 4x^3 + \ldots$

(*iv*) $1 + x + x^2 + x^3 + \ldots$

(*v*) $1 + \frac{13}{6}x + \frac{55}{72}x^2 \ldots$

2. (*i*) $\frac{208}{9}a^{14}$ (*ii*) $\frac{1155}{8}x^{12}$

(*iii*) $\dfrac{2^{3/2}\left[\frac{3}{2}\left(\frac{3}{2}-1\right)\left(\frac{3}{2}-2\right)....\left(\frac{3}{2}-r+1\right)\right]}{\lfloor r}, \left(\frac{3}{2}x\right)$

(*iv*) $(-1)^r \dfrac{1.3.5...(2r-1)}{\lfloor r} x^r$, $|x| < 1/2$.

3. (*i*) $\dfrac{-27.23...3}{8.1}\left(\dfrac{3}{4}\right)^8 x^8$ (*ii*) $\dfrac{-208}{9}x^7$.

(*iii*) $\dfrac{-1463}{233}x^6$.

4. 0.2527 5. 5.00997

6. $4n$ 7. $\dfrac{(-1)^{n+1}}{9.2^{n-1}} - \dfrac{n+1}{3} - \dfrac{4}{9}$

8. 7 or − 4 9. $1 - 3x + 3x^2 - 2x^3 - 9x^4 + \ldots$

10. $x^n\left[1+n.\dfrac{y}{\tau}+\dfrac{n(n-1)}{\lfloor 2}\dfrac{y^2}{x^2}+\dfrac{n(n-1)(n-2)}{\lfloor 3}\dfrac{y^3}{x^3}+...\right]$ when $|x| > |y|$.

$y^n\left[1+n.\dfrac{x}{y}+\dfrac{n(n-1)}{\lfloor 2}\dfrac{x^2}{y^3}+\dfrac{n(n-1)(n-2)}{\lfloor 3}.\dfrac{x^3}{y^3}+...\right]$ when $|x| < |y|$.

11. 1. 13. 5.0133.

A APPLICATION OF BINOMIAL THEOREM

We illustrate below by means of examples how the Binomial Theorem is used in the summation of infinite series and certain other types of problems.

Example 25:

Sum the series

$$1+\frac{3}{4}+\frac{3.5}{4.8}+\frac{3.5.7}{4.8.12}+...\infty$$

Solution:

Comparing the given series with the expansion

$$(1+x)^n = 1+nx+\frac{n(n-1)}{\lfloor 2}x^2+\frac{n(n-1)(n-2)}{\lfloor 3}x^3+...$$

we find $nx = \dfrac{3}{4}$ ...(*i*)

$$\frac{n(n-1}{2}x^2 = \frac{3.5}{4.8} \qquad ...(ii)$$

(*i*) gives $x = \frac{3}{4n}$

Putting in (*ii*), we get

$$\frac{n(n-1}{2}.\frac{9}{16n^2} = \frac{3.5}{4.8}$$

$$\Rightarrow \left(1 - \frac{1}{n}\right)9 = 3.5 = 15$$

$$\Rightarrow n = -\frac{3}{2}$$

and, therefore, $\qquad x = -\frac{3}{4} \times \frac{2}{3} = -\frac{1}{2}$

Hence the given series is equal to

$$(1 - 1/2)^{-3/2} = (1/2)^{-3/2} = 2^{3/2} = 2\sqrt{2}.$$

Example 26:

Sum the series

$$\frac{5}{3.6} + \frac{5.7}{3.6.9} + \frac{5.7.9}{3.6.9.12} + \ldots$$

Solution:

The given series can be written as

$$= \frac{1}{3}\left[\frac{3.5}{3.6} + \frac{3.5.7}{3.6.9} + \frac{3.5.7.9}{3.6.9.12} + \ldots\right]$$

$$= \frac{1}{3}\left[2 + \frac{3.5}{3.6} + \frac{3.5.7}{3.6.9} + \ldots \qquad - 2\right]$$

$$= \frac{2}{3} + \frac{1}{3}\left[1 + 1 + \frac{3.5}{3.6} + \frac{3.5.7}{3.6.9} + \ldots\right]$$

The series inside the bracket will be summed up by means of Binomial theorem for any index.

Comparing thes series with the expansion

$$(1+x)^n = 1 + nx + \frac{n(n-1)}{2\,!}x^2 + \ldots,$$

we get $\qquad nx = 1 \qquad$ or $\quad x = \frac{1}{n}$

and $\frac{n(n-1)x^2}{2} = \frac{3.5}{3.6} = \frac{5}{6}$

$\Rightarrow \frac{n(n-1)}{2}\frac{1}{n^2} = \frac{5}{6} \qquad \Rightarrow n = -\frac{3}{2}$

This implies that $x = -2/3$

Consequently the series inside the bracket

$$= \left(1 - \frac{2}{3}\right)^{-3/2}$$

$$= \left(\frac{1}{3}\right)^{-3/2} = 3^{3/2} = 3\sqrt{3}.$$

Hence the given series $= -\frac{2}{3} + \frac{1}{3}(3\sqrt{3})$

$$= \sqrt{3} - \frac{2}{3}.$$

Example 27:

If $\quad y = \frac{1}{3} + \frac{1.3}{3.6} + \frac{1.3.5}{3.6.9} + \ldots$

prove that $\quad y^2 + 2y - 2 = 0$

Solution:

We have

$$y + 1 = 1 + \frac{1}{3} + \frac{1.3}{3.6} + \frac{1.3.5}{3.6.9} + \ldots$$

Comparing the series on the RHS with the binomial expansion

$$(1+x)^n = 1 + nx + \frac{n(n-1)}{\underline{|2}}x^2 + \ldots$$

we get $\quad nx = 1/3$

$$\frac{n(n-1)}{\underline{|2}}x^2 = \frac{1.3}{3.6}$$

giving thereby, $\quad n = -1/2 \quad$ and $\quad x = -2/3$

Thus $\quad y + 1 = \left(1 - \frac{2}{3}\right)^{-1/2} = \left(\frac{1}{3}\right)^{-1/2} = 3^{1/2}$

Squaring both sides, we get

$$y^2 + 1 + 2y = 3$$
$$\Rightarrow y^2 + 2y - 2 = 0.$$

Example 28:

If $p = q$ nearly and $n > 1$, show that

$$\frac{(n+1)p+(n-1)q}{(n-1)p+(n-1)q} = \left(\frac{p}{q}\right)^{1/n}$$

Apply this rule to find approximately the seventh root of $\frac{131}{133}$.

Solution:

It is given that $p = q$, nearly,

so let $q = p + h$ where h is very small.

Then LHS $= \frac{(n+1)p+(n-1)(P+h)}{(n-1)p+(n+1)(p+h)}$

$$= \frac{2np\left[1+\frac{n-1}{2n}.\frac{h}{p}\right]}{2np\left[1+\frac{n+1}{2n}.\frac{h}{p}\right]}$$

$$= \left(1+\frac{n-1}{2n}.\frac{h}{p}\right)\left(1+\frac{n+1}{2n}.\frac{h}{p}\right)^{-1}$$

$$= \left(1+\frac{n-1}{2n}.\frac{h}{p}\right)\left(1-\frac{n+1}{2n}.\frac{h}{p}\right), \text{ nearly}$$

$$= 1+\left(\frac{n-1}{2n}-\frac{n+1}{2n}\right)\frac{h}{p}+...$$

$$= 1-\frac{h}{np} \text{ approx.}$$

Again LHS $= \left(\frac{h}{p+h}\right)^{1/n} = \frac{1}{(1+h/p)^{1/n}}$

$$= \left(1+\frac{h}{p}\right)^{1/n} = 1-\frac{h}{np} \text{ approx.}$$

$\Rightarrow$LHS = RHS.

To do the second par, let $p = 131$, $q = 133$, $n = 7$.

Now $\left(\frac{131}{133}\right)^{1/7} = \left(\frac{p}{q}\right)^{1/7}$

$$= \frac{(7+1)131+(7-1)133}{(7-1)131+(7+1)133}$$

$$= \frac{1048+798}{786+1064}$$

$$= \frac{1846}{1850}$$

$$= \frac{923}{925} = .9978 \text{ nearly.}$$

EXERCISES

1. Sum the series

(*i*) $1-\frac{3}{4}+\frac{3.5}{4.8}-\frac{3.5.7}{4.8.12}+\ldots\infty.$

(*ii*) $1+\frac{1}{3}+\frac{1.3}{2\,!}\left(\frac{1}{3}\right)^2+\frac{1.3.5}{3\,!}\frac{3.5.7}{4.8.12}\left(\frac{1}{3}\right)^2+\ldots\infty.$

(*iii*) $1+\frac{1}{4}+\frac{1.4}{4.8}+\frac{1.4.7}{4.8.12}+\ldots\infty.$

(*iv*) $1+\frac{15}{8}+\frac{15.21}{8.16}+\frac{15.21.27}{8.16.24}+\ldots\infty.$

2. Prove that

$$\frac{1}{4}+\frac{1.3}{4.6}+\frac{1.3.5}{4.6.8}+\ldots=1.$$

3. Show that

$$1+\frac{1}{3}+\frac{2.5}{6.12}+\frac{2.5.8}{4.12.18}+\ldots\infty=2^{2/3}$$

4. Show that

$$\frac{7}{5}\left(1+\frac{1}{10^2}+\frac{1.3}{1.2}.\frac{1}{10^4}+\frac{1.3.5}{1.2.3}.\frac{1}{10^6}+\ldots.\right)=\sqrt{2}$$

5. Identify with a binomial expansion and find the sum of the series

$$2+\frac{5}{3\lfloor 2}+\frac{5.7}{3^2\lfloor 3}=\frac{5.7.9}{3^3\lfloor 4}+\ldots.$$

6. If x is so small, that its squares and higher powers can be neglected, then show that

(i) $$\frac{(1-x)^{1/3}+(1-5x)^2}{\sqrt[4]{16-x}}=1-\frac{989}{192}x.$$

(ii) $$\frac{(1+x)^{1/2}+(1+x)^{2/3}}{(1+x)^{2/3}+(1-x)^{1/2}}=1-\frac{1}{6}x.$$

(iii) $$\frac{\left(1+\frac{2x}{3}\right)^{-6}(4+3x)^{1/2}}{(4+x)^{3/2}}=\frac{1}{4}-\left(1-\frac{10x}{3}\right).$$